The Biochemical Mode of Action
of Pesticides

The Biochemical Mode of Action of Pesticides

Second Edition

J. R. CORBETT, K. WRIGHT
and A. C. BAILLIE

FBC Limited
Chesterford Park Research Station
Saffron Walden, Essex, England

Foreword by
The Rt. Hon. Lord Todd, OM, FRS

ACADEMIC PRESS 1984
Harcourt Brace Jovanovich, Publishers

London Orlando San Diego New York Austin
Boston Tokyo Sydney Toronto

ACADEMIC PRESS INC. (LONDON) LTD
24/28 Oval Road
London NW1 7DX

United States Edition published by
ACADEMIC PRESS INC.
Orlando, Florida 32887

British Library Cataloguing in Publication Data
Corbett, J. R.
    The biochemical mode of action of pesticides—
    2nd ed.
    1. Pesticides—Physiological effect
    I. Title   II. Wright, K.   III. Baillie, A. C.
    574.19′2   QP82.2.P4

    ISBN 0-12-187860-0
    LCCCN 73-9455

Typeset by Bath Typesetting Ltd, UK
and printed in Great Britain by Thomson Litho Ltd, East Kilbride, Scotland

Preface to the Second Edition

The second edition covers the new information that has become available since the first edition appeared in 1974, and has been almost completely rewritten. Key developments in this period include insights into the mechanism of action of: Hill reaction inhibitors; pyrethroid, cyclodiene and formamidine insecticides; insect hormone analogues; benzimidazole fungicides; fungicides inhibiting ergosterol biosynthesis; and the herbicide glyphosate. In addition we have included new or much expanded sections on insect pheromones, herbicide safeners, and the degree of predictability which is now possible in pesticide discovery.

Our objective with the second edition is the same as that of the first, namely to explain on a molecular basis what is known of the way in which pesticides interfere with the biochemistry of living organisms. We have attempted to be straightforward in our descriptions, and to provide suitable background information for those whose knowledge of biochemistry is limited. We have followed the same outline as before, where the emphasis was on biochemical similarities of action, so that compounds affecting the same basic biochemical mechanism will be considered together, even though their patterns of use are different.

In an attempt to avoid over-burdening the text with references we have in some cases referred to reviews rather than original papers. Some readers may regret that historical precedent is occasionally obscured by this procedure but it will be evident from the reviews and we hope that our account is easier to read as a result.

Throughout we have searched for information that might be of value to chemists concerned with introducing greater predictability into the search for new pesticides. This concern has led us to emphasise action in molecular terms, since this is the language of the chemist. We do not discuss the practical significance, the ecological implications, or other non-biochemical aspects of pesticides. Consequently certain pesticides may receive more attention than a practising agriculturist might have thought they warranted from their sales volume. Similarly, other topics, such as resistance, which are undoubtedly important in relation to field use of pesticides, are dealt with only in connection with the mode of action.

Despite this bias towards rationality in pesticide design, we have included the relevant information on all pesticides in the sixth edition of

The Pesticide Manual, edited by Charles Worthing in 1979, though we have done our best to include more recent compounds of interest. We are grateful to Dr Worthing for keeping us informed about potential new inclusions in the seventh edition of *The Pesticide Manual*, which was in preparation when our book was finalised. As in the first edition of this work, we have normally excluded compounds acting solely against bacteria and protozoa, repellents, fumigants and rodenticides.

In referring to pesticides we have in nearly all cases used the BSI (British Standards Institution) approved common name. In most cases this is the same as the ISO (International Standards Organisation) approved name. Otherwise we have used the chemical name, sometimes pointing out other common names in use.

The first edition was written by the first-named author alone, while this edition is the work of three. Every attempt has been made to write the book together and with continuity of thought, and not as a multi-authored treatise. We therefore jointly take responsibility for any errors or omissions. We are grateful to workers in other research institutes who have sent us reprints and other information, and to Professor N. Amrhein for his comments on a section of Chapter 6. Also we should like collectively to thank our colleagues in FBC Limited who have helped us, particularly Dr David Evans for the pheromone section, Mr Brian Wright, who was our adviser on kinetic matters, and the staff of our Information and Word Processing Departments.

We are also grateful to the Board of FBC Limited for permission to publish the book. Finally, we should like to acknowledge our debt to Lord Todd for writing the Foreword to this second edition.

J. R. Corbett, K. Wright and A. C. Baillie

Foreword

The relationship between chemical constitution and physiological activity has been the subject of intensive study and speculation since the earliest days of organic chemistry. This is perhaps not surprising since the initial stimulus to the science came from medicine and from the need to find a wider range of drugs for the cure of disease. This stimulus remains, but with the passage of time and under the spur of rising world population the control of pests in agriculture and in animal husbandry have also become problems of urgency and a massive development of chemical industry has taken place in endeavouring to provide solutions to them. The triumphs achieved in some areas of pest control have been many and striking, yet a precise theoretical basis for the design of pesticides still eludes us just as it does in the search for chemotherapeutic agents. Ehrlich's lock and key analogy for selective toxicity may well be valid, but it is of limited practical value unless one knows what kind of lock is on the physiological door which it is desired to open or shut. Put in a more scientific way, unless and until we understand the enzyme systems involved in vital processes and their vulnerability in different organisms, rational development of specific pesticides will be well-nigh impossible. Given such an understanding we could begin to design chemical compounds which will interfere with vital enzyme systems and which, if problems of stability and transport in the organism can be surmounted, will provide effective and specific pesticides. To the search for such a rational approach to pesticide design Dr Corbett has devoted his own researches and it forms the theme of this book.

It is now about forty years since the discovery that organophosphorus and carbamate insecticides are powerful inhibitors of acetylcholinesterase and the general acceptance of the view that this property is the essential basis of their insecticidal activity. Since then much work has been done on the biochemical effects of pesticides as a basis for understanding their activity. In a foreword to the first edition of this book I drew attention to the wide dispersal of work in this field in the scientific and technical literature and to the need for it to be collated and reviewed. That need Dr Corbett's book was designed to meet and it did so with marked success. The ten years that have passed since the publication of that edition, however, have seen an ever-increasing growth in the amount of research carried out on the biochemical mode of action of pesticides. These advances alone would warrant a second edition, and the authors are to be congratulated on the manner in which they have reviewed them and have at the same time provided a wealth of information on individual pesticides and discussed such important issues as resistance—an ever-present problem in the pesticide field.

This is an up-to-date and well-written source book for all chemists and biologists interested in the problems of pest control. Like all worthwhile reviews, this second edition stimulates as well as informs the reader and it should be studied by all who wish to have up-to-date knowledge of a rapidly expanding field or who seek to develop a rational approach to the design and synthesis of pesticides with predictable activity. Despite the large amount of work reviewed by Dr Corbett and his colleagues the goal of rational design seems far off, but at least progress to date encourages one to believe that success will ultimately crown the efforts being made. And that success will be needed if we are to meet the nutritional requirements of an ever-growing population and at the same time avoid indiscriminate attack on our environment.

Todd

Contents

1 | Pesticides Interfering with Respiration

I. BIOCHEMICAL BACKGROUND

All higher organisms rely on respiration to break down organic molecules, thereby providing energy and raw materials necessary for their continued survival. Consequently, all are, in principle, vulnerable to the action of inhibitors of this process.

Respiration falls naturally into a number of phases. Organic foodstuffs contain carbohydrates, fats or proteins, and these are first broken down to their constituent monomers which are, respectively, sugars, acetyl CoA (acetate bound in an 'activated' form) and amino acids. The carbon skeletons of the monomers are subsequently oxidised through glycolysis ('splitting of sugar') and/or the tricarboxylic acid (or Krebs) cycle in which the two-carbon units of acetyl CoA are degraded to CO_2 (Fig. 1.1). For glucose the process is described by the equation:

$$C_6H_{12}O_6 + 6O_2 \rightarrow 6CO_2 + 6H_2O$$

During the oxidative reactions of these procedures nicotinamide adenine dinucleotide (NAD^+) is reduced to NADH thus conserving the 'reducing power' of the substrates. The reducing equivalents are in turn relinquished through a chain of electron carriers during respiratory chain electron transport, so that the NAD^+ is regenerated, and electron flow through the chain culminates in the reduction of oxygen to water. This electron transport is coupled to the synthesis of the high-energy intermediate, adenosine-5'-triphosphate (ATP) from the diphosphate (ADP) and inorganic phos-

1

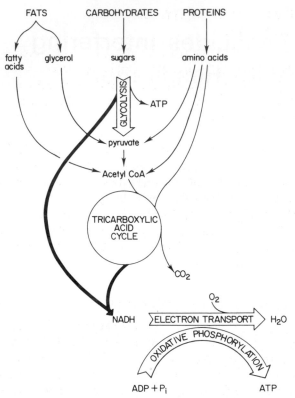

Fig. 1.1 Outline of the reactions of respiration. For abbreviations, see text

phate (P_i) by a process called oxidative phosphorylation (Fig. 1.1). A certain amount of ATP is also formed directly during glycolysis (substrate-level phosphorylation; see Fig. 1.2).

The terminal phosphodiester bond of ATP retains the energy that is required for many of the synthetic reactions carried out by living cells. Although ATP is an important end product of respiration, many of the intermediates in glycolysis and the tricarboxylic acid cycle are also important because they are the starting compounds for other cellular biosyntheses. Consequently a pesticide which inhibits respiration will deprive its target organism of both ATP and certain key metabolites.

The foregoing is a brief summary of the overall process. In the following sections of the chapter more detailed accounts will be given of those aspects with which pesticides are known to interfere. The effects of fungicidal inhibitors of intermediary metabolism and energy production have been reviewed by Lyr (1977) and, as part of a general review on the mode of action of fungicides, by Kaars Sijpesteijn (1977).

II. THE FORMATION OF ACETYL CoA

A. Glycolysis (Fig. 1.2)

Most organisms inhabit an environment containing oxygen, and under these conditions glycolysis may be defined as the conversion of glucose or glucose-1-phosphate (released from a store of starch or glycogen) into pyruvate. When oxygen is absent glycolysis would also include the further reduction of pyruvate to lactate (in mammals) or through acetaldehyde, to ethanol (alcoholic fermentation, occurring for instance in yeasts). These reductions use some of the accumulating NADH (which cannot be re-oxidised via the respiratory chain in the absence of oxygen) and the NAD^+ so regenerated allows glycolysis to proceed further and make available more ATP at the substrate level.

The reactions of glycolysis take place in the soluble phase of the cell, the cytoplasm. The pyruvate is further metabolised inside the mitochondria (see below) by a reductive decarboxylation which yields acetyl CoA. We shall return to this shortly.

B. Transfer of compounds into mitochondria

The only organelle to concern us in this chapter is the mitochondrion, in which most of the cell's energy is produced by oxidative phosphorylation, and in which fatty acid degradation and the later stages of carbohydrate breakdown take place. Mitochondria are bounded by a double membrane, the outer of which is freely permeable to many metabolites and the inner of which is convoluted inwards to form finger-like cristae. The inner membrane does not permit unrestricted access of metabolites to the interior, where enzymes of the tricarboxylic acid cycle are located. The components of electron transport are arranged in the inner membrane and, according to the chemiosmotic theory (p. 18), protons are pumped across this membrane during electron transfer.

Since the mitochondrial inner membrane is selectively permeable and, at least in animals (as opposed to many plants: Moore and Rich, 1980) cannot be penetrated by NADH, reducing power is transferred between the organelle and the surrounding cytoplasm by a shuttle. For example, in insects, dihydroxyacetone phosphate is reduced in the cytoplasm by NADH to form glycerol-3-phosphate, which moves into the mitochondrion and becomes re-oxidised, so delivering the reducing equivalents of NADH across the membrane. O'Brien et al. (1965) made a brave but unsuccessful attempt to design insecticides that would not harm mammals by synthesising inhibitors for the glycerol-3-phosphate shuttle, which was thought to be uniquely important to insects, and less so to mammals.

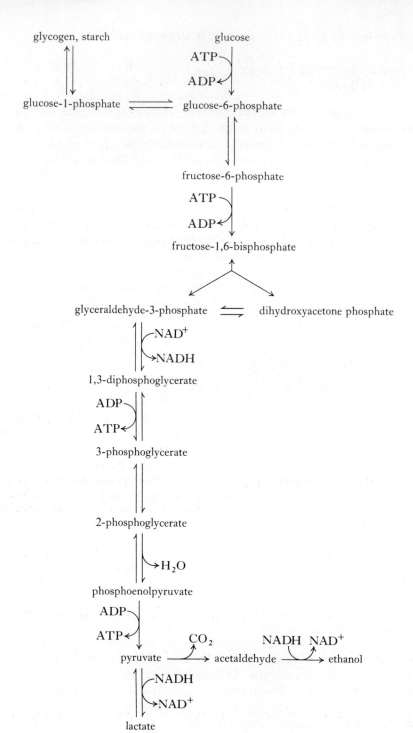

Fig. 1.2 The reactions of glycolysis. For abbreviations, see text

In a post-mortem on the failure of these compounds as insecticides O'Brien (1967) said 'we may never know whether the reason was that the shuttle is not vital, or that it was inadequately inhibited because the compounds were not potent enough, or that the compounds were rapidly metabolised'. If the possibility of lack of penetration of unmetabolised compound to the site of action is included, this serves as a salutary list of the possible ways in which compounds may fail in practice.

C. Conversion of pyruvate to acetyl CoA

Once inside the mitochondria, pyruvate is converted to acetyl CoA (a thioester with the sulphur-containing coenzyme A; acetyl CoA takes part

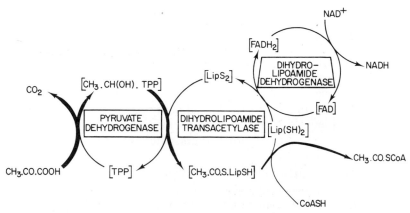

Fig. 1.3 Reaction sequence of the pyruvate dehydrogenase complex. Square brackets indicate that the intermediate is bound to the enzyme. TPP, thiamine pyrophosphate; $LipS_2$ etc., lipoamide moiety and its reduced forms; CoASH, coenzyme A; FAD, flavin adenine dinucleotide

in a variety of acetyl transfer reactions). The conversion of pyruvate to acetyl CoA is accomplished by the pyruvate dehydrogenase complex (EC 1.2.4.1)* on which two classes of commercial pesticide are thought to work.

The three enzymes which constitute the complex are shown in Fig. 1.3 and the overall reaction is:

$$CH_3.CO.COOH + NAD^+ + CoASH \rightarrow CH_3.CO.SCoA + NADH + H^+ + CO_2$$
pyruvate   acetyl CoA

*This is the Enzyme Commission classification which groups enzymes according to the reactions they catalyse.

The complex is huge with a molecular mass of over 7 million daltons in mammals, and it contains various proportions of the three enzymes according to source (Hucho, 1975).

There is an enzyme in the tricarboxylic acid cycle (see Fig. 1.1), α-oxoglutarate dehydrogenase (EC 1.2.4.2) which proceeds by an analogous mechanism.

Although the following account emphasises pyruvate dehydrogenase as the target for some arsenic and probably all copper fungicides, it is likely that α-oxoglutarate dehydrogenase is also affected. As we shall see, the proposed mechanism of action of the pesticides is through an interaction with the -SH groups of lipoic acid. The evidence is not conclusive however, and it seems likely that the action of the compounds also involves combination with other important thiol groups within the cell.

1. *Arsenic-containing pesticides*

The arsenic-containing pesticides listed in Table 1.1 have been in use for many years, and there has been little recent work on their mechanism of action. It seems likely that the first five compounds act by producing toxic inorganic arsenic ions that are soluble in water (Martin, 1964). The first three compounds produce arsenite ions (ions of arsenious acid, $As(OH)_3$), while the next two produce arsenate ions (ions of arsenic acid, $O=As(OH)_3$).

Although it is frequently stated that arsenates are biologically effective by reduction to arsenites, the evidence for this, at least in regard to pesticidal usage of arsenic containing compounds, cannot be regarded as conclusive.

In an exhaustive review of the inhibitory effects of arsenicals on living systems, Webb (1966) describes work showing that arsenates and arsenites may be inter-converted in living systems. Arsenates and arsenites are said to be equally toxic to insects (Brown, 1963) which may indicate that interconversion occurs. On the other hand, arsenite is a more effective herbicide than arsenate, though it does not seem that arsenate is extensively reduced in plant tissue to give arsenite (Sachs and Michael, 1971).

Arsenate itself has a known biochemical effect since it can act as an uncoupler (p. 19) of both oxidative phosphorylation and the substrate-level phosphorylation associated with glycolysis (Slater, 1963). It does this by mimicking the phosphate ion and being incorporated into key high-energy-containing intermediates which rapidly break down if they contain arsenate in place of the normal phosphate.

The arsenite ions produced directly by the first three compounds in Table 1.1 probably kill organisms by inhibiting either or both of the

Table 1.1   Arsenic compounds used as pesticides

Name* and use	Structure
Arsenous oxide (rodenticide)	As_2O_3
Paris green (insecticide)	$(MeCOO)_2Cu.3Cu(AsO_2)_2$
Sodium arsenite (insecticide and herbicide)	solid solution of NaOH and As_2O_3
Calcium arsenate (insecticide)	basic calcium arsenates of the type $3Ca_3(AsO_4)_2 . Ca(OH)_2$
Lead arsenate (insecticide)	$PbHAsO_4$
Methylarsonic acid as disodium salt as monosodium salt (herbicides)	$MeAsO(OH)_2$ $MeAsO(ONa)_2$ $MeAsO(OH).ONa$
Dimethylarsinic acid (herbicide)	$Me_2AsO.OH$

*Pesticides usually have at least three names: the common or trivial name, which is used most frequently; the proprietary trade name, which is not used in this book; and the chemical name, defining the structure. Sometimes, as in this table, the chemical name is used as the common name. The common and chemical names are given in *The Pesticide Manual* (Worthing, 1979).

pyruvate and α-oxoglutarate dehydrogenase complexes. The evidence for this may be considered under four headings.

(a) *Inhibition of the enzyme* in vitro

Webb (1966) lists numerous examples of the *in vitro* inhibition of pyruvate and α-oxoglutarate oxidising systems by arsenite concentrations of 100 μM and less. For instance, Hoskins *et al.* (1956) found that the oxidation of α-oxoglutarate by honey-bee mitochondria was 38% inhibited by 50 μM arsenite. However, Webb (1966) lists some 15 other enzymes that are at least 30% inhibited by a concentration of 100 μM or less, so the effect is not specific to the two dehydrogenases.

(b) *Accumulation of oxo-acids in poisoned organisms*

Under certain conditions inhibition of the dehydrogenases should lead to the accumulation of the relevant oxo-acid, and Webb (1966) lists a variety of organisms and tissues where arsenite has caused such accumulation. These are of fungal, plant, insect and animal origin.

(c) Inhibition of respiration in vivo

Treatment of insects with arsenite may cause inhibition of respiration (Brown, 1963).

(d) Symptoms of poisoning

Insects poisoned with arsenite show symptoms which resemble those caused by the respiratory inhibitor rotenone (p. 21) but are unlike those caused by neuroactive insecticides (O'Brien, 1967). Thus death is caused by progressive inactivity, and does not involve convulsions.

In summary, there is reasonable evidence that arsenite kills higher animals mainly by interference with oxo-acid oxidation, though this is probably not its sole locus of action. By extension this mode of action is also likely in other organisms and the limited evidence available for insects supports this.

Turning now to the exact molecular mode of action, it is generally believed that arsenite inhibits keto-acid oxidation by combining with the two thiol groups of the lipoamide component of the dihydrolipoamide transacetylase (Fig. 1.3):

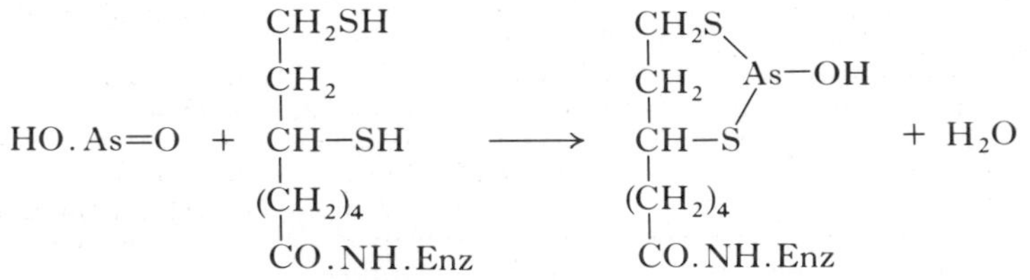

$$
\begin{array}{ccc}
 & \begin{array}{l} CH_2SH \\ | \\ CH_2 \\ | \\ HO.As{=}O \;+\; CH{-}SH \\ | \\ (CH_2)_4 \\ | \\ CO.NH.Enz \end{array} & \longrightarrow \quad \begin{array}{l} CH_2S \\ \;\;\;\;\;\;\;\;\backslash \\ CH_2 \;\;\;As{-}OH \\ \;\;\;\;\;\;\;\;/ \\ CH{-}S \\ | \\ (CH_2)_4 \\ | \\ CO.NH.Enz \end{array} \quad +\; H_2O \\
\end{array}
$$

arsenite   lipoamide on
(in solution)   dihydrolipoamide
transacetylase

Webb (1966) states that trivalent arsenicals do not react with any group of biological importance other than thiols, and Aldridge and Cremer (1955) demonstrated that arsenite did not combine with glutathione (a monothiol) but reacted avidly with the dithiol dimercaptopropanol ($CH_2SH.CHSH.CH_2OH$). However, there does not appear to be any direct evidence that lipoamide is indeed the actual site of inhibition, and some data (Webb 1966) suggests that arsenite may react with thiol groups on the surface of dihydrolipoamide dehydrogenase, the final enzyme in the oxo-acid oxidation enzyme system (Fig. 1.3).

The pentavalent arsenical herbicides methylarsonic acid (and its salts) and dimethylarsinic acid have not been studied much from a biochemical point of view. Dimethylarsinic acid is probably not reduced in the plant to a trivalent arsenical, and appears to be metabolically stable (Sachs and Michael, 1971), though it will react chemically with thiol groups (Jacobson *et al.*, 1972).

2. *Copper compounds and copper carriers used as fungicides* (*Table 1.2*)

(*a*) *Inorganic copper compounds*

Copper compounds have been widely used as fungicides since the discovery in 1885 of the fungicidal action of Bordeaux mixture (McCallan, 1967), which is made from copper sulphate and calcium oxide. The copper compounds listed in Table 1.2 probably all act via the cupric ion Cu^{2+} (Martin, 1969), which means that cuprous oxide must be oxidised before becoming effective, though the mechanism, enzymatic or otherwise, by which this is done is apparently not known.

Table 1.2 Antifungal compounds containing or working via copper

Name	Structure
Copper containing compounds	
Copper naphthenates ('Naphthenic acids')	soluble copper salts of naphthenic acids from petroleum
Copper oxychloride	$3Cu(OH)_2 . CuCl_2$ or $Cu_2(OH)_3Cl$
Copper sulphate	$CuSO_4 . 5H_2O$ (as a component of Bordeaux mixture and Cheshunt compound)
Cuprous oxide	Cu_2O
Copper carrying compounds Oxine–copper	
Ferbam	$(Me_2N . CS . S)_3Fe$
Phenylmercury dimethyldithiocarbamate	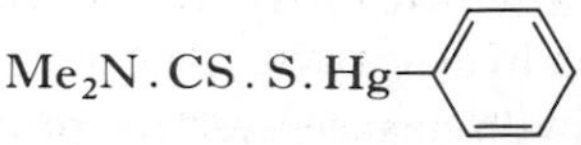 $Me_2N . CS . S . Hg-$
Thiram	$Me_2N . CS . S . S . CS . NMe_2$
Ziram	$(Me_2N . CS . S)_2 Zn$

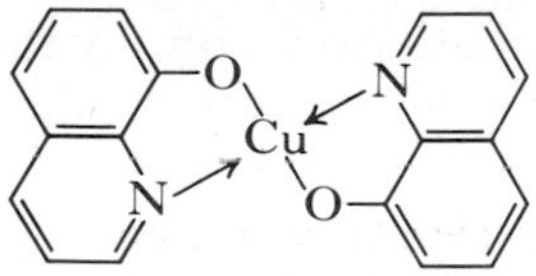

Like nearly all of the early fungicides, inorganic copper compounds prevent the spread of the infection but will not eradicate an existing one (Evans, 1968). This is consistent with the view that their site of action is the fungal spore. It has long been known that fungal spores will concentrate copper ions from the surrounding medium (Martin, 1969), and Somers (1963) used radioactive copper to show that fungal conidia may concentrate copper by up to about 100-fold over that in the immediate environment, depending on the conditions. It is not unreasonable to assume that in a field situation the spores might contact a spray solution containing 1 ppm copper, so the ion might be concentrated to 100 ppm in the spore, which gives a final concentration of 1.5 millimoles of copper kg^{-1} of spore wet weight, or approximately 1.5 mM. This high concentration of copper inside the spore is consistent with the generally held view that the fungitoxic activity of copper ions is due to a non-specific affinity for various groups in the cell, such as imidazole, carboxyl, phosphate or thiol, resulting in non-specific denaturation of protein and enzymes (Somers, 1961; Lyr, 1977).

However, Kaars Sijpesteijn (1970), in an elegant review of the mode of action of agricultural fungicides, suggests that copper ions might be fungitoxic through interference with the pyruvate dehydrogenase system, though she does not exclude other additional effects.

This view is based on work done with isolated dihydrolipoamide dehydrogenase (Fig. 1.3). Using a yeast enzyme, Wren and Massey (1966) showed that, at a concentration of 10 cupric ions to one molecule of enzyme flavin, inhibition occurred. The enzyme from pig heart was affected more strongly, and it is thought that the same mechanism of inhibition is involved. Casola et al. (1966) showed that the inhibition of pure pig dihydrolipoamide dehydrogenase was not caused by irreversible binding of copper to the enzyme, but by the oxidation of two thiol groups on the enzyme protein, followed by further damaging oxidative effects.

Although copper sulphate has been shown to inhibit oxygen uptake by some fungal spores at concentrations that just prevent germination (McCallan et al., 1954), there do not appear to be any studies on whether pyruvate or α-oxoglutarate accumulates in fungi treated with inorganic copper. Consequently, we may conclude only that the inhibition of dihydrolipoamide dehydrogenase, and hence of the pyruvate (and possibly α-oxoglutarate) dehydrogenase system, provides a possible explanation of the effect of cupric ions on fungi, and that other mechanisms may operate as well.

If copper does work by inhibiting oxo-acid oxidation, one would expect copper salts to be generally biocidal. However, possibly due to lack of uptake by higher plants and insects, inorganic copper salts are not used widely to control pests other than fungi.

(b) Copper carriers

The mode of action of oxine copper and the dialkylthiocarbamates (Table 1.2) has been reviewed by Albert (1979) and Kaars Sijpesteijn (1970). In brief there is evidence that at least part of the anti-fungal activity of these compounds is due to their ability to chelate copper and carry it into the fungal cell, where it may interfere with oxo-acid oxidation.

(i) *Oxine-copper*. The anti-fungal compound oxine-copper is based on 8-hydroxyquinoline:

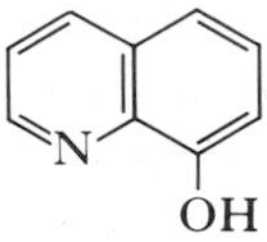

This compound is used in chemical analysis because of its ability to form lipid soluble metal chelates, and it has been known as a disinfectant and antiseptic agent since 1895. The chemistry and mode of action have been reviewed by Lukens (1969).

The importance of copper in the toxic action of 8-hydroxyquinoline sulphate, formerly used as a fungicide, was shown by Kaars Sijpesteijn *et al.* (1957a). The minimum concentration necessary to inhibit growth of *Aspergillus niger* was 10 ppm in the total absence of copper, but only 0.1 ppm if 0.03 ppm $CuSO_4.5H_2O$ was present. However, if the concentration of copper sulphate was increased to 3 ppm the minimum inhibitory level of 8-hydroxyquinoline rose to 0.5 ppm. The interpretation of this surprising result is that the toxicity at the 0.1 ppm level was due to formation of the 1 : 2 copper to 8-hydroxyquinoline complex (oxine–copper), while the decrease in toxic effect when more copper was added was due to formation of the 1 : 1 complex.

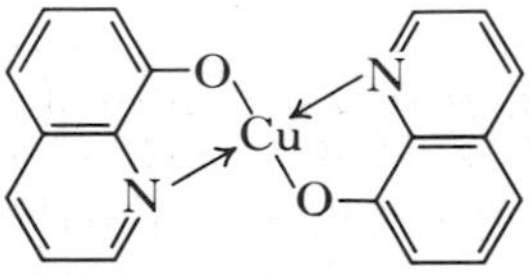

1 : 2 complex 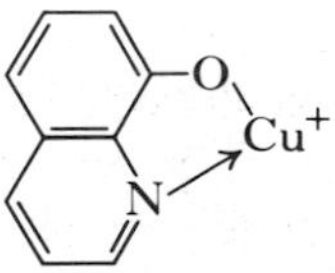   1 : 1 complex

The relative lack of toxicity of the 1 : 1 complex is probably caused by its water solubility preventing it from penetrating the lipid layers of the fungal cell (Block, 1956).

One might expect the 1 : 1 complex to be the actual toxic agent since it possesses residual combining power available for reaction with cell components. On this theory the 1 : 2 complex would penetrate the cell and equilibrate to form some of the toxic 1 : 1 complex. There are two lines of indirect evidence supporting this view.

Firstly excess 8-hydroxyquinoline, which is known to be able to penetrate the fungal cell (Lukens, 1969), reverses the toxic effect of oxine-copper (Block, 1956). Presumably this occurs because the 1 : 1 complex inside the cell is converted to the 1 : 2 complex. A similar situation occurs with 8-hydroxyquinoline on bacteria, except that ferric iron replaces copper (Albert, 1979).

Secondly, McNew and Gershon (1969) synthesised copper chelates of the type A–Cu–B, where A was a substituted or unsubstituted 8-hydroxy-quinoline and B was a substituted arylhydroxy acid. Fungitoxicity tests showed that A–Cu–B was essentially equivalent in activity to compounds of the type A–Cu–A. If the mixed chelate dissociated in the fungal cell, the possible products are, according to McNew and Gershon, B and Cu–B (neither of which proved to be particularly fungitoxic) and A–Cu, which, if the quinoline were unsubstituted, is the 1 : 1 complex of copper with 8-hydroxyquinoline. Since A–Cu is the only common dissociation product of A–Cu–A and A–Cu–B, it seems very likely that the toxic agent is the 1 : 1 complex.

The mechanism by which the 1 : 1 complex of copper with 8-hydroxy-quinoline and B was a substituted arylhydroxy acid. Fungitoxicity tests to accumulate in treated fungal cells (Kaars Sijpesteijn, 1970) so the pyruvate dehydrogenase complex could be one site of action. If this is correct one might speculate that thiol groups of this system, either of lipoamide itself or of the enzyme protein, might be the actual sites of reaction, since oxidation of thiols by cupric ions is enhanced by 8-hydroxy-quinoline (Lukens, 1969), and we have already seen that cupric ion itself may catalyse the oxidation of thiol groups of dihydrolipoamide dehydrogenase.

(*ii*) *Dialkyldithiocarbamates*. Although the mechanism of action of the dialkyldithiocarbamates resembles that of 8-hydroxyquinoline, it differs in detail. The mode of anti-fungal action of the dialkyldithiocarbamates has been reviewed by Janssen and Kaars Sijpesteijn (1961) and Albert (1979).

The dialkyldithiocarbamates are derivatives of the unstable dithio-carbamic acid $NH_2 . CS . SH$, and should be distinguished from the related alkylenebisdithiocarbamates which are salts of acids having the general formula $HS . CS . NH–X–NH . CS . SH$, where X is an alkylene group, typically $-CH_2.CH_2-$. The alkylene bisdithiocarbamates have a different mode of action (p. 300).

Phenylmercury dimethyldithiocarbamate is likely to have a dual mode of action since the phenylmercury moiety is toxic in its own right, reacting with single thiol groups in a non-specific manner (p. 300).

With increasing concentration of sodium dimethyldithiocarbamate in the culture medium of *Aspergillus niger*, effects on fungal growth fell into three phases (Janssen and Kaars Sijpesteijn, 1961). Initially, at low concentrations, inhibition occurred but only when the copper concentration was reasonably high (say 10 ppm, with about 0.5 ppm of the dithiocarbamate); this was considered to be due to the formation of the 1 : 1 copper to dithiocarbamate complex:

$$\text{Me} \diagdown \atop \text{Me} \diagup \text{N-C} {\diagup \text{S} \diagdown \atop \diagdown \text{S} \diagup} \text{Cu}^+$$

Secondly, as the concentration of the dimethyldithiocarbamate was increased, fungitoxicity was diminished as the 1 : 2 complex formed:

$$\text{Me} \diagdown \atop \text{Me} \diagup \text{N-C} {\diagup \text{S} \diagdown \atop \diagdown \text{S} \diagup} \text{Cu} {\diagup \text{S} \diagdown \atop \diagdown \text{S} \diagup} \text{C-N} {\diagup \text{Me} \atop \diagdown \text{Me}}$$

This contrasts with the situation found with 8-hydroxyquinoline where the 1 : 2 complex is toxic. This anomaly has been explained by Kaars Sijpesteijn and Janssen (1958), who showed that the water solubility of the 1 : 2 copper complex with dimethyldithiocarbamate was very low (0.01 ppm), while that of the corresponding complexes with 8-hydroxy-quinoline was 1 ppm. Presumably there is not enough of the 1 : 2 dimethyl-dithiocarbamate complex in solution to cause fungitoxicity, unless the fungus is very sensitive.

Finally, at high (about 50 ppm) concentrations of dimethyldithio-carbamate, a second inhibition level occurred. This required no copper and may be attributed to an effect of the free dithiocarbamate ion.

The situation need not be identical in all fungi. For instance, in the study mentioned (Janssen and Kaars Sijpesteijn, 1961) *Fusarium oxy-sporum* seemed to be insensitive to the 1 : 1 complex, and *Glomerella cingulata* seemed so sensitive that even the small amounts of the 1 : 2 complex in solution were sufficient to inhibit the growth.

To summarise, copper is required to achieve high fungitoxicity with low concentrations of dimethyldithiocarbamates. The 1 : 1 complex will penetrate the cell, and is presumed to be the toxic agent, while the 1 : 2 complex is generally too insoluble in water to allow penetration to occur. This contrasts with 8-hydroxyquinoline, where the 1 : 1 complex, although still the toxic agent, cannot penetrate the cell, while the 1 : 2 complex can.

That the dialkyldithiocarbamates inhibit pyruvate dehydrogenase is supported by the fact that pyruvate accumulates after treatment in, e.g., *Aspergillus niger* (Kaars Sijpesteijn *et al.*, 1957b). The likelihood exists therefore that the two classes of copper chelators, like the inorganic copper fungicides, act principally through their effect on the lipoamide-containing dehydrogenases.

This conclusion is supported by the demonstration that various dithiocarbamate derivatives inhibited pyruvate dehydrogenase and α-oxoglutarate dehydrogenase from mammalian tissues (DuBois *et al.*, 1961) though in this work succinate dehydrogenase was found to be inhibited as well. This was also true in a detailed study of the mechanism of action of ziram on yeast respiratory processes (Briquet *et al.*, 1976). Work on whole yeast, isolated mitochondria, and extracted enzymes led to the conclusion that pyruvate, α-oxoglutarate and succinate dehydrogenases, all of which were 50% inhibited at less than 10 μM, were the principal targets. The inhibition of succinate dehydrogenase was investigated in some detail. Zinc ions were not the important component of ziram since 30 μM ZnCl$_2$ had no effect on the enzyme, which was protected from ziram inhibition by reduced thiol compounds (e.g. 2-mercaptoethanol and glutathione). In addition the number of free-SH groups available in yeast mitochondria was diminished by ziram and indications were obtained that thiol groups of two levels of reactivity were involved. In this context, the reactivity of thiram and related compounds with nucleophiles has recently been reemphasised (Jen and Cava, 1982).

Briquet *et al.* (1976) concluded that succinate, pyruvate and α-oxoglutarate dehydrogenases were the primary targets for ziram in yeast. Copper ions were not implicated at any stage and the authors do not discuss their results in relation to the older work considered above. We have already mentioned that the free dithiocarbamate ion appears to be fungitoxic at high concentrations and that the sensitivity of fungi to the dialkyldithiocarbamates varies. It is possible that the results obtained by Briquet *et al.* (1976) apply only to yeast, but further investigations seem to be required.

3. s-*Butylamine*

Control of *Penicillium* spp. on citrus fruits after harvest can be achieved by the fungistatic action of s-butylamine (Eckert, 1969):

$$\begin{array}{c} \text{Et} \\ \diagdown \\ \text{CH-NH}_2 \\ \diagup \\ \text{Me} \end{array}$$

A systematic investigation by Yoshikawa and Eckert (1976) focused attention on pyruvate oxidation as the site of action of the compound and this was followed up by a more detailed investigation (Yoshikawa et al., 1976). Pyruvate dehydrogenase was suggested as the site of action since:

(a) the compound inhibited oxidation of pyruvate but not citrate, isocitrate, succinate, malate, acetyl CoA or NADH in isolated Penicillium mitochondria;

(b) oxidative phosphorylation itself (see below) was not affected.

(c) isolated pyruvate dehydrogenase was strongly inhibited by 20 mM s-butylamine though tricarboxylic acid cycle enzymes were hardly affected.

(d) s-butylamine inhibited the isolated enzyme in a competitive fashion, with a K_i of 13.8 mM (cf. K_m of 0.228 mM).*

These kinetic constants show that s-butylamine is a purely reversible inhibitor binding less well than the substrate, and at first sight it seems unlikely that pyruvate dehydrogenase is the site of action. However Eckert et al. (1975) had previously shown that the fungus could concentrate s-butylamine from 0.5 mM in the medium to 15 mM (assuming uniform distribution throughout the cell) in 30 minutes. It is possible that mitochondria also concentrate the compound, so an inhibitory concentration within the mitochondrial matrix cannot be ruled out. In this

---

*The dissociation constant (K_i) between any inhibitor (I) and a receptor such as an enzyme (E) which associate reversibly as follows:

$$E + I \rightleftharpoons EI$$

is given by the expression

$$K_i = \frac{[E][I]}{[EI]}$$

The square brackets denote concentrations, so that K_i is expressed as a concentration; K_i measures the extent of dissociation and will be large when dissociation is high, and small when dissociation is low. Therefore compounds that bind very tightly have low dissociation constants. In practice a dissociation constant of 1 μM or less indicates tight binding.

For the simple enzyme system:

$$E + S \underset{k_{-1}}{\overset{k_1}{\rightleftharpoons}} ES \overset{k_2}{\to} E + P$$

K_m is defined as $\dfrac{k_{-1} + k_2}{k_1}$ (the Michaelis-Menten constant)

and is also equal to the substrate concentration which gives half-maximal velocity of the overall reaction. It is therefore a measure of the affinity of the enzyme for its substrate. It is not the same as K_s, the dissociation constant for the ES complex (analogous to K_i) unless $k_{-1} \gg k_2$.

case α-oxoglutarate dehydrogenase was probably not affected since mito-chondrial oxidation of the tricarboxylic acid cycle substrates citrate and isocitrate was unimpaired.

Secondary-butylamine is highly specific in that certain *Penicillium* species are unaffected by it. The specificity is not due to lack of uptake since insensitive species also accumulate *s*-butylamine and related amines. (Eckert *et al.*, 1975).

D. Acetyl CoA formation from fats

A unique mode of action, interference with fat oxidation, has been suggested for the experimental miticide hexadecyl cyclopropanecarboxylate (cyclo-prate). The oxidation of fatty acids as fuel takes place inside the mito-chondria, and is thus separated physically from the cytoplasmic synthesis of the fatty acids that the cell requires. The mitochondrial inner membrane is, however, impermeable to fatty acids, and they are conveyed across the membrane as an ester of the carrier molecule carnitine:

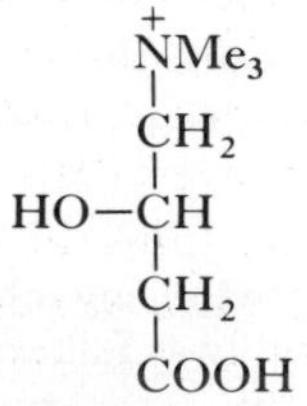

$$\overset{+}{N}Me_3$$
$$|$$
$$CH_2$$
$$|$$
$$HO-CH$$
$$|$$
$$CH_2$$
$$|$$
$$COOH$$

On the basis of metabolism work in mammalian tissues it was suggested (Quistad *et al.*, 1979) that cycloprate:

$$\triangleright\!-COOC_{16}H_{33}$$

which selectively controls phytophagous mites, owes its activity to the complete sequestration of carnitine following hydrolysis to cyclopropane carboxylic acid.

cycloprate $\xrightarrow{\text{hydrolysis}}$ $\triangleright\!-COOH$ $\xrightarrow{+ \text{ carnitine}}$ $\triangleright\!-CO-O-\overset{\displaystyle CH_2}{\underset{\displaystyle CH_2}{CH}}$

$$\overset{+}{N}Me_3$$
$$|$$
$$CH_2$$
$$|$$
$$\triangleright\!-CO-O-CH$$
$$|$$
$$CH_2$$
$$|$$
$$COOH$$

cyclopropane
carboxylic acid

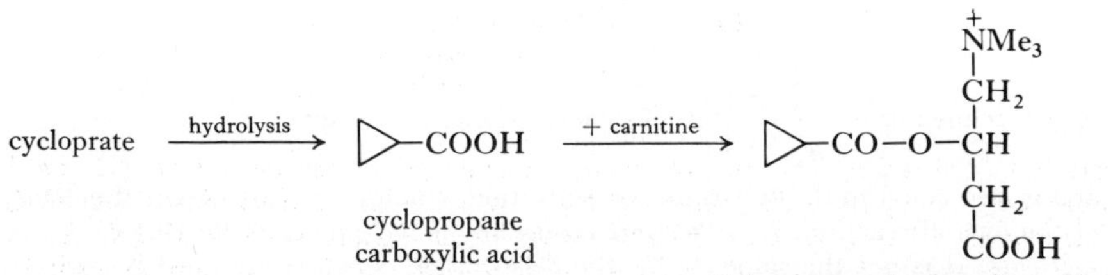

O-(cyclopropylcarbonyl)carnitine

Further study showed that O-(cyclopropylcarbonyl)carnitine is a significant metabolite of the acaricide in mite eggs, but not in other stages which are less susceptible to the compound. The authors are careful to point out that this evidence is somewhat circumstantial and they do not rule out other detrimental effects of cyclopropane carboxylic acid.

III. RESPIRATORY CHAIN PHOSPHORYLATION
A. Biochemical background
1. *Electron transport*

During a number of steps in glycolysis and the tricarboxylic acid cycle the cofactor NAD$^+$ becomes reduced to NADH. The flavoprotein component (flavin adenine dinucleotide; FAD) of the tricarboxylic acid cycle enzyme succinate dehydrogenase complex (EC 1.3.99.1) is also reduced, forming FADH$_2$. These two cofactors are re-oxidised by relinquishing their electrons through a chain of respiratory carriers, each of successively lower redox potential (p. 51), and the electrons are passed eventually to oxygen, reducing it to H$_2$O. A schematic drawing of the carriers is shown in Fig. 1.4. Many of them are haem-containing cytochromes, but non-haem iron proteins are also involved. Since it is difficult to ascertain the exact redox potential of carriers within the mitochondrial inner membrane the sequence of those which have closely similar redox potentials as measured *in vitro* cannot be certain. However the scheme in Fig. 1.4 would be acceptable to most workers.

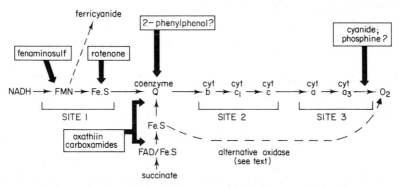

Fig. 1.4 The respiratory electron transport chain. The bracketed sites 1, 2 and 3 represent the steps which are thought to provide the driving force for ATP formation. NADH, reduced nicotinamide adenine dinucleotide; FMN, flavin mononucleotide; FAD, flavin adenine dinucleotide; Fe.S, iron-sulphur protein (non-haem); cyt, cytochrome

2. *Oxidative phosphorylation*

During the flow of electrons along the respiratory electron transport chain, energy is trapped by the synthesis of ATP from ADP and inorganic phosphate. The elucidation of the mechanism of the coupling of electron transport to this oxidative phosphorylation of ADP continues to occupy many research biochemists. Ideas proposed have been variants on one of three theories, which contend that, after release during a redox reaction and before use for ATP synthesis the energy is stored as either (*a*) a 'high-energy' chemical intermediate (there is a precedent for this since high-energy compounds occur in glycolysis and are capable of driving substrate-level phosphorylation); (*b*) a conformational change in the membrane (analogous to squashing a spring), or (*c*) a proton electrochemical potential generated by the movement of protons across the inner mitochondrial membrane and a measure of the extent to which the proton gradient across the membrane is removed from equilibrium (Boyer *et al.*, 1977; Nicholls, 1982).

Most biochemists now favour mechanism (*c*), the 'chemiosmotic' theory first proposed by Mitchell (1961). Stated simply, this is that, during the passage of electrons down the chain of respiratory carriers, H^+ ions are transported across the membrane, from inside to out, thus leading to the generation of a proton electrochemical potential which can drive ATP synthesis via an ATPase enzyme spanning the inner membrane. (The ATPase is so called because it can hydrolyse ATP in the reverse direction.) Supporters of other theories do not dispute that a proton gradient occurs, but regard it not as the primary event, but as an inevitable consequence of some other primary process.

The development and current aspects of the chemiosmotic theory are considered in a monograph by Nicholls (1982).

3. *Measurement of respiratory chain electron transport with the oxygen electrode*

Electron transport and oxidative phosphorylation are conveniently measured by suspending isolated mitochondria in an oxygen electrode, which continuously determines the concentration of oxygen in solution. The output from the electrode is coupled to a recorder to give traces like the ones shown in Fig. 1.5. Fig. 1.5*a* shows that intact mitochondria incubated with an oxidisable substrate take up oxygen only slowly until ADP is added. Now that the mitochondria have a substrate to phosphorylate, oxygen uptake is rapid until all the ADP has been converted to ATP, when the oxygen consumption falls again. For convenient reference in the literature, mitochondria consuming small amounts of oxygen in the absence

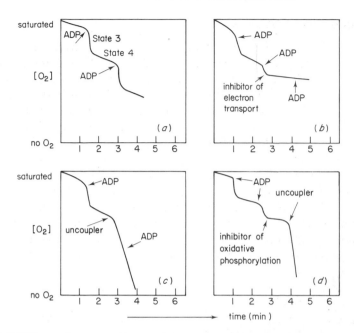

Fig. 1.5 Effects of compounds interfering with respiratory chain phosphorylation (a) pulses of oxygen uptake stimulated by ADP in coupled mitochondria (b) inhibition of oxygen uptake by an electron transport inhibitor (c) removal of respiratory control by an uncoupler (d) inhibition of oxidative phosphorylation and relief by an uncoupler. In all cases oxygen consumption can only be stimulated by ADP if there is sufficient inorganic phosphate present

of ADP are said to be in 'State 4', while in the presence of ADP the mitochondria are in 'State 3'. Mitochondria responding as shown in Fig. 1.5a are said to have 'respiratory control' or to be 'coupled', since electron transport only occurs when ADP is present (the slow background oxygen consumption is due to damaged mitochondria and can be ignored).

If an *inhibitor of electron transport* is added, electrons cannot be passed down the respiratory chain to oxygen, and oxygen consumption ceases (Fig. 1.5b). If, instead, an *uncoupler* of oxidative phosphorylation is added, the effect is to remove the respiratory control and allow maximal oxygen uptake whether or not ADP is added (Fig. 1.5c). Finally an *inhibitor of oxidative phosphorylation* (but not electron transport) will prevent oxygen uptake by coupled mitochondria, but once they are uncoupled, electron transport, and hence oxygen uptake, can proceed unimpaired (Fig. 1.5d).

Pesticides are known to act as inhibitors of electron transport and oxidative phosphorylation, and as uncouplers of oxidative phosphorylation. In the first case one would like to know the precise sites of interaction with the electron transport chain and to be convinced that compounds interfere at sufficiently low concentrations for this to be a plausible mode of action *in vivo*. In the case of uncouplers one usually measures the minimum uncoupling concentration, i.e. the concentration required to just release the mitochondria from respiratory control so that subsequent addition of ADP has no effect on the rate of oxygen uptake. An alternative, used particularly with manometric methods for measuring oxygen uptake employed in earlier studies, is to express the results as the concentration required to bring about 50% uncoupling. Measures of uncoupling may vary because of differences in the amount and quality of mitochondria used. These difficulties should be borne in mind when we later attempt to compare data from the literature.

B. Action of pesticides
1. *On electron transport*

A variety of pesticides has been shown to interfere with electron transport. For clarity it will be preferable to take them in order of the components in the electron transfer chain with which they are proposed to interfere (Fig. 1.4). The succinate dehydrogenase side branch will be left until the end.

(*a*) *Fenaminosulf*

Tolmsoff (1962) originally showed that the fungicide fenaminosulf:

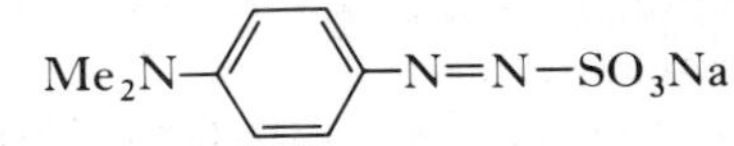

$$Me_2N-\langle\!\!\!\bigcirc\!\!\!\rangle-N{=}N{-}SO_3Na$$

was an 'effective inhibitor' of NADH but not succinate oxidation in the fungus *Pythium*. This suggested that the site of action was in the NADH dehydrogenase complex (EC 1.6.99.3) of the respiratory chain, i.e. that part before the branch point at coenzyme Q (also known as ubiquinone) (Fig. 1.4). This conclusion has been confirmed and extended by Schewe and colleagues working with electron-transporting membrane vesicles from the inner membrane of beef heart mitochondria, in which NADH (but not succinate) oxidation was inhibited with an I_{50}* of 1.4 μM (Halangk and Schewe, 1975; Schewe *et al.*, 1975). In the presence of NADH,

---

*I_{50}; I_x is the concentration of inhibitor that inhibits a process by x%. Related terms are also used in evaluating the effects of chemicals:
    LC_x; the concentration of a compound required to kill x% of the organisms under test.
    LD_x; the dose of compound required to kill x% of the organisms under test.

electrons can be passed directly from the FMN (flavin mononucleotide)-containing flavoprotein of the complex to ferricyanide (Fig. 1.4) and this activity was found to be inhibited by fenaminosulf (but not rotenone – see below) (Halangk and Schewe, 1975). Subsequent results with this NADH-ferricyanide redox system have shown that the inhibition is time dependent and irreversible, is enhanced by pre-incubation with NADH, takes place without any change being detectable in the FMN spectrum after acid extraction, and is decreased by both NAD^+ and FMN. The compound does not react chemically with NADH or FMN (Müller and Schewe, 1977). Together this evidence suggests that the principal site of action of fenaminosulf is the flavin-reducing part of NADH dehydrogenase. It has been suggested that fenaminosulf could form an acid-labile complex with FMN at the active site of the enzyme (Schewe and Müller, 1979).

Much of this work on the mode of action of a fungicide has been done using mammalian mitochondria, since it is much easier to make good mitochondria from mammalian tissues than, for example, from plant or fungal sources which have tougher cell walls that are only broken under conditions which tend to damage mitochondrial integrity. Although it is obviously preferable to use mitochondria from an appropriate source where possible, it is likely that mitochondria are sufficiently similar from different sources to allow extrapolation of modes of action.

(b) Rotenone

Information on the insecticidal use of rotenone was reviewed by O'Brien (1967), while Yamamoto (1970) and Storey (1981) have summarised information on its biochemical mode of action.

As demonstrated by competition experiments (Horgan et al., 1968) rotenone acts at the same site between NADH and coenzyme Q as piericidin A, an insecticidal compound produced by fungi (Takahashi et al., 1965; Yoshida et al., 1980). Rotenone reduced oxygen consumption by rat liver mitochondria to the extent of 95% at 0.6 μM (Lindahl and Oberg, 1960). Piericidin A binds and inhibits on the oxygen side of FMN, possibily, as shown in Fig. 1.4, to the iron-sulphur protein (Van Dam and Meyer, 1971).

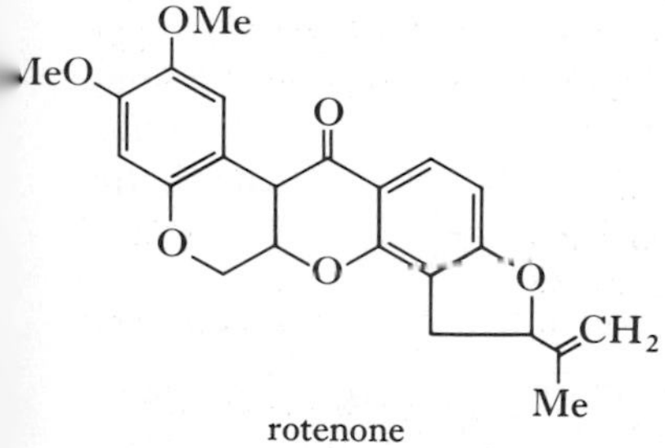

rotenone 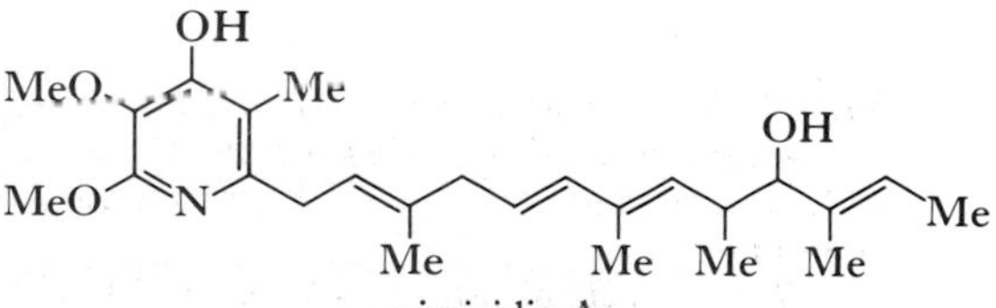   piericidin A

It should be noted that both of these compounds, but particularly rotenone, will interact with other components of the respiratory chain (see Horgan *et al.*, 1968) though they do not bind to these so tightly.

The symptoms of insects poisoned with rotenone differ from those produced by insecticides that act on the nervous system (Chapters 3 and 4), and include slowing of the heart beat, depression of respiratory movement, reduction in oxygen uptake, and an eventual limp paralysis (Brown, 1963).

Although these symptoms seem to fit well with a site of action in electron transport, recent work (Marshall and Himes, 1978) indicates that rotenone may also have an effect on microtubule assembly. Microtubules have various cellular roles and are assembled when required from protein subunits termed tubulin. Their most studied role is in mitotic spindle formation during cell division. We shall return to this in more detail in Chapter 6. Since rotenone inhibited tubulin self-assembly *in vitro* at concentrations in the range 2–10 μM (Marshall and Himes, 1978) this action may contribute to its toxic role *in vivo*.

(c) 2-Phenylphenol

This disinfectant and fungicide was found to inhibit respiration in the bacterium *Rhodospirillum rubrum* by acting on the respiratory chain in the region of coenzyme Q (Oelze and Kamen, 1975). In a subsequent paper

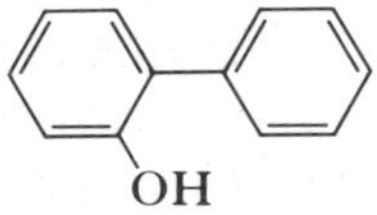

2-phenylphenol

OH

Oelze *et al.* (1978) found that NADH oxidation by membrane fractions from *R. rubrum* were half-inhibited by 80 μM 2-phenylphenol, though 360 μM compound was needed for the same degree of inhibition of succinate oxidation. On the basis that NADH and succinate oxidation contribute 70% and 30% respectively to the total oxidation capacity of the membranes, Oelze *et al.* (1978) calculated that 50% inhibition of cellular oxygen uptake would result from 130 μM compound simultaneously inhibiting both oxidations; this value is rather high to be certain that this is the primary mode of action. 2-Phenylphenol can inhibit carotenoid synthesis (see, e.g., Bucholtz *et al.*, 1977) but effects on this process could clearly only contribute to toxic effects in the case of photosynthetic organisms.

(d) Tridemorph

Tridemorph is an eradicant, systemic fungicide for use on cereal mildews (Worthing, 1979). The structure is given in Table 6.4. Current evidence no longer favours inhibition of electron transport as its mode of action, but the development of the story is recounted below as a cautionary tale.

Bergmann *et al.* (1975) found that *Torulopsis candida* respiration was 30% inhibited by 100 μM tridemorph, and Müller and Schewe (1976) conducted experiments with beef heart sub-mitochondrial particles which led to the conclusion that the site of action was, like rotenone, between FMN and coenzyme Q. The I_{50} at this site was 3.4 μM. They also found inhibition further down the electron transport chain in the region of site 2 (Fig. 1.4), though this was less sensitive with an I_{50} of 24 μM. Some artificial electron by-pass reactions were inhibited as well.

One might have expected more than 30% inhibition of fungal respiration by 100 μM tridemorph if its principal action was on a system to which it bound with an I_{50} in the micromolar region. The explanation could have been a species difference, but work by Kerkenaar and Kaars Sijpesteijn (1979) has thrown some light on the matter. These workers showed that unformulated tridemorph, even at high concentrations, had no effect on the respiration of several fungi. A surfactant used in the formulation of the commercial product did however inhibit respiration, in a manner which turned out to be synergised by tridemorph. Even so, inhibition of fungal growth could occur at concentrations of formulated product which did not affect respiration.

Hence we should not regard tridemorph as an inhibitor of respiration; it is in fact thought to interfere with sterol biosynthesis and it will therefore be discussed in Chapter 6. The work of Kerkenaar and Kaars Sijpesteijn (1979) serves as a reminder that pure active ingredient, rather than formulated product, should be investigated where possible.

(e) Dibutylchloromethyl tin (DBCT)

This compound is not used as a pesticide itself but is related to the trialkyl

Bu
\
Bu—SnCl   dibutylchloromethyl tin
/
ClCH$_2$

tins which act principally by inhibition of oxidative phosphorylation (see p. 40). However Moore *et al.* (1979) have published data indicating that DBCT behaves as an electron transport inhibitor in mung bean mitochondria, causing approximately 80% inhibition of, for example, succinate oxidation at 0.2 μM. Since uncoupled electron transport was equally sensitive this cannot be ascribed to inhibition of the phosphorylation process. The properties of the inhibition were consistent with action at the coenzyme Q level, possibly by inhibiting the reoxidation of a portion of the coenzyme Q pool by the cytochrome b-c_1 complex (Fig. 1.4).

As we shall see later, the evidence that the trialkyl tins act by inhibition of the ATPase responsible for ATP synthesis is good. However the work of Moore et al. (1979) suggests that this may not be the complete explanation.

(f) Cyanide

Hydrogen cyanide (HCN) and calcium cyanide ($Ca(CN)_2$), which decomposes to HCN in the presence of water, are used to control insects in enclosed areas by fumigation (Worthing, 1979).

Cytochrome oxidase (EC 1.9.3.1; cyt a/a_3) catalyses the terminal step in the electron transport chain, the oxidation of reduced cytochrome c by oxygen (Fig. 1.4), and it is reversibly inhibited to the extent of 50% by 0.01 μM cyanide (Dixon and Webb, 1964). Since cyanide combines readily with heavy metals to form stable complexes it is thought to inhibit cytochrome oxidase by combining with essential iron and copper atoms which occur in the enzyme and the kinetics of this interaction in intact mitochondria have been investigated (Wilson et al., 1972). The detailed nature of the inhibition varies with the experimental conditions and the redox state of the cytochrome oxidase (see Erecińska and Wilson, 1981).

Cyanide is not a specific poison, and Dixon and Webb (1964) list forty enzymes that are inhibited by it, though cytochrome oxidase is by far the most sensitive. Since cytochrome oxidase, not to mention the other less susceptible enzymes, is of widespread occurrence, it is not surprising that cyanide is generally toxic.

Hydrogen cyanide reduces the respiratory movements of insects, and cytochrome oxidase has been shown to be inhibited in insects treated in vivo (Brown, 1963).

(g) Phosphine (PH_3)

Aluminium phosphide is a compound which decomposes when moist to give phosphine, PH_3. This gas is spontaneously flammable but the product can be formulated so that other gases are also evolved in a non-flammable mixture. Phosphine is highly toxic to mammals but its toxicity to insects and its volatility make it suitable for use as a fumigant for stored grain.

The precise reasons for the toxicity of phosphine, in particular the fact that it is toxic only in the presence of oxygen (Bond et al., 1967) have been explored in detail by Chefurka et al. (1976) who examined its effects on mitochondria from mouse liver, housefly and granary weevil. The oxygen electrode was used to measure the activities of portions of the respiratory chain functionally isolated by the use of electron donors, acceptors and inhibitors of known properties.

The simplest conclusion from these experiments is that phosphine is a non-competitive inhibitor of cytochrome oxidase. K_i values were determined as 16 μM (mouse liver), 280 μM (granary weevil) and 72 μM (beef heart). In general these results were in proportion with the amount of inhibitor required to effectively inhibit mitochondrial respiration (I_{50} about 200 μM for state 3 respiration of mouse liver mitochondria). These concentrations are high for the interaction of a pesticide with its primary target, but may be realistic for a fumigant. No other more sensitive biochemical target has been identified.

(h) Oxathiin carboxanilides and related compounds (Table 1.3)

These compounds are systemic fungicides useful against basidiomycetes. Carboxin was the first to be subjected to detailed biochemical examination. Early work revealed that it inhibited oxidation of glucose and, to a greater extent, acetate by intact fungi (Mathre, 1970; Ragsdale and Sisler, 1970). Work with mitochondria from the sensitive fungus *Ustilago maydis* showed that NADH oxidation was unaffected, but that succinate oxidation was inhibited with a K_i in the region of 0.3–0.5 μM; mitochondria from other sources, including a non-susceptible fungus, were less affected (Mathre, 1971; White, 1971). Together with the finding that succinate accumulates in carboxin-treated cells of fungi (Mathre, 1970), this evidence strongly suggests that inhibition of succinate oxidation is the primary site of action of carboxin.

The enzyme succinate dehydrogenase occurs as part of a respiratory chain complex known as 'succinate dehydrogenase complex' or 'complex II'. It uses a flavoprotein (FAD) rather than a pyridine nucleotide as electron acceptor and also contains non-haem iron-sulphur proteins which mediate the transfer of electrons from reduced FAD ($FADH_2$) to coenzyme Q. In addition there are associated peptides and lipids. Since the earlier experiments referred to above, more recent work has confirmed that the site of action of the oxathiin carboxanilides is between FAD and coenzyme Q (Fig. 1.4).

Much of the work has been done on beef heart mitochondria; this is justified by the studies of Müller *et al.* (1977) who found that succinate-cytochrome c reductase of beef heart electron-transport particles had properties resembling succinate dehydrogenase of the carboxin-sensitive fungus *Trametes versicolor*, but not of the tolerant *Trichoderma viride*. However it should be noted that results vary somewhat with the type of preparation, the electron acceptor used in the *in vitro* system, and the carboxanilide tested (Coles *et al.*, 1978).

Table 1.3   Oxathiin carboxanilides and a related compound which act by inhibition of succinate oxidation

Name	Structure	$I_{50}(\mu M)$*
Carboxin	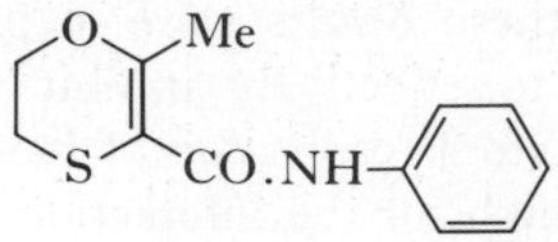	0.5
Oxycarboxin	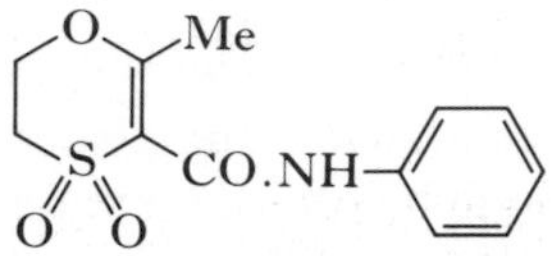	8.0
Pyracarbolid	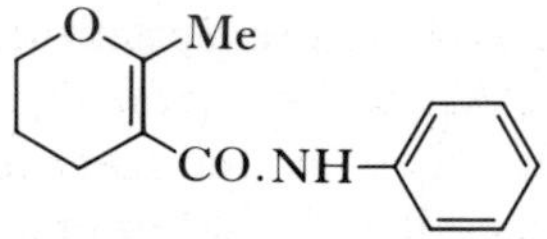	5.5
Fenfuram	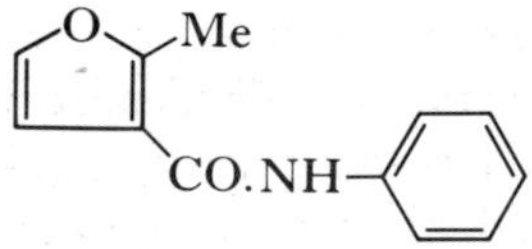	4.2
Benodanil	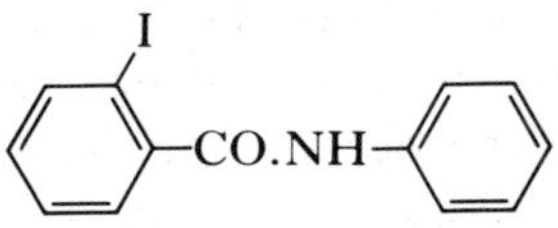	7.5
3'-Isopropoxy-2-methylbenzanilide (Mepronil proposed)	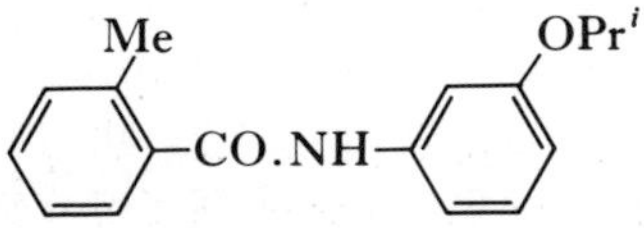	

*I_{50} for inhibition of succinate oxidation by *Ustilago maydis* mitochondria, using 2,6-dichlorophenolindophenol as electron acceptor (White and Thorn, 1975).

Although succinate oxidation is sensitive to carboxin *in situ*, solubilised succinate dehydrogenase is not inhibited by the fungicide, showing that other components present in the complex are required for inhibition to occur. This was shown in a study in which a complex II preparation exhibiting succinate-coenzyme Q reductase activity was isolated from beef heart mitochondria. The activity was sensitive, even under non-illuminating conditions, to inhibition by the photoreactive* compound azido-carboxin. The chemical also carried a general tritium label (Ramsay *et al.*, 1981).

azido-carboxin

When succinate dehydrogenase was extracted from complex II it was insensitive to azido-carboxin, which did not bind to any of the complex II components that had been removed. However, when the system was reconstituted by recombining the components, it regained sensitivity to azido-carboxin. When the inhibitor was added to complex II and irradiated to stimulate covalent bond formation, most of the specifically bound inhibitor was found in the low molecular mass peptides and phospholipid components of the complex. The binding of carboxin therefore requires the presence of other components of the complex as well as succinate dehydrogenase itself (Ramsay *et al.*, 1981).

A noteworthy point is that many fungi possess an 'alternative oxidase' by which electrons can be passed directly from the last iron-sulphur protein of complex II to oxygen (Fig. 1.4). This pathway can drive the synthesis of only a small amount of ATP, and it is not clear whether it has an important role (e.g. in heat production) or whether it represents an 'evolutionary relic' (see Lyr, 1977). However, it is sensitive to inhibition by carboxin (Sherald and Sisler, 1972; Lyr, 1977).

A range of carboxin analogues has been tested as inhibitors of succinate oxidation in electron-transfer particles (using phenazine methosulphate as electron acceptor), with both I_{50} and K_i values emerging in the 0.02–3 μM range (Mowery *et al.*, 1977). A much wider range of compounds was examined by White and Thorn (1975) who conducted a structure-activity study correlating growth inhibition of fungal cultures with inhibition of

---

*Photoreactive compounds are those containing groups which can be converted to a reactive species by light. Azides (R–N_3), which are the most commonly used compounds with this property, decompose on irradiation with ultraviolet or visible light to yield nitrogen gas and a highly reactive nitrene (R–$\overset{..}{N}$). Compounds with photoreactive groups are known as 'photoaffinity labels'.

succinate oxidation in mitochondria from two sensitive fungi (*Ustilago maydis* and *Cryptococcus laurentii*). These studies included most of the compounds shown in Table 1.3, where the results are given. With few exceptions (out of 93 compounds) the *in vitro/in vivo* activity correlations were good. All the I_{50} values in Table 1.3 are less than 10 μM and it therefore seems safe to conclude that inhibition of the succinate dehydrogenase complex is the principal biochemical site of action of these fungicides.

White and Thorn (1975) also considered the basic structure required for activity and were able to broadly support requirements for *in vivo* activity suggested earlier (ten Haken and Dunn, 1971):

$$\begin{array}{c} R \diagdown \quad Me \\ \diagup C = C \diagdown \\ R' \quad CONH - \langle \text{benzene ring} \rangle - X_n \end{array}$$

White and Thorn found that active inhibitors of succinate oxidation in mitochondria were amides of α,β-unsaturated carboxylic acids where the R and R' substituents on the double bond were part of a benzene ring or linked in a structure in which the double bond was adjacent to a nitrogen, oxygen or sulphur atom. The group on the amide nitrogen did not have to be phenyl (cyclohexyl was satisfactory), but compounds with alkyl groups were inactive. If the group was phenyl, various substituents (X_n) were possible. The carbonyl group was essential and an H atom on the amide nitrogen was highly desirable. Methyl could be replaced in certain conservative modifications, with the size of the group seeming more important than its electronic properties; I was the best, followed by Et > Br > Me > Cl > OH > F = H.

Further detailed structure-activity correlations for a series of related thiophene carboxamides are reported by White and Thorn (1980).

Also on the basis of structure-activity work, Lyr and colleagues (Lyr, 1977; Schewe *et al.*, 1979) have put forward a model of the oxathiin carboxanilide binding site in which two molecules of, for example, carboxin were considered to complex with the iron and sulphur atoms at the centre of the Fe.S protein. The methyl and phenyl groups were proposed to participate in hydrophobic interactions with the protein.

Genetic and resistance studies have also supported succinate dehydrogenase complex as the site of action of the oxathiin carboxanilides. Correlation between negligible activity of the complex and oxathiin resistance was noticed first by Georgopoulos and colleagues (Georgopoulos *et al.*, 1972; Georgopoulos and Vomvoyanni, 1972) and resistance to carboxin has more recently been ascribed to changes in the enzyme protein itself in *Aspergillus* (Gunatilleke *et al.*, 1976) and in both haploid and diploid strains of *Ustilago maydis* (Georgopoulos *et al.*, 1975). In addition, altered enzyme was partly responsible for carboxin-resistance in the laboratory-induced mutants from *U. hordei* investigated by Ben-Yephet *et al.* (1975).

White *et al.* (1978) and White and Thorn (1980) have shown that certain oxathiin and thiophene carboxamide fungicides are highly active against carboxin-insensitive succinate dehydrogenase complexes from carboxin-resistant fungal mutants. The implication of this work is that small structural modifications to the enzyme are involved in the development of resistance, and this simultaneously confers sensitivity to compounds of related structure.

(i) Miscellaneous compounds

Although the dinitroaniline herbicides will be dealt with in Chapter 5 as inhibitors of cell division, Moreland *et al.* (1972a) found that these compounds interfered with mitochondrial respiration as well as with photosynthesis at concentrations up to 100 μM. However, studies on respiration in tissue pieces showed that respiration was only inhibited to a large extent in areas where there was extensive cell division (Moreland *et al.*, 1972b). Hence cell division remains the most likely primary target of the dinitroanilines.

2. Uncouplers of oxidative phosphorylation

(a) Background

An introduction to oxidative phosphorylation and the methods by which it can be investigated has been given earlier (p. 18). An uncoupler of oxidative phosphorylation allows electron transport to proceed without ATP synthesis; indeed, relieved of the load of oxidative phosphorylation, the electron transport will normally proceed faster in the presence of an uncoupler.

Table 1.4 lists all the currently used pesticides that are thought to act by uncoupling oxidative phosphorylation. Comparison of the structures shows that all the compounds are, or could easily give rise to, lipid soluble acids. According to the chemical-coupling hypothesis such compounds act by catalysing general acid or base hydrolysis of hypothetical high-energy-containing chemical intermediates (Wilson *et al.*, 1971). On the chemiosmotic theory lipid soluble acids uncouple mitochondria by destroying the proton electrochemical potential produced during electron transport and required for the formation of ATP (Mitchell, 1966; Mitchell and Moyle, 1967). The protonated acid enters the mitochondrion and dissociates. Since the resulting anion is able to delocalise its charge over the aromatic nucleus, it is deemed able to return across the membrane, where it can pick up another proton and repeat the cycle. A minor variant

Table 1.4 Pesticides thought to act by uncoupling oxidative phosphorylation

Name and use	Structure

A. Dinitrophenol derivatives

Binapacryl
(acaricide, fungicide)

O_2N — benzene ring with O.CO.CH=CMe$_2$, CH(Et)(Me), NO$_2$ substituents

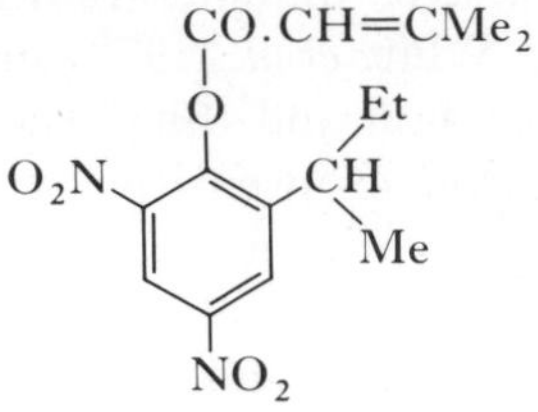

Dinobuton
(acaricide, fungicide)

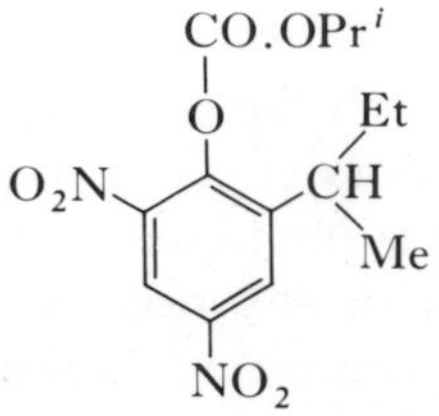

O_2N — benzene ring with O.CO.OPr^i, CH(Et)(Me), NO$_2$ substituents

Dinocap
(mixture of compounds,
where R is
1-methylheptyl,
1-ethylhexyl,
1-propylpentyl)
(acaricide, fungicide)

O_2N — benzene ring with O.CO.CH=CHMe, R, NO$_2$ substituents; and O_2N — benzene ring with O.CO.CH=CHMe, NO$_2$, R substituents

Dinoseb
(herbicide)

O_2N — benzene ring with OH, CH(Et)(Me), NO$_2$ substituents

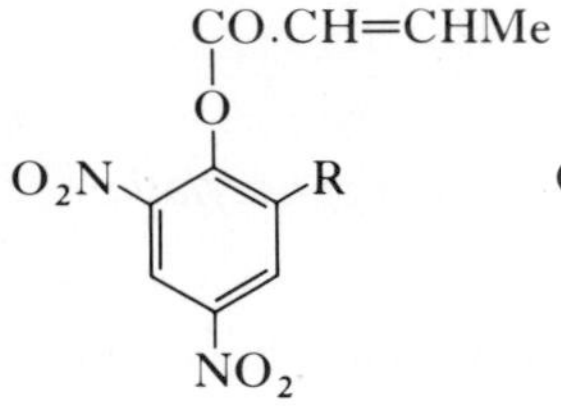

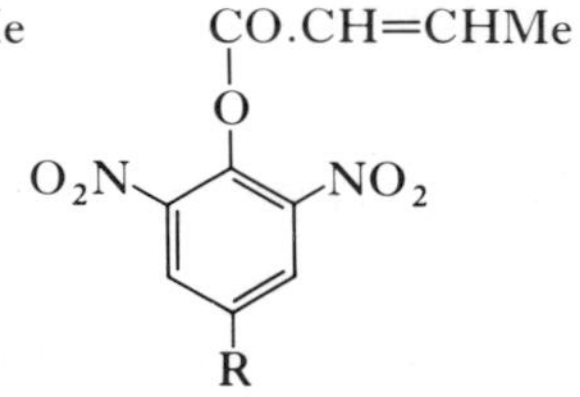

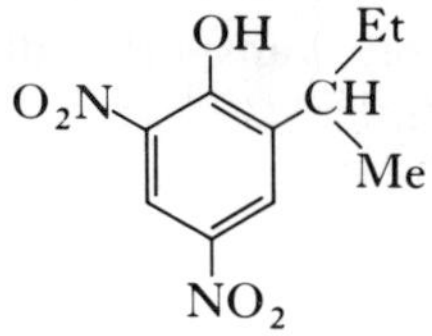

Dinoseb acetate
(herbicide)

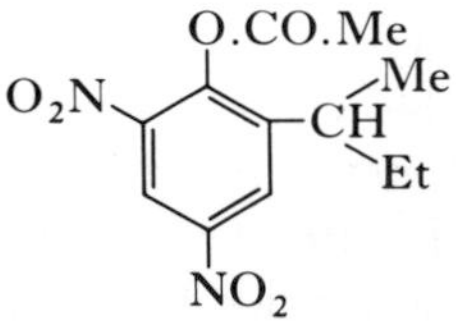

O_2N — benzene ring with O.CO.Me, CH(Me)(Et), NO$_2$ substituents

Name and use	Structure
Dinoterb (herbicide)	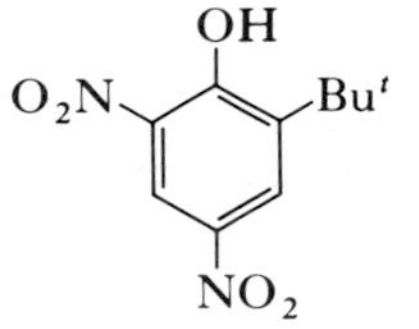
DNOC (insecticide, herbicide)	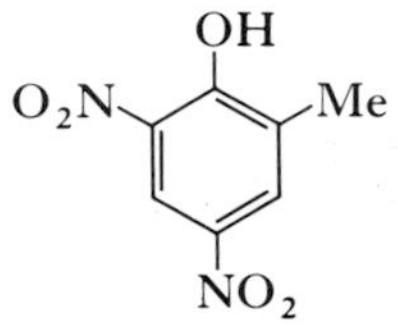

B. *Hydroxybenzonitriles*

Bromoxynil (herbicide)	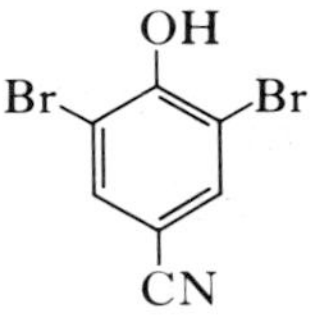
Bromoxynil octanoate (herbicide)	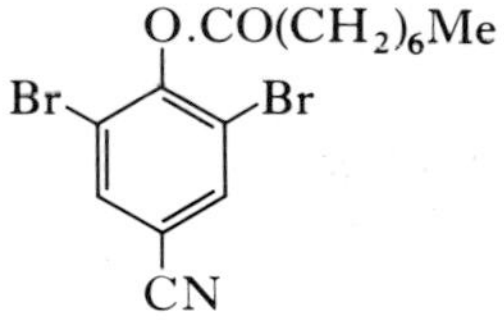
Ioxynil (herbicide)	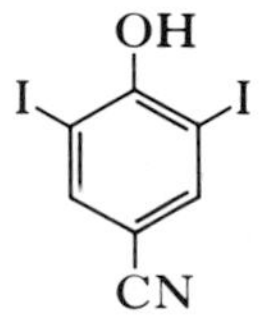

Name and use	Structure

Ioxynil octanoate
(herbicide)

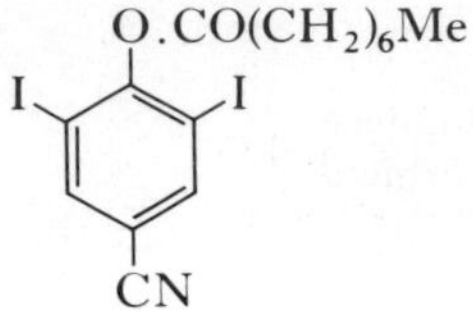

$O.CO(CH_2)_6Me$ with ring bearing two I substituents and CN

C. Miscellaneous compounds

Bromofenoxim
(herbicide)

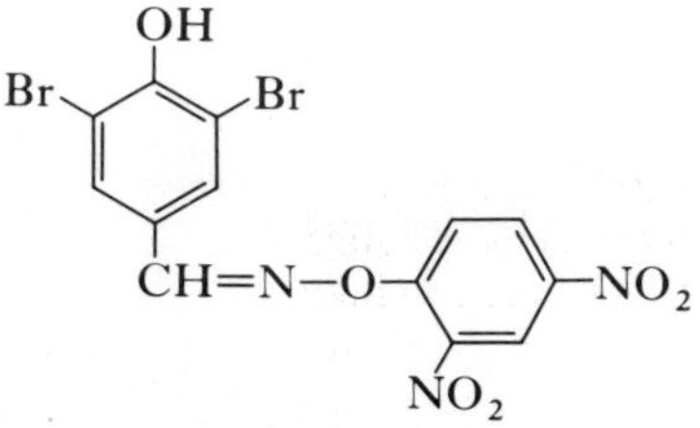

OH, Br, Br ring with $CH=N-O-$ ring bearing NO_2 and NO_2

Dichlorophen
(fungicide, bactericide)

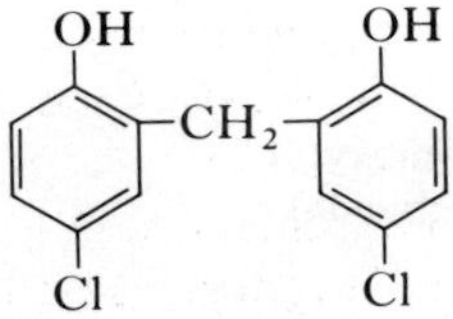

Two OH, Cl-substituted rings joined by CH_2

Drazoxolon
(fungicide)

Me, $N-N-H$ ring with Cl, and isoxazolone ($N-O-$, O)

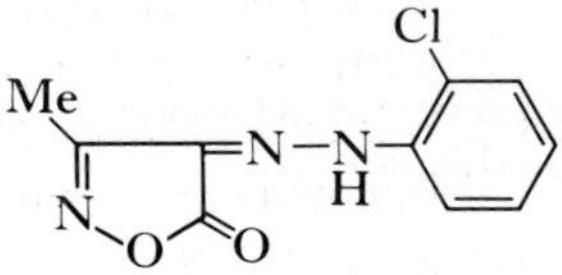

Malonoben
(acaricide)

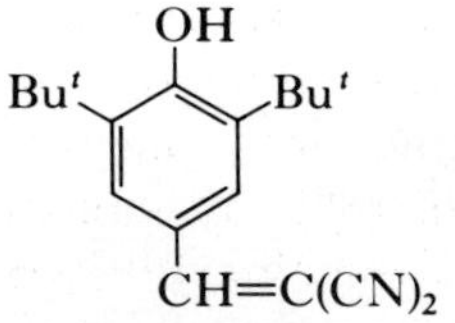

OH, Bu^t, Bu^t ring with $CH=C(CN)_2$

Name and use	Structure

Niclosamide
(molluscicide)

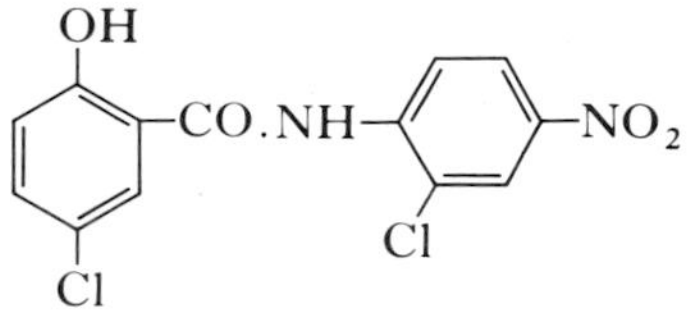

Pentachlorophenol
(insecticide, fungicide,
defoliant, herbicide)

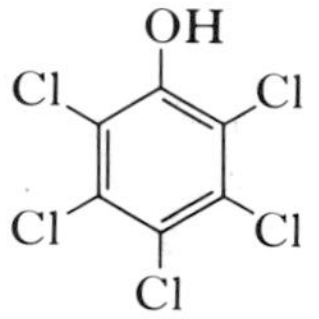

has the anion returning across the membrane in complex with a cation so that the complex has no net charge (Kessler *et al.*, 1977; Green and Vande Zande, 1981).

The chemiosmotic explanation is promulgated by most general biochemistry text books on the basis of a body of work correlating uncoupling activity with lipophilicity. It has been supported more recently by Cunarro and Weiner (1975), who found a close correlation between uncoupling activity in respiring mitochondria and ability to promote H^+ transport into mitochondria. Consistent results have also been obtained with artificial membrane preparations, but the results have been disputed (see Cunarro and Weiner, 1975, for original references).

This convenient and convincing view has been called into question recently, particularly by Hanstein and Hatefi (see, e.g., Hanstein, 1976a,b; Hatefi, 1980), who support a discrete uncoupler binding site, an idea which had arisen earlier (Williamson and Metcalf, 1967; Weinbach and Garbus, 1969). Hanstein and Hatefi (1974a) showed that 2,4,6-trinitrophenol (picric acid), unlike 2,4-dinitrophenol, did not uncouple intact mitochondria, but only caused uncoupling in sub-mitochondrial particles, which have their membranes turned inside out during preparation. Although this can be accommodated by the chemiosmotic theory (see McLaughlin, 1981; Hanstein and Kiehl, 1981), it raised the possibility of a binding site accessible from only one side of the membrane. This idea was also investigated in work with uncouplers modified by the inclusion of reactive (or photoreactive) groups which are able to covalently react with the sites to which they bind. This follows the principle of active-site-directed irreversible inhibitors of enzymes (Baker, 1967), also known as 'affinity labels'. Compounds which have been used in this way are collated in Table 1.5. As a result of this work an uncoupler binding site on the matrix side of the inner membrane is envisaged.

Table 1.5 Compounds used as affinity labels of a presumed uncoupler binding site

Structure	Reference
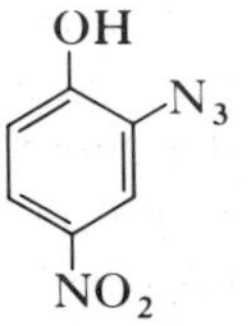	Hanstein (1976a,b)
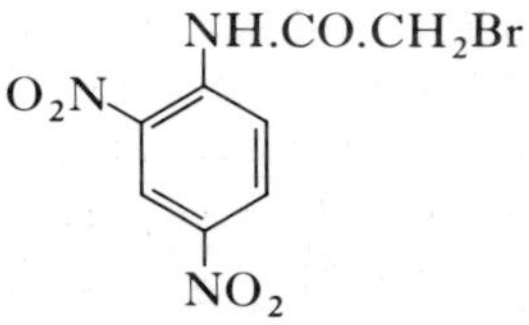	Chen *et al.* (1975)
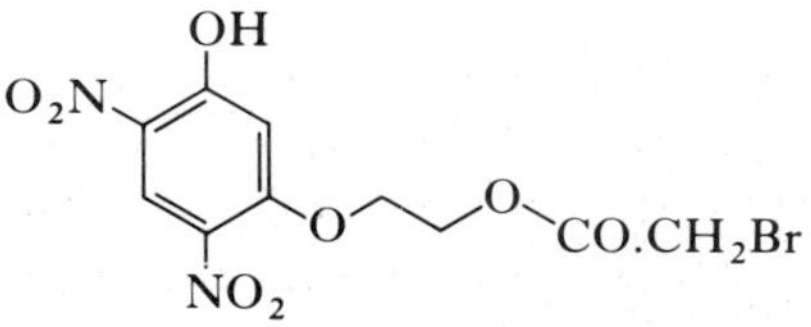	Wang *et al.* (1973)
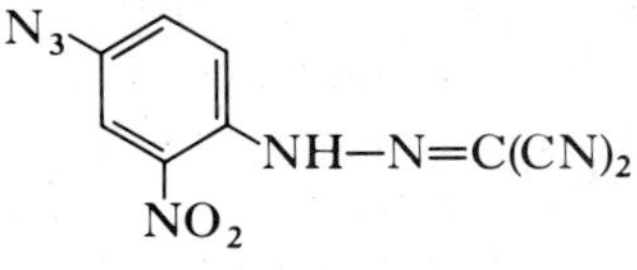	Katre and Wilson (1978)

The characteristics of this binding site were summarised by Hanstein (1976b).

(*i*) It is specific for uncouplers of the anionic aromatic type.

(*ii*) Binding of such compounds is unaffected by other types of uncoupler (e.g. ion-conducting compounds or arsenate).

(*iii*) It is independent of the energy-state of the mitochondria.

(*iv*) The number of binding sites (0.3–0.6 nmoles/mg protein) is comparable to that of other components of the respiratory chain.

(*v*) It is present in mitochondria from a wide variety of sources.

The binding of tritiated 2-azido-4-nitrophenol was studied under conditions where it was not rendered reactive by irradiation and therefore bound reversibly (Hanstein and Hatefi, 1974b) and the dissociation constant was found to be about 6 μM at 3°C. Other, unlabelled uncouplers of a similar type showed competitive inhibition with respect to 2-azido-4-nitrophenol, as demonstrated by double-reciprocal plots. This type of analysis is exactly analogous to that being applied to the binding site for inhibitors of photosynthetic electron transport as will be described in the next chapter.

The properties of the proteins with which covalent bonds were formed by compounds in Table 1.5 were not sufficiently similar to allow generalisation about their role. Hanstein (1976b) envisaged a binding site close to the subunit of the ATPase which carries out the phosphorylation of ADP. Indeed, a comparison of the properties of particular polypeptides of the ATPase complex with those of the proteins which become photo-labelled by 2-azido-4-nitrophenol does suggest that uncouplers bind to a component of the ATPase (Blondin, 1980; Hatefi, 1980). Katre and Wilson (1978) also propose an uncoupler binding site. They discuss previous work in the context of their results, including work by Draber et al. (1972) whose structure-activity correlations on a series of α-acyl-α-cyanocarbonyl-phenylhydrazones made as potential pesticides could not be reconciled solely on the basis of lipophilicity and acidity.

The existence of an uncoupler binding site is supported by work on mutants. Griffiths et al. (1974) mention, but do not discuss, an uncoupler-resistant mutant of yeast, and such a mutant from the bacterium Bacillus megaterium (Decker and Lang, 1977) is probably a single point mutation and may be due to altered ATPase (Guffanti et al., 1981). (Bacteria do not have mitochondria but they do carry out oxidative phosphorylation by a mechanism thought to be very similar to the one operating in higher organisms.) Further, of 21 strains of the bacterium Escherichia coli selected for resistance to tributyl tin (which is an ATPase inhibitor—see p. 40), 19 were also resistant to uncouplers of oxidative phosphorylation (Ito and Ohnishi, 1981). For this reason (inter alia) the authors conclude that these strains have a modified ATPase. Such a specific modification represents a more likely basis for resistance to uncouplers than wholesale alteration to the nature of the membrane, as seems to be required according to the chemiosmotic view.

In a useful summary of much of this work Criddle et al. (1979) accept the suggestion of a mode of uncoupling associated with the ATPase. However, they stress that this need not rule out important direct effects of uncouplers on membrane permeabilities. This seems a prudent conclusion to reach in a field with a history of changing ideas. It is not immediately reconcilable with the majority-supported chemiosmotic view of oxidative phosphorylation, though a suggestion as to how this may be done has been put forward by Phelps and Hanstein (1977). According to this idea an uncoupler anion could first bind to the uncoupler binding site according to its dissociation constant, and block energy transfer. The anion would then be protonated from the cytoplasm side, be released from the binding site because of its now reduced affinity, and dissociate to release its proton on the matrix side to complete the process.

(b) Dinitrophenol derivatives

As shown in Table 1.4, dinitrophenol derivatives are used to control fungi, insects and weeds. The pesticidal use of dinitroalkylphenols has been reviewed by Kirby (1966).

The dinitrophenol derivatives, with the exception of dinocap (which is a mixture of isomers) are based on the following phenols:

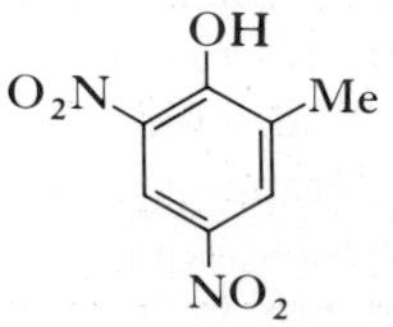

DNOC 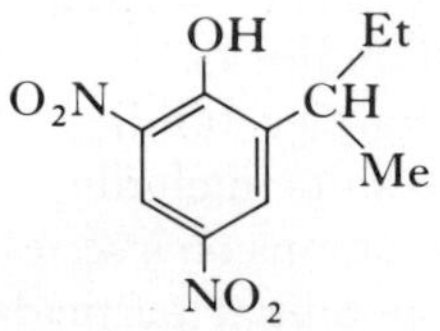   dinoseb 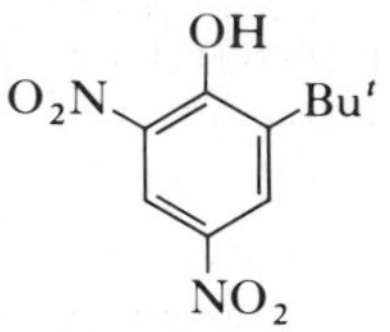   2-t-butyl-4,6-
                                               dinitrophenol

The esterified compounds are mainly used for the control of mites and powdery mildew infections of plants. Both pests occur superficially on the plant and the esterified compounds are probably preferable to the free phenols because they are less likely to be phytotoxic and more likely to persist in the surface layers of the plant than the free phenols. Thus, binapacryl, used for the control of mites and powdery mildew, is an ester of the herbicide, dinoseb. Once the ester has been taken up by the pest it is presumed to generate the phenol.

Esterified nitophenols tend to be unstable to chemical hydrolysis and could presumably be hydrolysed in the organism either chemically or enzymatically to yield the toxic dinitrophenol (Kirby and Frick, 1958). The esterified compounds used as herbicides must also undergo hydrolysis, either in the plant or the soil, before they can be effective as uncouplers. They presumably have some advantage of uptake or persistence not found with the free phenol to which they give rise.

Table 1.6 shows the uncoupling activity of two of the parent phenols, as well as that of 2,4-dinitrophenol itself, on mitochondria isolated from a variety of sources. The uncoupling activity of the other parent phenols

Table 1.6 Uncoupling activity of pesticides on mitochondria isolated from various sources. (See footnote for abbreviations. Note that when comparing activities the limitations discussed in the text must be considered)

Compound	Uncoupling activity (μM) in mitochondria isolated from:		
	Mammals	Insects; mites	Plants; fungi
Dinitrophenol derivatives			
2,4-Dinitrophenol	30; RL; U_{50}[a]	50; FT; muc[d]	40–50; sweet potato; ms[c]
	20; RL; muc[b]	97; FT; U_{50}[h]	$\leqslant$130; *Aspergillus niger*; ms[e]
	50; ML; muc[d]	50; BT; muc[d]	200; *Aspergillus oryzae*; U_{50}[k]
	1; MB; muc[d]		
DNOC	20; ML; muc[d]	20; FT; muc[d]	20–30; pea shoots; U_{50}[f]
	20; MB; muc[d]	19; FT; U_{50}[h]	
		30; BT; muc[d]	
Dinoseb	1; ML; muc[d]	0.8; FT; muc[d]	
	0.5; MB; muc[d]	0.8; BT; muc[d]	
Hydroxybenzonitriles			
Bromoxynil	3.2–5; RL; U_{50}[a]		50; pea shoots; U_{50}[f]
			60; white potato; ms[g]
Ioxynil	1.0–1.3; RL; U_{50}[a]		25–30; pea shoots; U_{50}[f]
			10; white potato; ms[g]
Miscellaneous compounds			
Bromofenoxim			1.6; white; potato; muc[l]
Drazoxolon	3; RL; U_{50}[j]		
Malonoben	0.006; RL; muc[m]		
Niclosamide		0.3; FT; U_{50}[h]	
Pentachlorophenol	5; RL; U_{50}[i]		

Abbreviations: RL, rat liver; ML, mouse liver; MB, mouse brain; FT, housefly thorax; BT, honeybee thorax; U_{50}, concentration causing 50% uncoupling, usually determined with a Warburg manometer; muc, minimum uncoupling concentration; ms, concentration giving maximum stimulation of oxygen uptake.

References: [a]Parker (1965); [b]Chappell (1964); [c]Wiskich and Bonner (1963); [d]Ilivicky and Casida (1969); [e]Watson and Smith (1967); [f]Kerr and Wain (1964); [g]Ferrari and Moreland (1969); [h]Williamson and Metcalf (1967); [i]Weinbach and Garbus (1965); [j]Parker and Summers (1970); [k]Kawakita (1970); [l]Moreland and Blackmon (1970); [m]Nishizawa *et al.* (1974).

seems not to have been reported, but might be expected to be similar. The level of *in vitro* activity quoted for DNOC and dinoseb falls in the range 0.5 to 30 μM, which is probably sufficiently powerful to explain their biological activity.

If the dinitrophenol-based pesticides do kill by uncoupling oxidative phosphorylation, an increase in respiration followed by death would be expected. Although 2,4-dinitrophenol is not used as a pesticide we shall include it here since it is the archetypal uncoupler and has consequently been the subject of much study. It has been shown to stimulate the respiration of man, various insects, yeast, and higher plant tissues (Simon, 1953; Gaur and Beevers, 1959).

Insecticides based on dinitrophenols cause an initial stimulation of insect respiration (Metcalf, 1955; Brown, 1963) followed by a decline and death. To take an example, Harvey and Brown (1951) showed that 10 μg of DNOC injected into a German cockroach stimulated oxygen uptake to a level some three times that of normal within about 75 minutes after application. Respiration started to decline after 100 minutes, and death ensued at approximately 500 minutes after injection.

Effects of uncouplers on respiration in plant tissues have been examined both *in vitro* and *in vivo*. Biochemical studies show that 2 to 10 μM dinoseb stimulates respiration and inhibits phosphate uptake and ATP formation by tomato leaf discs kept in the dark so that photosynthesis could not occur (Wojtaszek *et al.*, 1966). Wojtaszek (1966) reasoned that if dinoseb works by uncoupling oxidative phosphorylation, those plants which inherently maintain a high level of ATP should be relatively tolerant. Using leaf discs from thirteen species of plants picked for their varying susceptibility to dinoseb, he showed that the ability to accumulate phosphate from solution in either the light or the dark was correlated with tolerance to the herbicide. Since the uptake of phosphate is also correlated with the amount of ATP formed (Wojtaszek *et al.*, 1966), the hypothesis that dinoseb kills by inhibiting the formation of ATP is supported.

During a 24 h period following treatment, more damage occurred to intact plants treated with 2,4-dinitrophenol if they were maintained in darkness, than if they were kept in the light. Further, the wavelengths of light that were most effective in alleviating injury were those that are most effective in promoting photosynthesis (Mellor and Salisbury, 1965). These results are explicable if the compound reduced the level of ATP, which was then replaced by photosynthetic phosphorylation (Chapter 2). For this explanation it is necessary to assume that any effect of the nitrophenol on photosynthesis (see below and p. 82) had not occurred during this period. Wojtaszek (1966) also found the effect of dinoseb to be reduced if a light period followed treatment, though the opposite was noted by Desmoras and Jaquet (1964).

Nitrophenols can also affect photosynthetic electron transport and this will be taken up again in Chapter 2 (p. 81). However, dinoseb applied to variegated plants had a similar desiccant action on green, white or variegated leaves (Bovey and Miller, 1968) so plant tissue does not need to be photosynthetic to be killed by the compound.

Also, it is known that nitrophenol herbicides (and hydroxybenzonitriles, discussed below) can inhibit mitochondrial electron transport, though at higher concentrations than are required to cause uncoupling. On this basis they are classified by Moreland (1980) as 'inhibitory uncouplers'.

A compound not included in the *Pesticide Manual* (Worthing, 1979) since it is not a pesticide in the conventional sense is 3-trifluoromethyl-4-nitrophenol. Niblett and Ballantyne (1976) report that the compound, used as a lamprey larvicide, is, predictably, an uncoupler of oxidative phosphorylation, causing 50% uncoupling of rat liver mitochondria at nanomolar concentrations.

Before leaving the dinitrophenol-based pesticides it should be noted that dinocap will give, on hydrolysis, a 2,6-dinitrophenol with an alkyl group in the 4 position. Ilivicky and Casida (1969) found that 4-s-butyl-2,6-dinitrophenol was not an uncoupler, but a rather active (at 0.1 μM) inhibitor of oxidative phosphorylation, so that administration of dinocap might, after hydrolysis bring about both uncoupling and inhibition of oxidative phosphorylation.

(c) Hydroxybenzonitriles

Of the hydroxybenzonitrile herbicides listed in Table 1.4 ioxynil is probably a sufficiently powerful uncoupler (10 to 30 μM; Table 1.6) for this to explain its herbicidal action. However, the activity of bromoxynil (50 to 60 μM) seems rather low for uncoupling of oxidative phosphorylation to be the sole cause of herbicidal activity. Both compounds inhibit photosynthetic electron transport in chloroplasts; the relative contributions of effects on respiration and photosynthesis to the phytotoxicity of these herbicides is discussed on p. 82.

(d) Miscellaneous compounds

The molluscicide niclosamide stimulated the uptake of oxygen by snail tissue maximally at 0.1 nM with succinate as substrate (Ishak *et al.*, 1970), strongly suggesting an uncoupling action. This has been shown for another salicylanilide 5-chloro-3-*t*-butyl 2′ chloro-4′-nitrosalicylanilide (not used as a pesticide) which uncoupled housefly thorax mitochondria at 5 nM (Williamson and Metcalf, 1967).

Pentachlorophenol is a biocide. Action as an uncoupler is supported by the demonstration that it stimulates oxygen uptake of snail tissue, with a maximal effect at 0.1 μM on glutamate oxidation (Ishak *et al.*, 1970). As might be expected from an uncoupler it is primarily a contact herbicide (Fryer and Evans, 1968) though it has also been found to inhibit photosynthetic electron transport and phosphorylation at micromolar concentrations (Krogmann *et al.*, 1959).

Drazoxolon will uncouple rat liver mitochondria (Table 1.6) so that its fungicidal action could be due to this effect.

The miticide and fungicide malonoben gave complete uncoupling of rat liver mitochondria at a concentration of 6 nM (Nishizawa *et al.*, 1974). Because of its potency as an uncoupler, malonoben has been investigated further (Terada and Van Dam, 1975; Terada, 1975). Spectral and other investigations of the effect of the compound on intact mitochondria and model membrane systems led Terada (1975) to conclude that malonoben can cross the mitochondrial inner membrane, but remains localised in a certain region of the membrane during the uncoupling process. This seems to fit with the idea of a discrete uncoupler binding site, as referred to above.

Although it now seems likely that dichlobenil acts by inhibiting cellulose biosynthesis (p. 270), Moreland *et al.* (1974) showed that the 3- and 4-hydroxylated derivatives of dichlobenil were capable of uncoupling mung bean mitochondria, as well as inhibiting electron transport at higher concentrations and interfering with chloroplast functions. Ring hydroxylation and conjugation are known for this compound so it seemed possible that such metabolic products could contribute to the biological activity (Moreland *et al.*, 1974). However, as these authors point out, Verloop (1972) states that some plants (e.g., garden cress) are very sensitive to dichlobenil though hydroxylation is not known to occur to any great extent.

3. *Inhibitors of oxidative phosphorylation*

The distinction between the properties of inhibitors and uncouplers of oxidative phosphorylation is explained on p. 19. Table 1.7 lists those pesticides thought to work by inhibiting oxidative phosphorylation. The first group consists of trisubstituted tins, whose fungicidal activity and mode of action have been reviewed by Kaars Sijpesteijn *et al.* (1969). As already mentioned (p. 39) there is a possibility that the pesticidal effect of dinocap may be partially or wholly due to inhibition, rather than uncoupling, of oxidative phosphorylation.

(a) *Trisubstituted tin compounds*

The hypothesis that the trisubstituted tin pesticides (Table 1.7) work by inhibiting oxidative phosphorylation is based on evidence originally obtained by Aldridge and his group that such compounds are active

Table 1.7 Pesticides thought to act by inhibition of oxidative phosphorylation

Name and use	Structure
Bis(tributyltin) oxide (biocide; timber protection)	$(Bu_3Sn)_2O$
Cyhexatin (acaricide)	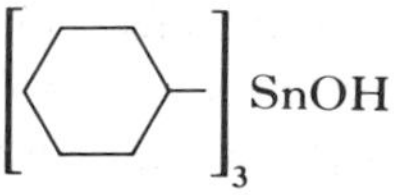 SnOH
Fenbutatin oxide (acaricide)	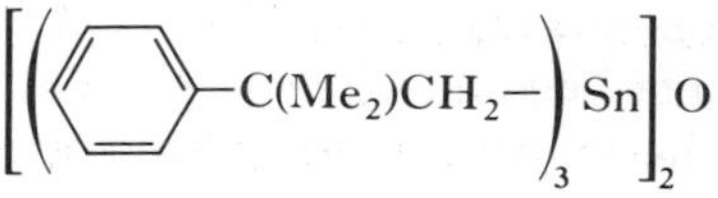
Fentin acetate (fungicide)	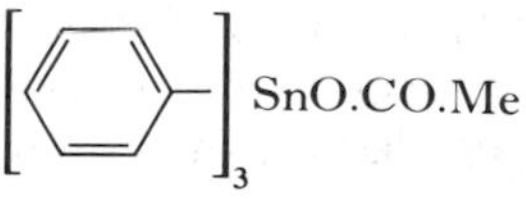 SnO.CO.Me
Fentin hydroxide (fungicide)	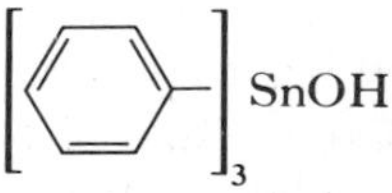 SnOH
Chlorfenson (acaricide)	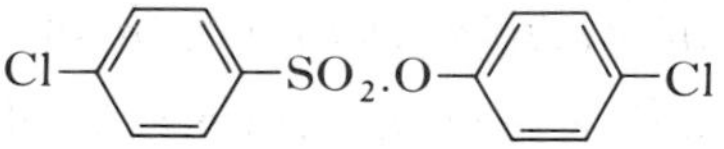
Tetradifon (acaricide)	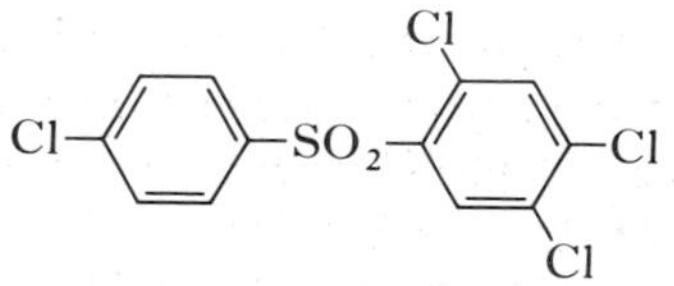
Tetrasul (acaricide)	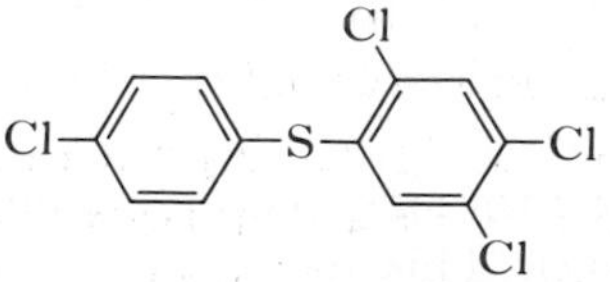

inhibitors of oxidative phosphorylation in isolated mammalian mitochondria. Fentin acetate is easily hydrolysed in dilute aqueous solution
(Kaars Sijpesteijn *et al.*, 1969) to fentin hydroxide, which ionises to the
triphenyl tin cation; fentin hydroxide causes marked inhibition of oxidative
phosphorylation in rat liver mitochondria at 3 μM (Stockdale *et al.*, 1970)
and an ATPase preparation from houseflies was also susceptible to
inhibition by micromolar concentrations of triphenyltin (chloride) (Pieper
and Casida, 1965). It seems likely that such *in vitro* activity is sufficient
to explain the *in vivo* effect. The ATPase is not susceptible when solubilised,
and it is therefore thought that the membrane-located component of the
ATPase is the target (Selwyn, 1976). According to the chemiosmotic view,
this component contains the pore through which protons flow during ATP
synthesis. The molecular nature of the interaction of trialkyl tins with the
ATPase has not been established with certainty.

Trialkyl tins in general have been found to act on mitochondria in
three ways (see Aldridge *et al.*, 1977; Cain and Griffiths, 1977):

(*i*) By mediating exchange of Cl^- and OH^- across the mitochondrial
membrane in Cl^--containing media. Under appropriate conditions this
can be observed as a stimulation of ATPase activity. Whether this contributes to the toxic effect is unclear.

(*ii*) By inhibiting the ATPase as mentioned above (Stockdale *et al.*,
1970).

(*iii*) By causing gross swelling of the mitochondria.

These effects have been recognised mainly by work on the lower alkyl
tins which are not in fact used as pesticides. In these cases ATPase
stimulation is seen at low concentrations, with the inhibition becoming
evident at higher levels. In the case of higher molecular weight compounds
such as triphenyl tin, ATPase inhibition occurs at particularly low concentrations (in the range 1–10 μM in rat liver mitochondria) so that no
stimulatory effect is seen (Aldridge *et al.*, 1977).

In the case of cyhexatin, the activity against the mitochondrial ATPase
in homogenates of the two-spotted spider mite has been investigated
(Desaiah *et al.*, 1973). The low I_{50} of 0.62 nM fully supports the proposed
mode of action.

Trialkyl tins have also been tested on chloroplast functions. For instance,
triphenyl tin chloride is a powerful inhibitor of ATP synthesis and coupled
electron transport in spinach chloroplasts ($I_{50} = 1$ μM) (Gould, 1976).
Such interference with the same function in this different organelle tends
to support the proposed mode of action, but the compounds are not used
as herbicides.

At high concentrations, fentin acetate is a caterpillar feeding deterrent
(Worthing, 1979). This has been attributed to the direct inhibition of

digestive enzymes (Ishaaya *et al.*, 1977) and is unlikely to be connected with the principal mode of action, discussed above.

(b) Sulphur-containing acaricides

The mitochondrial ATPase may also be the site of action of the sulphur-containing acaricides shown in Table 1.7.

Chlorfenson and tetradifon were found to inhibit fish brain enzyme with I_{50} values of 2.7 μM and *c.* 0.05 μM respectively (Cutcomp *et al.*, 1972; Desaiah *et al.*, 1972). The same site of action was found in rat liver mitochondria, in which the I_{50} was found to be in the region of 5–25 μM, depending on the way the inhibition was quantified (Bustamante and Pedersen, 1973).

There does not seem to be any information on the mode of action of tetrasul, but in view of its structure it seems likely that it is active after oxidation to tetradifon.

C. Etridiazole

It is difficult to classify this fungicide since two modes of action have been proposed. However, since both are concerned with mitochondrial function, it seems reasonable to consider the compound in this chapter.

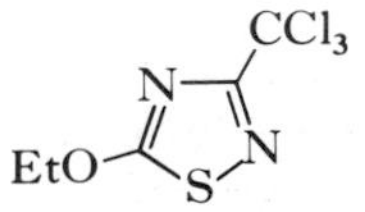

etridiazole

Halos and Huisman (1976a) found that concentrations of etridiazole which inhibited the growth of *Pythium* species also caused inhibition of respiration. Experiments on mitochondria from *P. ultimum* revealed that, e.g., succinate-driven electron transport was half-inhibited by 18 μM etridiazole due to an action between the b and c cytochromes. However, higher concentrations (up to 400 μM) gave only the same degree of inhibition, which presumably has to be explained by poor penetration of the compound into the mitochondria.

An action on electron-transfer reactions was supported by the further demonstration that fungal lines which had become relatively insensitive to etridiazole, accumulated compounds tentatively identified as coenzyme Q derivatives which may have been able to take part in redox reactions so as to by-pass the site of inhibition (Halos and Huisman, 1976b).

Lyr *et al.* (1977) and Radzhun and Casperson (1979), using *Mucor mucedo*, agreed that mitochondrial function was disrupted by etridiazole but suggested that this was due to extensive membrane damage resulting from the release of phospholipases within the mitochondria (and perhaps elsewhere within the cell). The suggestion was supported by a number of lines of evidence.

(*i*) One of the early symptoms visible by electron microscopy was damage to the mitochondrial inner membrane.

(*ii*) Incubation of isolated mitochondria *in vitro* with 38 μM etridiazole led to lipid changes which could have been caused by phospholipases, i.e. there was cleavage of the ester link between the fatty acid and glycerol moieties of the phospholipids.

(*iii*) The growth-inhibitory effects and the mitochondrial damage caused by 38 μM etridiazole could be overcome by simultaneously exposing the fungus to compounds known to inhibit phospholipases.

Radzhun and Casperson (1979) could not support inhibition of electron transport as the primary effect of etridiazole, since respiration by isolated *Mucor* mitochondria was less sensitive to the fungicide than growth of the intact fungus.

From the above evidence we can conclude that the site of etridiazole action is probably within the mitochondrion but we cannot describe its effects with any greater precision.

REFERENCES

Albert, A. (1979). "Selective Toxicity". Chapman and Hall, London

Aldridge, W. N. and Cremer, J. E. (1955). *Biochem. J.* **61**, 406–418

Aldridge, W. N., Street, B. W. and Skilleter, D. N. (1977). *Biochem. J.* **168**, 353–364

Baker, B. R. (1967). "The Design of Active-Site Directed Irreversible Enzyme Inhibitors". John Wiley and Sons, New York

Ben-Yephet, Y., Dinoor, A. and Henis, Y. (1975). *Phytopathol.* **65**, 936–942

Bergmann, H., Lyr, H., Kluge, E. and Ritter, G. (1975). In "Systemfungizide" (H. Lyr and C. Polter, eds) pp. 183–188. Akademie-Verlag, Berlin

Block, S. (1956). *J. Ag. Food Chem.* **4**, 1042–1046

Blondin, G. A. (1980). *Biochem. Biophys. Res. Commun.* **96**, 587–594

Bond, E. J., Monro, H. A. U. and Buckland, C. T. (1967). *J. Stored Prod. Res.* **3**, 289–294

Bovey, R. W. and Miller, F. R. (1968). *Weed Res.* **8**, 128–135

Boyer, P. D., Chance, B., Ernster, L., Mitchell, P., Racker, E. and Slater E. C. (1977). *Ann. Rev. Biochem.* **46**, 955–1026

Briquet, M., Sabadie-Pialoux, N. and Goffeau, A. (1976). *Arch. Biochem. Biophys.* **174**, 684–694

Brown, A. W. A. (1963). In "Insect Pathology: An Advanced Treatise" (E. A. Steinhaus, ed.) Vol. 1, pp. 65–131. Academic Press, London and New York

Bucholtz, M. L., Maudinas, B. and Porter, J. W. (1977). *Chem.-Biol. Interact.* **17**, 359–362

Bustamante, E. and Pedersen, P. L. (1973). *Biochem. Biophys. Res. Commun.* **51**, 292–298

Cain, K. and Griffiths, D. E. (1977). *Biochem. J.* **162**, 575–580

Casola, L., Brumby, P. E. and Massey, V. (1966). *J. Biol. Chem.* **241**, 4977–4984

Chappell, J. B. (1964). *Biochem. J.* **90**, 237–248
Chefurka, W., Kashi, K. P. and Bond, E. J. (1976). *Pestic. Biochem. Physiol.* **6**, 65–84
Chen, C.-C., Yang, M., Durst, H. D., Saunders, D. R. and Wang, J. H. (1975). *Biochemistry* **14**, 4122–4126
Coles, C. J., Singer, T. P., White, G. A. and Thorn, G. D. (1978). *J. Biol. Chem.* **253**, 5573–5578
Criddle, R. S., Johnston, R. F. and Stack, R. J. (1979). In "Current Topics in Bioenergetics" (D. R. Sanadi, ed.) Vol. 9, pp. 89–145. Academic Press, London and New York
Cunarro, J. and Weiner, M. W. (1975). *Biochim. Biophys. Acta* **387**, 234–240
Cutcomp, L. K., Desaiah, D. and Koch, R. B. (1972). *Life Sci.* **11**, 1123–1133
Decker, S. J. and Lang, D. R. (1977). *J. Biol. Chem.* **252**, 5936–5938
Desaiah, D., Cutcomp, L. K., Koch, R. B. and Yap, H. H. (1972). *Life Sci.* **11**, 389–395
Desaiah, D., Cutkomp, L. K. and Koch, R. B. (1973). *Life Sci.* **13**, 1693–1703
Desmoras, J. and Jacquet, P. (1964). Proc. 16th Int. Symp. of Crop Protection, pp. 633–642
Dixon, M. and Webb, E. C. (1964). In "Enzymes". Longman, London
Draber, W., Büchel, K. H. and Schäffer, G. (1972). *Z. Naturforsch.* **27**, 159–171
DuBois, K. P., Raymund, A. B. and Hietbrink, B. E. (1961). *Toxicol. and Appl. Pharmacol.* **3**, 236–255
Eckert, J. W. (1969). *World Rev. Pest Control* **8**, 116–137
Eckert, J. W., Bretschneider, B. F. and Rahm, M. L. (1975). In Proc. 8th Int. Plant Prot. Conf., Moscow. Sect. III, "Chem. Control", p. 215
Erecińska, M. and Wilson D. F. (1981). In "Inhibitors of Mitochondrial Functions" (M. Erecińska and D. F. Wilson, eds) pp. 145–164. Pergamon Press, Oxford
Evans E. (1968). In "Plant Diseases and their Chemical Control". Blackwell, Oxford
Ferrari T. E. and Moreland D. E. (1969). *Plant Physiol.* **44** 429–434
Fryer J. D. and Evans S. A. (eds) (1968). In "Weed Control Handbook", 5th edn, Vol. 1. Blackwell, Oxford
Gaur, B. K. and Beevers, H. (1959). *Plant Physiol.* **34**, 427–432
Georgopoulos, S. G. and Vomvoyanni, V. (1972). In "Herbicides, Fungicides, Formulation Chemistry" (A. S. Tahori, ed.) pp. 337–346. Gordon and Breach, New York
Georgopoulos, S. G., Alexandri, E. and Chrysayi, M. (1972). *J. Bacteriol.* **110**, 809–817
Georgopoulos, S. G., Chrysayi, M. and White, G. A. (1975). *Pestic. Biochem. Physiol.* **5**, 543–551
Gould, J. M. (1976). *Eur. J. Biochem.* **62**, 567–575
Green, D. E. and Vande Zande, H. (1981). *Biochem. Biophys. Res. Commun.* **100**, 1017–1024
Griffiths, D. E., Houghton, R. I. and Lancashire, W. E. (1974). In "Genetics and Biogenesis of Chloroplasts and Mitochondria" (T. Bucher, W. Newpert, W. Sebald and S. Werner, eds) pp. 175–185. North-Holland, Amsterdam
Guffanti, A. A., Blumenfeld, H. and Krulwich, T. A. (1981). *J. Biol. Chem.* **256**, 8416–8421
Gunatilleke, I. A. U. N., Arst, H. N. and Scazzocchio, C. (1976). *Genet. Res.* **26**, 297–305

Halangk, W. and Schewe, T. (1975). In "Systemfungizide" (H. Lyr and C. Polter, eds) pp. 177–182. Akademie-Verlag, Berlin

Halos, P. M. and Huisman, O. C. (1976a). *Phytopathol.* **66**, 158–164

Halos, P. M. and Huisman, O. C. (1976b). *Phytopathol.* **66**, 152–157

Hanstein, W. G. (1976a). *Trends Biochem. Sci.* **1**, 65–67

Hanstein, W. G. (1976b). *Biochim. Biophys. Acta* **456**, 129–148

Hanstein, W. G. and Hatefi, Y. (1974a). *Proc. Natl. Acad. Sci. U.S.A.* **71**, 288–292

Hanstein, W. G. and Hatefi, Y. (1974b). *J. Biol. Chem.* **249**, 1356–1362

Hanstein, W. G. and Kiehl, R. (1981). *Biochem. Biophys. Res. Commun.* **100**, 1118–1125

Harvey, G. T. and Brown, A. W. A. (1951). *Can. J. Zool.* **29**, 42–53

Hatefi, Y. (1980). *Ann. N.Y. Acad. Sci.* **346**, 434–443

Horgan, D. J., Ohno, H., Singer, T. P. and Casida, J. E. (1968). *J. Biol. Chem.* **243**, 5967–5976

Hoskins, D. D., Cheldelin, V. H. and Newburgh, R. W. (1956). *J. Gen. Physiol.* **39**, 705–713.

Hucho, F. (1975). *Ang. Chem. Int. Ed. Engl.* **14**, 591–601

Ilivicky, J. and Casida, J. E. (1969). *Biochem. Pharmacol.* **18**, 1389–1401

Ishaaya, I., Holmstead, R. L. and Casida, J. E. (1977). *Pestic. Biochem. Physiol.* **7**, 573–577

Ishak, M. M., Sharaf, A. A., Mohamed, A. M. and Mousa, A. H. (1970). *Comp. Gen. Pharmacol.* **1**, 201–208

Ito, M. and Ohnishi, Y. (1981). *FEBS Lett.* **136**, 225–230

Jacobson, K. N., Murphy, J. B. and Das Sarma, B. (1972). *FEBS Lett.* **22**, 80–82

Janssen, M. J. and Kaars Sijpesteijn, A. (1961). In "Fungicides in Agriculture and Horticulture", Society of Chemical Industry Monograph **15**, pp. 40–51

Jen, K.-Y. and Cava, M. P. (1982). *Tetrahedron Lett.* **23**, 2001–2004

Kaars Sijpesteijn, A. (1970). *World Rev. Pest Control* **8**, 85–93

Kaars Sijpesteijn, A. (1977). In "Systemic Fungicides" (R. W. Marsh, ed.) pp. 131–159. Longman, London

Kaars Sijpesteijn, A. and Janssen, M. J. (1958). *Nature (Lond.)* **182**, 1313–1314

Kaars Sijpesteijn, A., Janssen, M. J. and Dekhuyzen, H. M. (1957a). *Nature (Lond.)* **180**, 505–506

Kaars Sijpesteijn, A., Janssen, M. J. and van der Kerk, G. J. M. (1957b). *Biochim. Biophys. Acta* **23**, 550–557

Kaars Sijpesteijn, A., Luijten, J. G. A. and van der Kerk, G. J. M. (1969). In "Fungicides: An Advanced Treatise" (D. C. Torgeson, ed.) Vol. 2, pp. 331–366. Academic Press, London and New York

Katre, N. V. and Wilson, D. F. (1978). *Arch. Biochem. Biophys.* **191**, 647–656

Kawakita, M. (1970). *J. Biochem. (Tokyo)* **68**, 625–631

Kerkenaar, A. and Kaars Sijpesteijn, A. (1979). *Pestic. Biochem. Physiol.* **12**, 124–129

Kerr, M. W. and Wain, R. L. (1964). *Ann. Appl. Biol.* **54**, 441–446

Kessler, R. J., Vande Zande, H., Tyson, C. A., Blondin, G. A., Fairfield, J., Glasser, P. and Green, D. E. (1977). *Proc. Natl. Acad. Sci. U.S.A.* **74**, 2241–2245

Kirby, A. H. M. (1966). *World Rev. Pest Control* **5**, 30–44

Kirby, A. H. M. and Frick, E. L. (1958). *Nature (Lond.)* **182**, 1445–1446

Krogmann, D. W., Jagendorf, A. T. and Avron, M. (1959). *Plant Physiol.* **34**, 272–277

Lindahl, P. E. and Oberg, K. E. (1960). *Nature (Lond.)* **187**, 784

Lukens, R. J. (1969). In "Fungicides: An Advanced Treatise" (D. C. Torgeson, ed.) Vol. 2, pp. 395–446. Academic Press, London and New York

Lyr, H. (1977). In "Antifungal Compounds" (M. R. Siegel and H. D. Sisler, eds) Vol. 2, pp. 301–332. Dekker, New York

Lyr, H., Casperson, G. and Laussmann, B. (1977). *Z. Allg. Microbiol.* **17**, 117–129

Marshall, L. E. and Himes, R. H. (1978). *Biochim. Biophys. Acta* **543**, 590–594

Martin, H. (1964). In "The Scientific Principles of Crop Protection". Edward Arnold, London

Martin, H. (1969). "Fungicides: An Advanced Treatise" (D. C. Torgeson, ed.) Vol. 2, pp. 102–119. Academic Press, London and New York

Mathre, D. E. (1970). *Phytopathol.* **60**, 671–676

Mathre, D. E. (1971). *Pestic. Biochem. Physiol.* **1**, 216–224

McCallan, S. E. A. (1967). In "Fungicides: An Advanced Treatise" (D. C. Torgeson, ed.) Vol. 1, pp. 1–37. Academic Press, London and New York

McCallan, S. E. A., Miller, L. P. and Weed, R. M. (1954). *Contrib. Boyce Thompson Inst.* **18**, 39–68

McLaughlin, S. (1981). In "Chemiosmotic Proton Circuits in Biological Membranes" (V. P. Skulachev and P. C. Hinkle, eds) pp. 601–609. Addison-Wesley, Reading, Mass.

McNew, G. L. and Gershon, H. (1969). *Residue Reviews* **25**, 107–122

Mellor, R. S. and Salisbury, F. B. (1965). *Plant Physiol.* **40**, 506–512

Metcalf, R. L. (1955). "Organic Insecticides". Interscience, New York

Mitchell, P. (1961). *Nature (Lond.)* **191**, 144–148

Mitchell, P. (1966). *Biol. Rev.* **41**, 445–502

Mitchell, P. and Moyle, J. (1967). *Biochem. J.* **104**, 588–600

Moore, A. L. and Rich, P. R. (1980). *Trends Biochem. Sci.* **5**, 284–288

Moore, A. L., Linnett, P. E. and Beechey, R. B. (1979). *Biochem. Soc. Trans.* **7**, 1120–1122

Moreland, D. E. (1980). *Ann. Rev. Plant Physiol.* **31**, 597–638

Moreland, D. E. and Blackmon, W. J. (1970). *Weed Sci.* **18**, 419–426

Moreland, D. E., Farmer, F. S. and Hussey, G. G. (1972a). *Pestic. Biochem. Physiol.* **2**, 342–353

Moreland, D. E., Farmer, F. S. and Hussey, G. G. (1972b). *Pestic. Biochem. Physiol.* **2**, 354–363

Moreland, D. E., Hussey, G. G. and Farmer, F. S. (1974). *Pestic. Biochem. Physiol.* **4**. 356–364

Mowery, P. C., Steenkamp, D. J., Ackrell, B. A. C., Singer, T. P. and White, G. A. (1977). *Arch. Biochem. Biophys.* **178**, 495–506.

Müller, W. and Schewe, T. (1976). *Acta. Biol. Med. Ger.* **35**, 693–707

Müller, W. and Schewe, T. (1977). *Acta. Biol. Med. Ger.* **36**, 967–980

Müller, W., Schewe, T., Lyr, H. and Zanke, D. (1977). *Z. Allg. Microbiol.* **17**, 359–372

Niblett, P. D. and Ballantyne, J. S. (1976). *Pestic. Biochem. Physiol.* **6**, 363–366

Nicholls, D. G. (1982). "Bioenergetics. An Introduction to the Chemiosmotic Theory". Academic Press, London and New York

Nishizawa, Y., Sumida, S. and Muraoka, S. (1974). In "Mechanism of Pesticide Action" (G. K. Kohn, ed.) pp. 169–175. American Chemical Society, Washington, D.C.

O'Brien, R. D. (1967). In "Insecticides: Action and Mechanism". Academic Press, London and New York

O'Brien, R. D., Cheung, L. and Kimmel, E. C. (1965). *J. Insect. Physiol.* **11**, 1241–1246

Oelze, J. and Kamen, M. D. (1975). *Biochim. Biophys. Acta* **387**, 1–11

Oelze, J., Fakoussa, R. M. and Hudewentz, J. (1978). *Arch. Microbiol.* **118**, 127–132
Parker, V. H. (1965). *Biochem. J.* **97**, 658–662
Parker, V. H. and Summers, L. A. (1970). *Biochem. Pharmacol.* **19**, 315–317
Phelps, D. C. and Hanstein, W. G. (1977). *Biochem. Biophys. Res. Commun.* **79**, 1245–1254
Pieper, G. R. and Casida, J. E. (1965). *J. Econ. Entomol.* **58**, 392–400
Quistad, G. B., Staiger, L. E., Schooley, D. A., Sparks, T. C. and Hammock, B. D. (1979). *Pestic. Biochem. Physiol.* **11**, 159–165
Radzhun, B. and Casperson, G. (1979). In "Systemfungizide" (H. Lyr and C. Polter, eds) pp. 195–295. Akademie-Verlag, Berlin
Ragsdale, N. N. and Sisler, H. D. (1970). *Phytopathol.* **60**, 1422–1427
Ramsay, R. R., Ackrell, B. A. C., Coles, C. J., Singer, T. P., White, G. A. and Thorn, G. D. (1981). *Proc. Natl. Acad. Sci. U.S.A.* **78**, 825–828
Sachs, R. M. and Michael, J. L. (1971). *Weed Sci.* **19**, 558–564
Schewe, T. and Müller, W. (1979). In "Systemfungizide" (H. Lyr and C. Polter, eds) pp. 317–324. Akademie-Verlag, Berlin
Schewe, T., Hiebsch, C. and Halangk, W. (1975). *Acta. Biol. Med. Ger.* **34**, 1767–1775
Schewe, T., Müller, W., Lyr, H. and Zanke, D. (1979). In "Systemfungizide" (H. Lyr and C. Polter, eds) pp. 241–251. Akademie-Verlag, Berlin
Selwyn, M. J. (1976). In "Organotin Compounds: New Chemistry and Applications" (J. J. Zuckerman, ed.) pp. 204–226. American Chemical Society, Washington, D.C
Sherald, J. L. and Sisler, H. D. (1972). *Plant Cell Physiol.* **13**, 1039–1052
Simon, E. W. (1953). *Biol. Rev.* **28**, 453–479
Slater, E. C. (1963). In "Metabolic Inhibitors" (R. M. Hochster and J. H. Quastel, eds) Vol. II, pp. 503–516. Academic Press, London and New York
Somers, E. (1961). *Ann. Appl. Biol.* **49**, 246–253
Somers, E. (1963). *Ann. Appl. Biol.* **51**, 425–437
Stockdale, M., Dawson, A. P. and Selwyn, M. J. (1970). *Eur. J. Biochem.* **15**, 342–351
Storey, B. T. (1981). In "Inhibitors of Mitochondrial Functions" (M. Erecińska and D. F. Wilson, eds) pp. 101–108. Pergamon Press, Oxford
Takahashi, N., Suzuki, A. and Tamura, S. (1965). *J. Am. Chem. Soc.* **87**, 2066–2068
ten Haken, P. and Dunn, C. L. (1971). *Proc. 6th Br. Insectic. Fungic. Conf.*, **2**, 453–462
Terada, H. (1975). *Biochim. Biophys. Acta* **387**, 519–532
Terada, H. and Van Dam, K. (1975). *Biochim. Biophys. Acta* **387**, 507–518
Tolmsoff, W. J. (1962). *Phytopathol.* **52**, 755
Van Dam, K. and Meyer, A. J. (1971). *Ann. Rev. Biochem.* **40**, 115–160
Verloop, A. (1972). *Residue Reviews* **43**, 55–103
Wang, J. H., Yamauchi, O., Tu, S-I., Wang, K., Saunders, D. R., Copeland, L. and Copeland, E. (1973). *Arch. Biochem. Biophys.* **159**, 785–791
Watson, K. and Smith, J. E. (1967). *Biochem. J.* **104**, 332–339
Webb, J. L. (1966). "Enzyme and Metabolic Inhibitors", Vol. III. Academic Press, London and New York
Weinbach, E. C. and Garbus, J. (1965). *J. Biol. Chem.* **240**, 1811–1819
Weinbach, E. C. and Garbus, J. (1969). *Nature (Lond.)* **221**, 1016–1018
White, G. A. (1971). *Biochem. Biophys. Res. Commun.* **44**, 1212–1219
White, G. A. and Thorn, G. D. (1975). *Pestic. Biochem. Physiol.* **5**, 380–395
White, G. A. and Thorn, G. D. (1980). *Pestic. Biochem. Physiol.* **14**, 26–40

White, G. A., Thorn, G. D. and Georgopoulos, S. G. (1978). *Pestic. Biochem. Physiol.* **9**, 165–182
Williamson, R. L. and Metcalf, R. L. (1967). *Science* **158**, 1694–1695
Wilson, D. F., Ting, H. P. and Koppelman, M. S. (1971). *Biochemistry* **10**, 2897–2902
Wilson, D. F., Erecińska, M. and Brocklehurst, E. S. (1972). *Arch. Biochem. Biophys.* **151**, 180–187
Wiskich, J. T. and Bonner, W. D. Jr. (1963). *Plant Physiol.* **38**, 594–604
Wojtaszek, T. (1966). *Weeds* **14**, 125–129
Wojtaszek, T., Cherry, J. H. and Warren, G. F. (1966). *Plant Physiol.* **41**, 34–38
Worthing, C. R. (1979). "The Pesticide Manual", 6th edn. British Crop Protection Council, London
Wren, A. and Massey, V. (1966). *Biochim. Biophys. Acta* **122**, 436–449
Yamamoto, I. (1970). *Ann. Rev. Entomol.* **15**, 257–272
Yoshida, S., Nagao, Y., Watanabe, A. and Takahashi, N. (1980). *Agric. Biol. Chem.* **44**, 2921–2924
Yoshikawa, M. and Eckert, J. W. (1976). *Pestic. Biochem. Physiol.* **6**, 471–481
Yoshikawa, M., Eckert, J. W. and Keen, N. T. (1976). *Pestic. Biochem. Physiol.* **6**, 482–490

2 | Herbicides Interfering with Photosynthesis

I. BIOCHEMICAL BACKGROUND

In the account that follows, reliance has been placed on the many good reviews that have been written about photosynthesis, for instance by Gregory (1977) and various contributors to the treatise edited by Trebst and Avron (1977). More recent progress is reviewed by Vermaas and Govindjee (1981a) and Kaplan and Arntzen (1982).

A particularly succinct review of photosynthesis as related to herbicide action has been provided by Pfister and Arntzen (1979).

A. Function of chloroplasts

The green parts of plants are so coloured because they contain large amounts of the green pigment chlorophyll, located in numerous chloroplasts, the cellular organelles in which photosynthesis occurs. The overall process consists of the conversion of carbon dioxide and water into carbohydrate – that is, the reverse of respiration – and is accomplished in two stages, via the so-called *light* and *dark* reactions.

Both sets of reactions occur in the chloroplast which is bounded by a double outer membrane and contains within it the chlorophyll-rich *thylakoid membranes*; these carry out the light reactions whereby the energy of sunlight is harvested and trapped in chemically usable form as ATP and the reduced form of the coenzyme nicotinamide adenine dinucleotide-2′-phosphate (NADPH). These two compounds represent the end products of the light reactions and they are used to reduce carbon dioxide in the so-called dark reactions which take place in the soluble part of the chloroplast, the *stroma*. Since there is no herbicide known to interfere primarily with the dark reactions we will not consider them further. Conversely over half of the herbicides in current use act on the light reactions of photosynthesis.

B. Light reactions of photosynthesis

The energy of light is absorbed by two discrete photosystems, each of which contains a chlorophyll molecule at the reaction centre where the primary redox reaction of photosynthesis takes place. This chlorophyll molecule is associated with about 300 others, and these, together with some other accessory pigments such as carotenoids, harvest light and transmit its energy to the reaction centre. The properties of the photosystems are such that together they enable the plant to efficiently absorb light throughout the visible spectrum. The light is used to excite electrons, which are then able to reduce the primary acceptor of each photosystem.

The strength of oxidising and reducing agents is measured by the redox potential scale, which quantifies the tendency which a compound has to gain or lose electrons. Compounds having the most negative values on this scale are those most likely to donate electrons, and are thus the best reducing agents; a compound can only be a net donor of electrons to a molecule with a less negative redox potential.

Figure 2.1 shows, in diagrammatic form, how the two photosystems (or light reactions as they are also known) function together to excite electrons to the energy level of the primary acceptor of photosystem I. The diagram is a modern version of the 'Z' scheme first proposed by Hill and Bendall (1960). The two photosystems operate in series so that the primary reductant of photosystem II passes electrons through a series of electron carriers of successively lower reducing power (analogous to those involved in oxidative phosphorylation) to the reaction centre of photosystem I. Following excitation in a second light reaction the electrons are eventually passed to $NADP^+$.

$$NADP^+ + 2e^- + H^+ \rightarrow NADPH$$

The electrons excited by photosystem II come ultimately from water. In four separate photochemical events, four electrons are excited from a

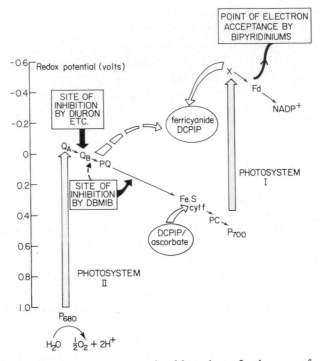

Fig. 2.1 Non-cyclic electron transport in chloroplasts. In the case of components interacting with the electron carriers, the major interaction is shown by a continuous arrow, and the minor interaction by a broken arrow. In this figure and in Fig. 2.2 we have included only components relevant to our consideration of herbicide action. DBMIB, dibromo-methyl-isopropyl-benzoquinone; DCPIP, 2,6-dichlorophenolindophenol: for abbreviations of electron transfer components, see text

manganese-containing 'water-splitting' complex at the reaction centre and passed (via an intermediate electron acceptor which may be pheophytin, a chlorophyll molecule lacking Mg^{++}; see Vermaas and Govindjee, 1981a) to the first stable acceptor, the quinone Q_A (see below). The complex accumulates the four positive charges remaining and then interacts with water, which provides four electrons to replace the four lost:

$$2H_2O \longrightarrow 4e^- + O_2 + 4H^+$$

Oxygen is a by-product of the reaction. The protons are released into the space contained within the thylakoid discs (see Fig. 2.2) and these protons, together with others conveyed from the stroma to the intra-thylakoid space during the redox reactions of plastoquinone (see Fig. 2.2 and the next section) form a gradient which is used to drive the synthesis of ATP; this constitutes photosynthetic phosphorylation.

C. Electron transfer components

Figure 2.1 shows the components between the two photosystems. The first stable acceptor of photosystem II (Q_A) quenches the fluorescence (p. 67) from the reaction centre chlorophyll (P_{680}*) which would occur if no electron acceptor was available. Q_A is thought to be a plastoquinone molecule in a special environment, and it acts as a 1-electron carrier.

Following Q_A is a second electron carrier Q_B also thought to be a plasto-quinone molecule in a particular environment. This acts as a 2-electron carrier, taking electrons one at a time from Q_A, and thus delivering them in pairs to the plastoquinone (PQ) pool in the form of the quinol PQH_2. From here the electrons pass to the reaction centre of photosystem I (P_{700}) through an iron-sulphur protein (Fe.S), cytochrome f (cyt f) and the copper protein plastocyanin (PC).

After further light energy has been absorbed by photosystem I, its primary acceptor (X or P_{430}) has sufficient reducing power to reduce ferredoxin (Fd). This in turn reduces $NADP^+$ via a flavoprotein (FP) dehydrogenase, so generating NADPH, one of the products of the two light reactions.

The process as we have described it so far is termed non-cyclic electron transport, but it is possible for electrons to be passed back from the reducing side of photosystem I through an additional cytochrome carrier to plastoquinone in a process known as cyclic electron transport. This may occur when supplementary ATP is required *in vivo* (Gimmler, 1977; Shahak *et al.*, 1981). (The mechanism for ATP synthesis is explained below.)

D. Membrane structure and vectorial electron flow

Figure 2.2 illustrates the approximate location of the components of the photosynthetic light reactions within the thylakoid membrane (see, e.g., Renger, 1979; Kaplan and Arntzen, 1982). The reaction centres are located towards the inside of the membrane and the quinones Q_A and Q_B towards the stromal side, but covered by a protein shield. Plasto-quinone is mobile within the membrane, and when it has received 2 electrons from photosystem II, it takes 2 protons from the stromal side of the membrane, becoming reduced to PQH_2. The PQH_2 releases its protons on the inside of the membrane while the electrons are passed to the iron-sulphur protein and cytochrome f. Thus, for every pair of

---

*This component, like many in photosynthesis, is named with respect to its spectral properties; P indicates that it is a pigment, and 680 is the absorption maximum in nanometres.

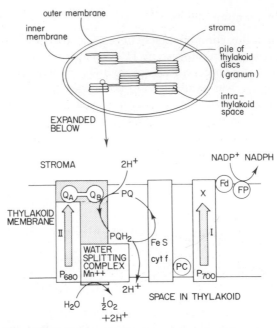

Fig. 2.2 Structure of the chloroplast and arrangement of major photosynthetic components in the thylakoid membrane. The rectangles which span the membrane depict the protein complexes which are able to move around within the lipid bilayer. For abbreviations of electron transfer components, see text

electrons passing through the photosystems, two protons are moved into the space inside the thylakoid.

A further two protons are produced inside the thylakoid when water is split and oxygen is formed (Fig. 2.2). Thus, during photosynthetic electron transport, net proton movement inwards takes place generating a proton gradient. According to present ideas this proton gradient, like the one generated across the mitochondrial inner membrane during oxidative phosphorylation, is used to drive the phosphorylation of ADP to ATP, the second product of the light reactions. (We will not consider the detail of photosynthetic phosphorylation since no herbicide is thought to have its primary effect here.)

The photosystems can be individually extracted from the thylakoid membranes in the form of protein complexes. Each of the two complexes comprises the light-harvesting apparatus and the immediate donors and acceptors for the photosystem. In addition the iron-sulphur protein and cytochrome f are part of another complex, as indicated in diagrammatic form in Fig. 2.2. The three complexes are able to move about in the fluid lipid membrane. Although it used to be assumed that each photosystem

II complex would be located near a photosystem I complex, current evidence suggests that the membranes which are pressed closely together in the centre of a granum (Fig. 2.2) contain mainly photosystem II complexes. Photosystem I is thought to be located mainly on the exposed regions of the membrane (i.e. on the outside of the grana and on the membrane extending across the stroma). In this model the plastoquinone is presumed to move laterally in the membrane, so that the photosystems are functionally, if not structurally, connected (Anderson, 1981). Alternatively it has been suggested that the two photosystems could function independently (Arnon et al., 1981).

E. Measurement of electron transport

In 1937 Hill showed that cell-free preparations of green plants could catalyse a light-dependent evolution of oxygen provided that an artificial electron acceptor, such as ferricyanide or a reducible dye, was present (see Hill, 1965). The resulting 'Hill reaction' can be written

$$2H_2O + 2A \xrightarrow[\text{chloroplasts}]{\text{light,}} 2AH_2 + O_2,$$

where A may be referred to as a 'Hill reagent' or 'Hill acceptor'. The Hill reaction has since been widely used in academic studies of photosynthesis and in work with herbicides.

The Hill reaction catalysed by isolated chloroplasts can be measured by the evolution of oxygen with either an oxygen electrode or, less conveniently, a Warburg manometer. However, most workers prefer the spectrophotometric method in which the formation of AH_2 is measured by a change in absorbance of light. The reduction of ferri- to ferrocyanide is particularly suitable from this point of view and ferricyanide is the acceptor used in many tests of the effects of herbicides on photosynthetic electron transport. In conventionally isolated chloroplasts under continuous illumination, ferricyanide is probably reduced predominantly by the primary acceptor of photosystem I (Izawa, 1980).

Acceptors other than ferricyanide are used in Hill reaction studies. However, it is satisfactory to compare measurements of inhibition obtained using different acceptors, with the obvious proviso that sensible results can only be obtained if the acceptor acts after the site of action of the inhibitor.

II. INHIBITORS OF PHOTOSYNTHETIC ELECTRON TRANSPORT
A. General introduction

Many of the herbicides interfering with photosynthesis act at a single site, and these compounds are divided into chemical classes in Table 2.1, which also contains references to work in which their effect on photosynthesis has been quantified. Some common structural requirements for activity have been pointed out (Moreland, 1969; Büchel, 1972; Draber et al., 1974; Trebst and Harth, 1974); inhibitors commonly have a lipophilic moiety in close association with the group –C(X)–N< where X = O or NH. One of the atoms attached to the nitrogen is often H, but this is not obligatory (Draber et al., 1974).

Table 2.1   Herbicides which act by inhibiting the Hill reaction

Name	Structure	I_{50} (μM)*
A. Ureas Benzthiazuron	—NH.CO.NHMe	
Buturon	 Cl—⟨⟩—NH.CO.N⟨ CH(Me)C≡CH / Me	
Chlorbromuron	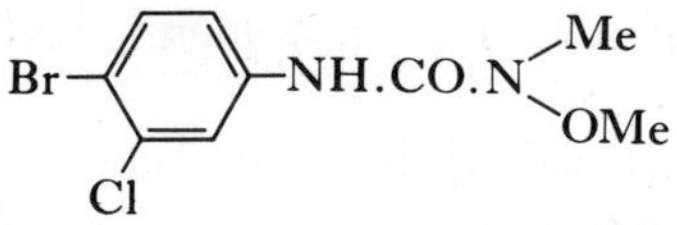 Br—⟨⟩—NH.CO.N⟨ Me / OMe (Cl)	0.06^a 0.22^b
Chloroxuron	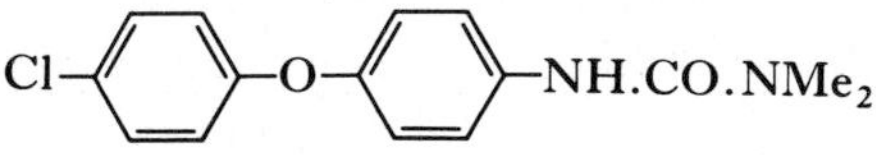 Cl—⟨⟩—O—⟨⟩—NH.CO.NMe$_2$	0.05^a 0.16^c
Chlortoluron	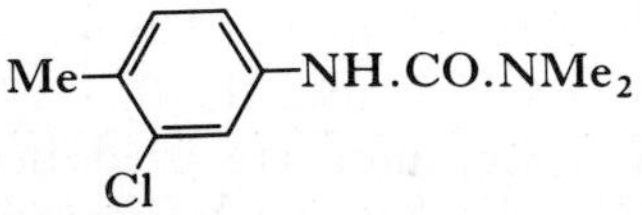 Me—⟨⟩—NH.CO.NMe$_2$ (Cl)	0.11^a

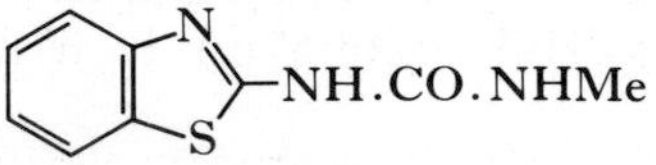

*I_{50} data are examples from the literature. They should be taken as a guide only, since variations in chloroplast source and test method can affect the result obtained. This is well illustrated by the comparative data collated by Dicks (1978).

Name	Structure	I_{50} (μM)
Cycluron	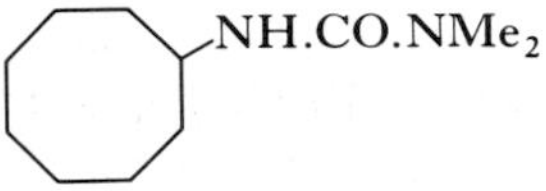 NH.CO.NMe$_2$	5.0^c
Difenoxuron	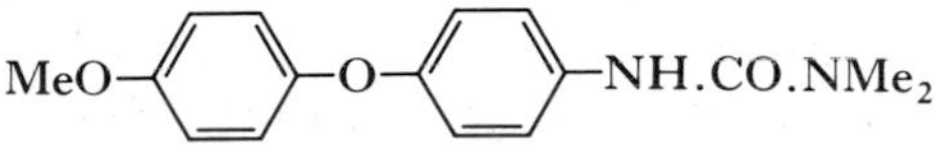 MeO—⟨⟩—O—⟨⟩—NH.CO.NMe$_2$	
Diuron	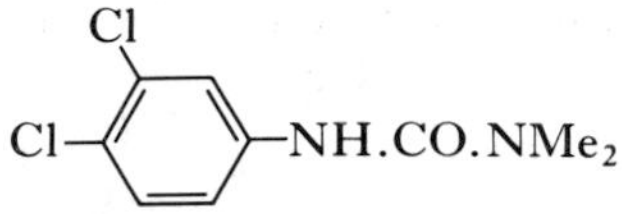 Cl, Cl—⟨⟩—NH.CO.NMe$_2$	0.04^a 0.18^c 0.03^d
Ethidimuron	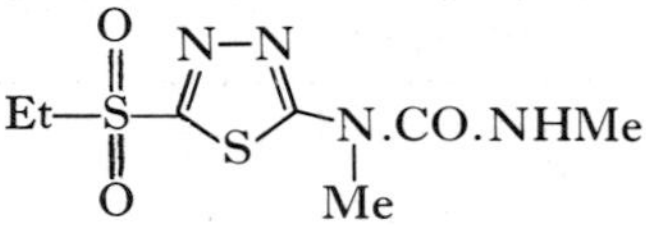 Et—S(=O)$_2$—⟨N–N / S⟩—N(Me).CO.NHMe	
Fenuron	 ⟨⟩—NH.CO.NMe$_2$	2.9^a 10^c 6.3^d
Fluometuron	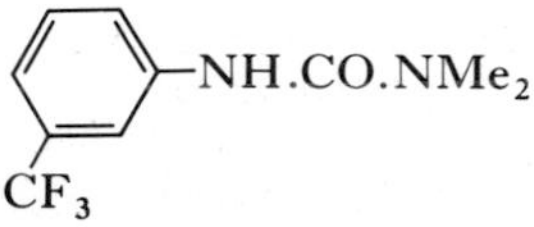 ⟨⟩—NH.CO.NMe$_2$, CF$_3$	1.1^a 6.0^e
Isoproturon	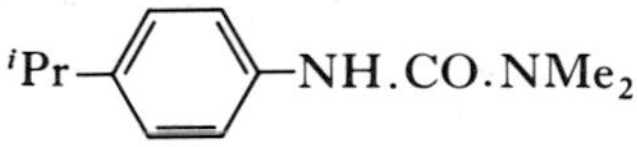 ^{i}Pr—⟨⟩—NH.CO.NMe$_2$	0.17^a
Linuron	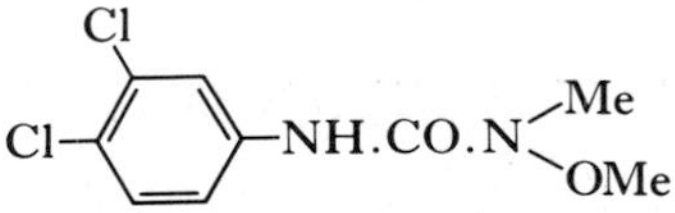 Cl, Cl—⟨⟩—NH.CO.N(Me)(OMe)	0.09^a 0.20^c 0.40^f
Methabenzthiazuron	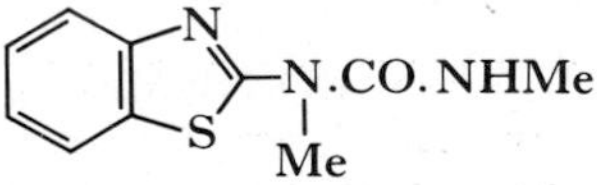 ⟨benzothiazole⟩—N(Me).CO.NHMe	$<1^g$ $c.1^h$

Name	Structure	I_{50} (μM)
Metobromuron	 Br—⬡—NH.CO.N(Me)(OMe)	0.94^a
Metoxuron	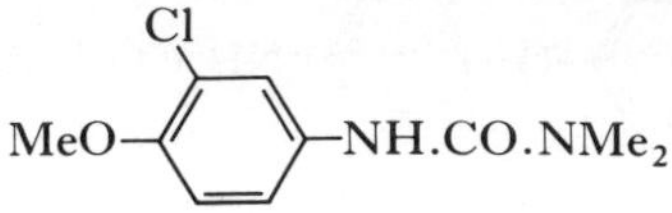 MeO—⬡(Cl)—NH.CO.NMe$_2$	0.26^a
Monolinuron	 Cl—⬡—NH.CO.N(Me)(OMe)	0.89^a
Monuron	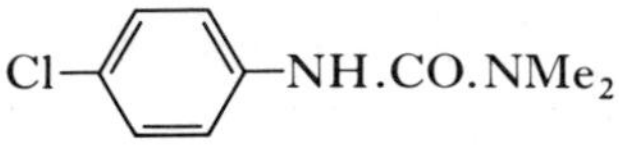 Cl—⬡—NH.CO.NMe$_2$	0.38^a 0.50^d
Neburon	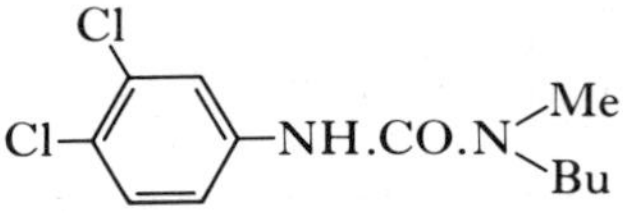 Cl,Cl—⬡—NH.CO.N(Me)(Bu)	0.14^c
Siduron	 ⬡—NH.CO.NH—cyclohexyl(Me)	$c.10$–60^l
Tebuthiuron	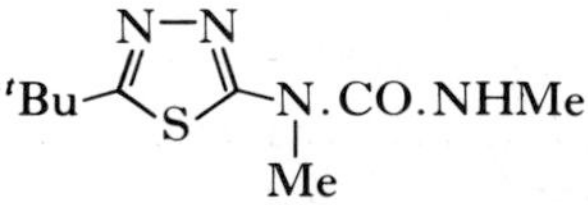 ^{t}Bu—[N-N,S thiadiazole]—N(Me).CO.NHMe	0.05–0.5^j
Thiazafluron	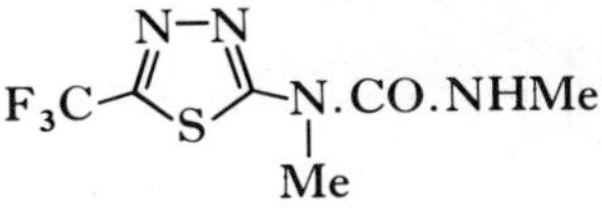 F$_3$C—[N-N,S thiadiazole]—N(Me).CO.NHMe	

Name	Structure	I_{50} (µM)
B. Cyclic ureas		
Buthidazole	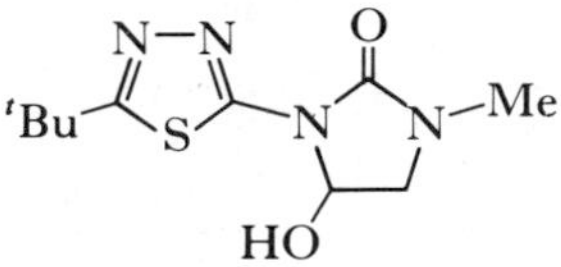	0.05–0.5[j]
Isocarbamid		
Methazole	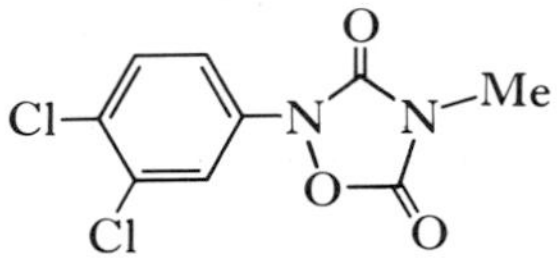	
C. Triazines		
Ametryne	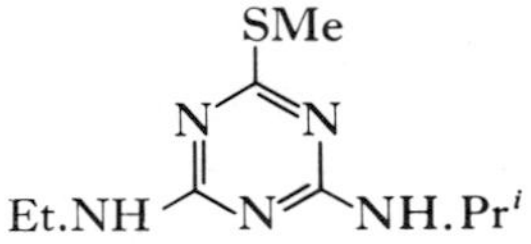	0.36[k]
Atrazine	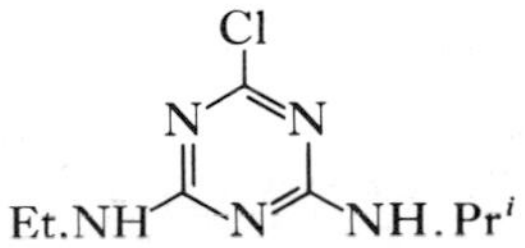	0.25[d] 1.51[k] 0.20[l]
Aziprotryne	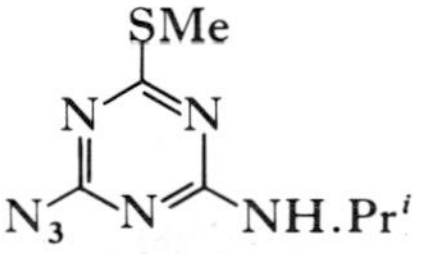	
Cyanazine	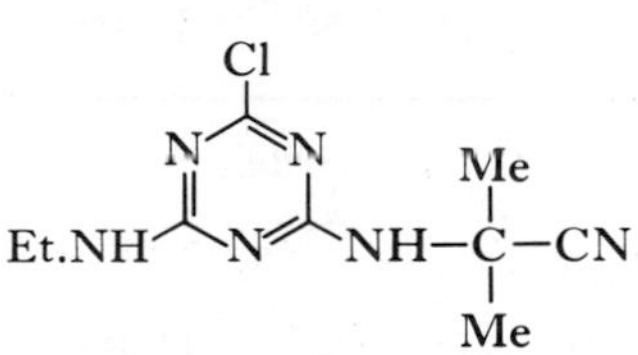	0.20[l]

Buthidazole:
^tBu ... $N{-}Me$, with HO substituent

Isocarbamid:
$^iPrCH_2NH.CO.N$ ... NH

Methazole:
Cl ... $N{-}Me$, Cl

Ametryne:
SMe; $Et.NH$ ... $NH.Pr^i$

Atrazine:
Cl; $Et.NH$ ... $NH.Pr^i$

Aziprotryne:
SMe; N_3 ... $NH.Pr^i$

Cyanazine:
Cl; $Et.NH$ ... $NH{-}\underset{Me}{\overset{Me}{C}}{-}CN$

Name	Structure	I_{50} (μM)
Desmetryne	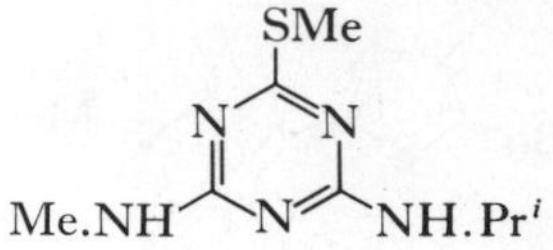 SMe; Me.NH, NH.Pr^i	1.41^k 0.17^m
Dimethametryn	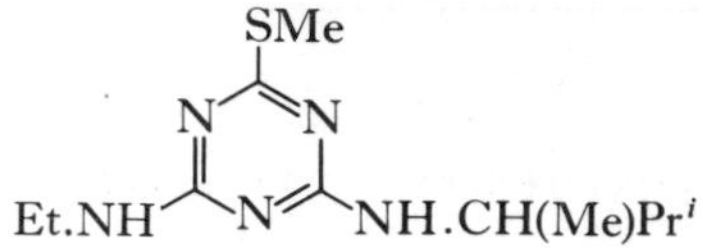 SMe; Et.NH, NH.CH(Me)Pr^i	
Dipropetryn	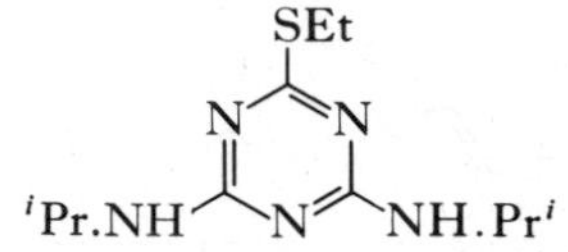 SEt; ^{i}Pr.NH, NH.Pr^i	
Eglinazine-ethyl	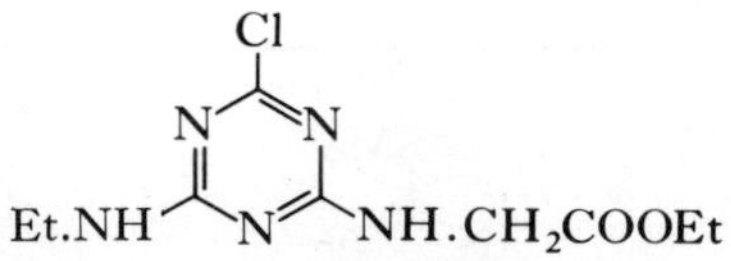 Cl; Et.NH, NH.CH$_2$COOEt	
Methoprotryne	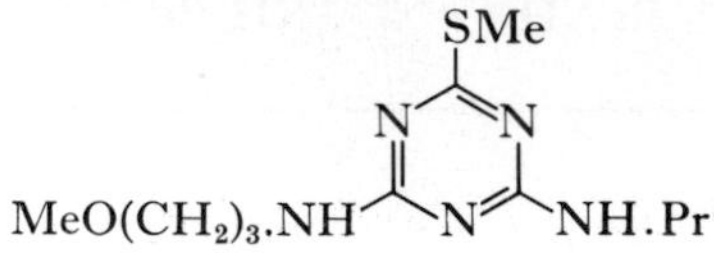 SMe; MeO(CH$_2$)$_3$.NH, NH.Pr^i	
Proglinazine-ethyl	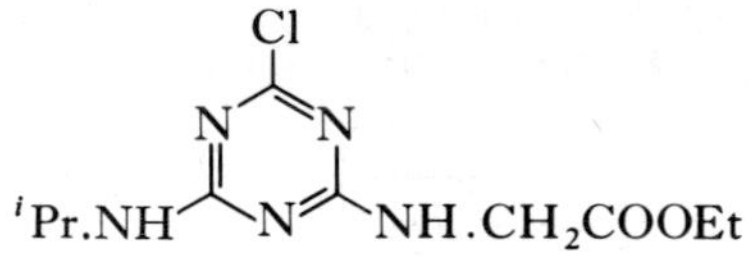 Cl; ^{i}Pr.NH, NH.CH$_2$COOEt	
Prometon	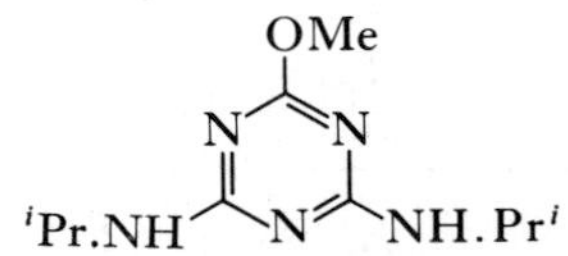 OMe; ^{i}Pr.NH, NH.Pr^i	2.0^d 1.3^m
Prometryne	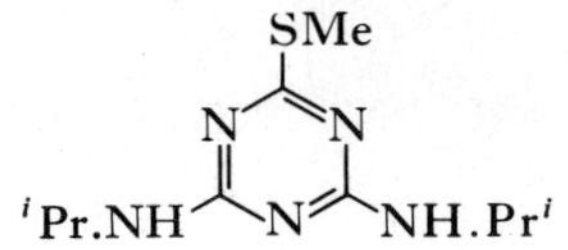 SMe; ^{i}Pr.NH, NH.Pr^i	0.21^k 0.09^m

Name	Structure	I_{50} (μM)
Propazine	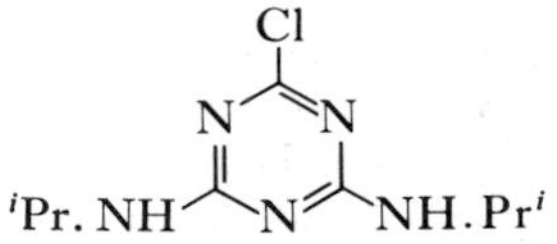	0.50[d] 0.35[m] 4.3[n]
Secbumeton	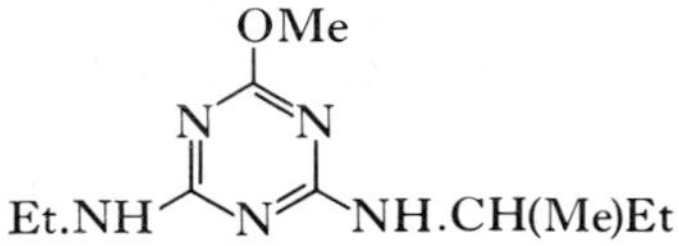	2.51[k]
Simazine	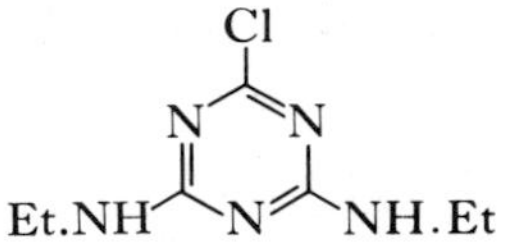	2.2[c] 0.4[d] 0.7[m] 5.9–8.4[n]
Simetryne	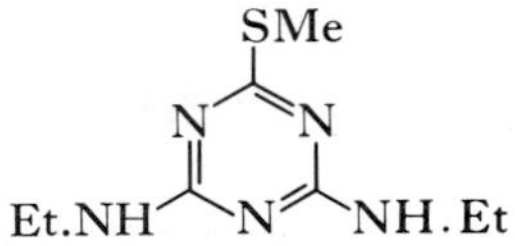	0.3[k]
Terbumeton	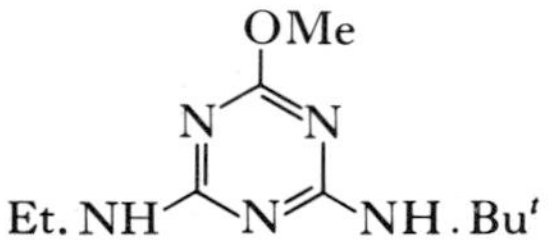	0.28[k]
Terbuthylazine	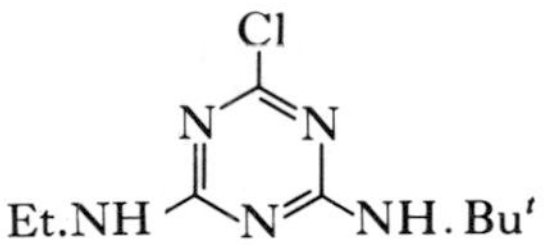	0.35[k]
Terbutryne	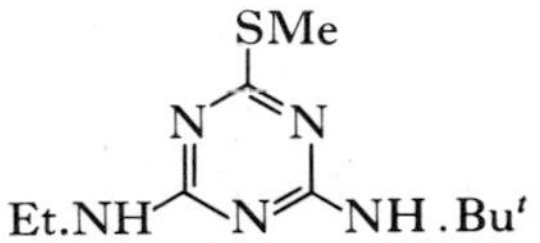	0.071[k]
Trietazine	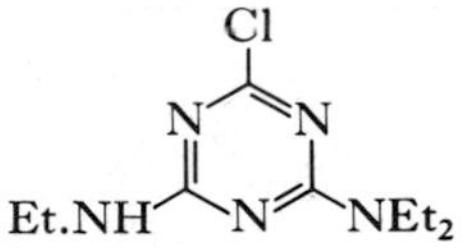	>1000[k] 270[n] c.100[o]

Name	Structure	I_{50} (μM)
D. Acylanilides Monalide	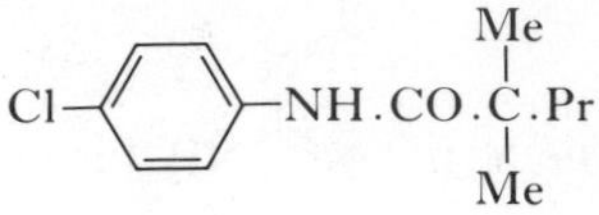	
Pentanochlor	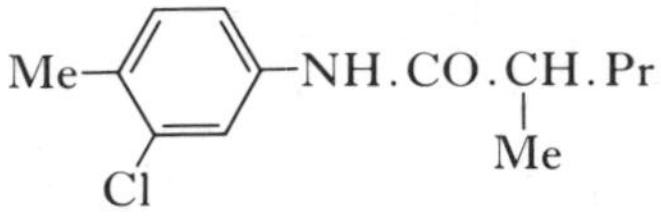	
Propanil	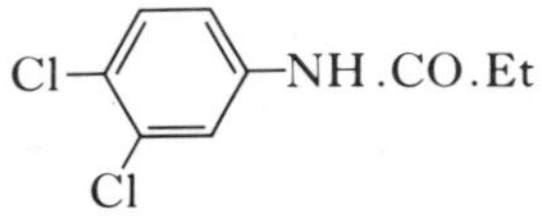	0.78^c 0.16^d
E. Uracils Bromacil	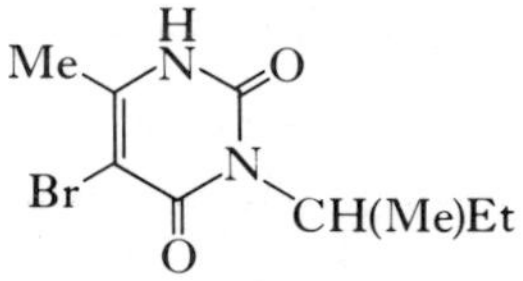	0.49^c 1.4^p 2.0^q
Lenacil	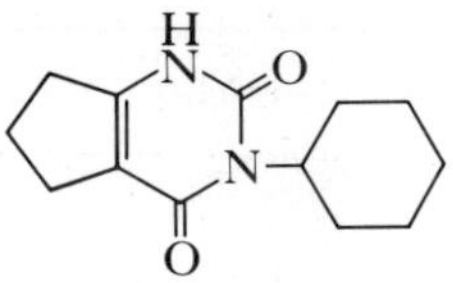	0.28^c 0.85^p
Terbacil	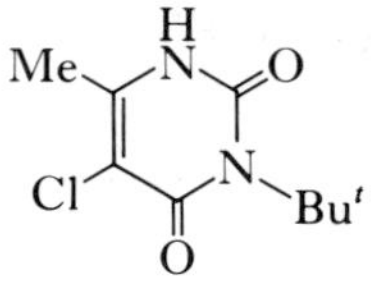	0.37^r
F. Phenylcarbamates Desmedipham	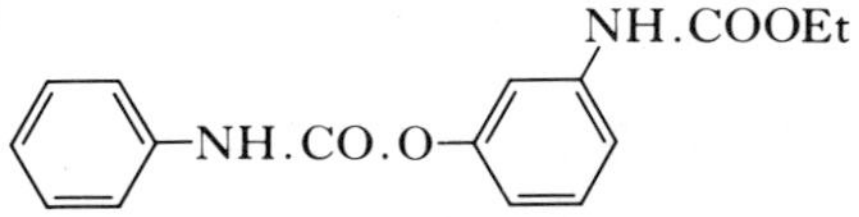	0.23^s
Phenisopham	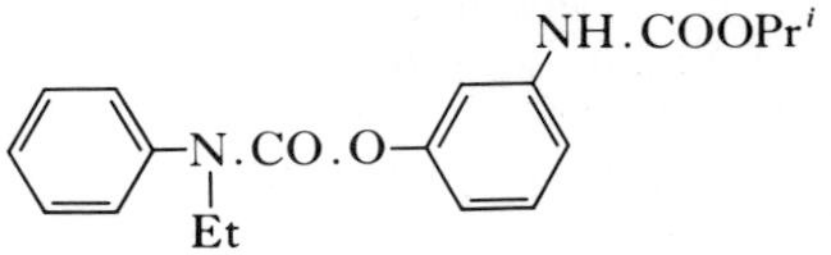	

Name	Structure	I_{50} (μM)
Phenmedipham		0.15[s]
G. Triazinones Isomethiozin		
Metamitron		0.8[t] 4[u]
Metribuzin		0.2[v] 0.23[w]
H. Miscellaneous compounds Bentazone		100[x] 40–48[y] 50[z] 20–150[aa]
Chloridazon		7[bb] 6[cc]
Hexazinone		

Name	Structure	I_{50} $^\backslash$ (μM)
Oxadiazon	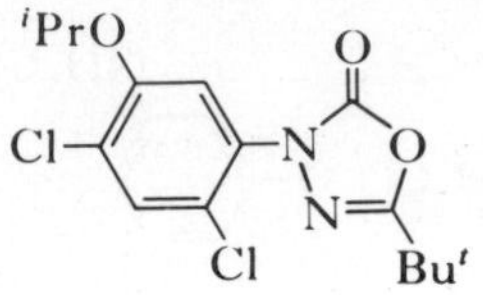 iPrO, Cl, N-O, N, Cl, Bu^t, O	
Quinonamid	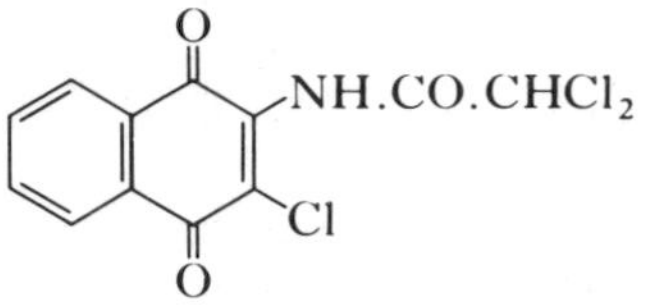 O, NH.CO.CHCl$_2$, Cl, O	

References: [a]Dicks (1978); [b]Green et al. (1966); [c]Moreland (1969); [d]Good (1961); [e]Wessels and van der Ween (1956); [f]Muschinek et al. (1979); [g]Fedtke (1973); [h]De Villiers et al. (1979); [i]De Mur et al. (1972); [j]Hatzios et al. (1980); [k]Ebert and Dumford (1976); [l]Brewer et al. (1979); [m]Gysin and Knuesli (1960); [n]Moreland and Hill (1962); [o]Dicks (1974); [p]Hilton et al. (1964); [q]Hoffman et al. (1964); [r]Hoffman (1972); [s]Schulz (1969); [t]Draber et al. (1974); [u]Schmidt and Fedtke (1977); [v]Trebst and Wietoska (1975); [w]Draber et al. (1969); [x]Retzlaff and Fischer (1973); [y]Mine and Matsunaka (1975); [z]Pfister and Arntzen (1979); [aa]Böger et al. (1977); [bb]Hilton et al. (1969); [cc]Frank and Switzer (1969).

All these compounds are thought to bind to a single site with which we shall be concerned in this section. We will first describe the evidence which indicated the point at which the herbicides interfere, and then examine the location of this site within the photosynthetic membranes and its properties. Specific points of interest related to individual classes of Hill reaction inhibitor will be considered at the end of the section.

As we describe various new experimental approaches that have been taken in the last few years it will become apparent that phenolic herbicides (Table 2.2), which can inhibit the Hill reaction (but may not have their primary action through this effect – see p. 82), do so in a slightly different way; the current view is that, while these compounds inhibit the Hill reaction at the same place in the electron transport chain as the chemicals whose general features are outlined above, they nevertheless do not seem to bind to the chloroplast membrane in exactly the same way.

The effects of herbicides on photosynthetic electron flow have been reviewed by, for instance, Moreland (1980), Trebst (1980) and Fedtke (1982).

Table 2.2 Phenolic herbicides that can inhibit the Hill reaction

Name	Structure	$I_{50}(\mu M)$

A. Hydroxybenzonitriles and a related compound

Bromofenoxim

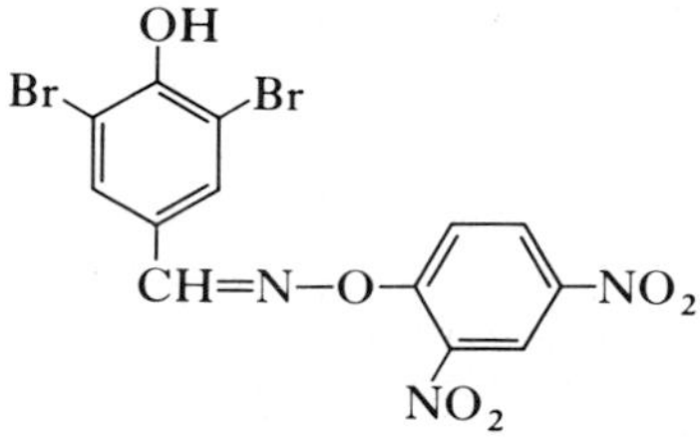

OH
Br — Br
CH=N—O— —NO$_2$
NO$_2$

4.6^a

Bromoxynil
(bromoxynil octanoate
is also in use)

OH
Br — Br
CN

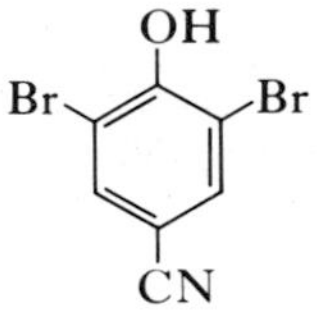

2.6^a
18^b
1^c

Ioxynil
(ioxynil octanoate
is also in use)

OH
I — I
CN

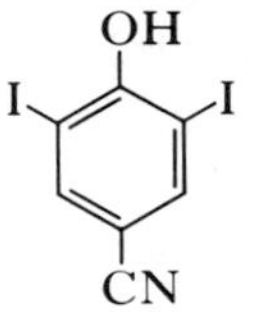

1.0^b
0.1^c
$0.5-10^d$
$0.07-0.5^e$
0.74^f

B. Phenols

Dinoseb
(dinoseb acetate
is also in use)

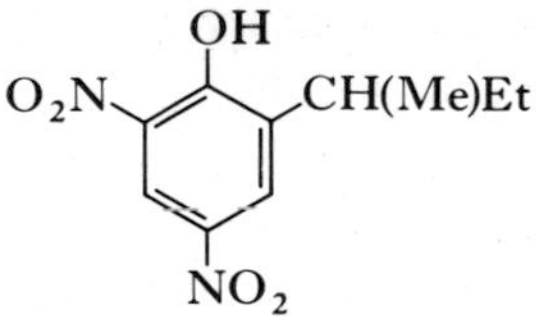

OH
O$_2$N — CH(Me)Et
NO$_2$

$< 10^g$
$13-17^h$

Dinoterb

OH
O$_2$N — Bu^t
NO$_2$

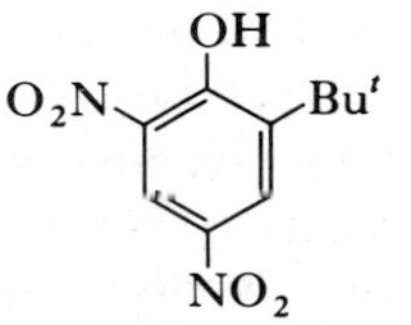

$c.\ 2^i$

Name	Structure	I_{50} (μM)
DNOC	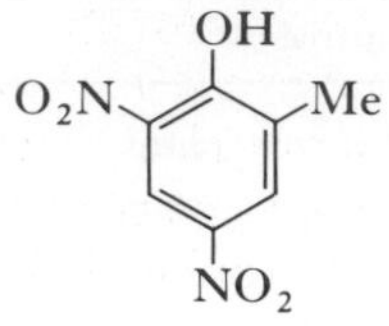	$1-15^j$

References: [a]Moreland and Blackmon (1970); [b]Kerr and Wain (1964); [c]Katoh (1972); [d]Desmoras and Jacquet (1964); [e]Gromet-Elhanan (1968); [f]Moreland (1969); [g]Younis and Mohanty (1980); [h]Moreland and Hill (1962); [i]Belbachir et al. (1980); [j]van Rensen et al. (1977).

B. Position of binding site within the electron transport chain

Some five years after the initial announcement of the discovery of herbicidal activity in the urea family (Bucha and Todd, 1951), Wessels and van der Veen (1956) showed that 0.2 μM diuron caused a 50% inhibition of electron transfer to the Hill acceptor, 2,6-dichlorophenolindophenol. This was one of the earliest demonstrations that a herbicide could act by inhibiting photosynthetic electron transfer, and many workers have since confirmed that diuron is an extremely potent inhibitor of the Hill reaction. It is the archetypal Hill reaction inhibitor and the site to which it and most of the other photosystem II herbicides bind is often referred to as the 'diuron site'. Wessels and van der Veen (1956) also showed that the closely related monuron caused 50% inhibition at 4 μM, and that its effect on isolated chloroplasts was reversible since it could be removed by washing.

This and much of the earlier work was reviewed by Moreland (1967). The outline account below illustrates the evidence that has been used to identify the site of herbicide action.

In the first place, photosystem I activity was found to remain unaffected in the presence of diuron. One way of demonstrating this employed a mutant of the alga Scenedesmus which could completely by-pass its photosystem II by using hydrogen directly as an electron source for photosystem I, so that although carbon dioxide was fixed there was no evolution of oxygen since water was not split by photosystem II. Bishop (1958) showed that, although diuron inhibited the normal photosynthetic production of oxygen from water by such a Scenedesmus mutant to the extent of 50% at 0.5 μM, the herbicide had no effect on the uptake of hydrogen when this was supplied to allow photoreduction to occur.

Another approach utilised artificial donors and acceptors to functionally isolate photosystem I and demonstrate its insensitivity to diuron. Chloroplasts supplied with 2,6-dichlorophenolindophenol, with ascorbate to maintain it in the reduced form, will take up electrons from the reduced indophenol (Fig. 2.1), so by-passing light reaction II (Vernon and Avron, 1965). Whereas a 10 μM solution of diuron causes more than 98% inhibition of normal photosynthetic reduction of $NADP^+$ by isolated chloroplasts, the addition of 2,6-dichlorophenolindophenol with ascorbate restores $NADP^+$ reduction to 94% of the original value (Vernon and Zaugg, 1960).

Further support for a lack of effect on photosystem I comes from systems in which only this reaction is operative. For instance, diuron does not affect photosynthetic reactions taking place in far-red light, when photosystem I is primarily involved (Moreland, 1967). In addition, photosynthetic bacteria, which lack photosystem II, are extremely insensitive to diuron (Hoffmann et al., 1964).

If diuron does not act on photosystem I, it must act either on photosystem II or on the electron transport chain. Further, the evidence already described above indicates that the site must be before the point at which reduced dichlorophenolindophenol donates electrons to the chain. This is confirmed by the influence of diuron on the oxidation and reduction of some of the electron transport components (see Moreland, 1967).

Further definition of the location of the diuron site has been sought by use of the electron acceptor silicomolybdate, which mediates a diuron-insensitive Hill reaction. However, there is still uncertainty as to the exact site at which this compound accepts electrons (see Böger, 1982, for discussion) so results are difficult to interpret.

Action of inhibitors in the vicinity of Q_A is supported by observations on the effect of diuron and related compounds on chloroplast fluorescence (fluorescence is the release of energy as light when an electron falls from an excited state back to the ground state). If Q_A is available to accept electrons produced by the water-splitting activity of photosystem II then no fluorescence is seen; it has been quenched by Q_A since this has accepted the electron in becoming reduced to Q_A^-. The rate of increase of fluorescence observed upon illumination of dark adapted chloroplasts is dramatically increased by diuron (see Pfister and Arntzen, 1979). This is attributed to the action of diuron in maintaining Q_A in its reduced form which can no longer accept energised electrons from the chlorophyll molecules; as a result, these electrons lose their energy via fluorescence.

The early experiments were carried out before the involvement of the quinone Q_B was suspected. In dark adapted chloroplasts, a single flash of light will result in the transfer of a single electron to Q_A, reducing it to Q_A^- (step 1, Fig. 2.3), and this will then transfer its electron to Q_B (step 2). A second flash of light allows Q_B to collect its second electron (steps 3

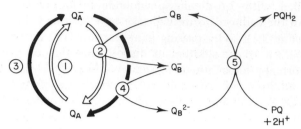

Fig. 2.3 Electron transfer between Q_A and plastoquinone. Steps 1 and 3 represent electrons excited by photosystem II. For explanation, see text

and 4), and the electron pair can then be passed to plastoquinone (step 5). (Alternatively, the Q_B^{2-} may exchange for a PQ molecule from the pool.) By the use of different numbers of light flashes, the effect of electron transport inhibitors on chlorophyll fluorescence can be examined, and this will yield information about the redox states of the components. Data obtained using this approach demonstrate that Q_A is functional as an electron carrier in chloroplasts treated with diuron, but that Q_B is not active as an electron acceptor under these conditions (see Pfister and Arntzen, 1979).

Thus the current view is that photosynthesis inhibitors bind to the chloroplast so as to prevent the correct function of the second quinone acceptor Q_B. This could occur by displacement of Q_B from its binding site (p. 76), by alteration of the redox potential of Q_B so as to make it harder to reduce, or by impairing the $Q_A \rightarrow Q_B$ reduction by an allosterically-induced shape change (see van Rensen, 1982).

C. Nature of the site
1. *Tryptic digestion of chloroplast membranes*

Incubation of chloroplasts with the proteolytic enzyme trypsin leads to inhibition of electron transport, presumably due to removal of part or all of a component of the chloroplast (Mantai, 1970). Chloroplasts treated with trypsin have the following properties:

(*a*) there is no effect on oxygen evolution, so that the oxidising side of photosystem II is probably left intact;

(*b*) the inhibitory effect of diuron, whether seen as inhibition of electron flow to ferricyanide, or as an increase in chlorophyll fluorescence, is diminished (Regitz and Ohad, 1976; Renger *et al.*, 1976; Renger, 1976). These and other experiments led Renger (1976) to postulate that a protein shield covered Q_A and Q_B and contained the diuron binding site (Fig. 2.2), which could be digested away by trypsin. Diuron was thought to inhibit electron flow by an allosteric effect, i.e. it bound at one site and exerted its effect at another part of the shield protein through an alteration in the shape of the protein. The physiological function of the protein is discussed below (p. 76). The proposal is consistent with the observation that diuron affords some protection against trypsin digestion of thylakoid membranes from the pondweed *Spirodela oligorrhiza* (Mattoo *et al.*, 1981).

The effect of trypsin digestion on the ability of diuron to inhibit electron transport could, on the basis of the observations described so far, be due to the removal of the binding site, or to digestion of part of the shield so that the ferricyanide could gain access to Q_A, thus short-circuiting the inhibitory action (Renger, 1979; Trebst, 1979; van Rensen and Kramer, 1979). Direct binding studies, which we shall discuss in the next section, do indeed demonstrate that the binding sites are being degraded, but it is also clear that a new site for ferricyanide reduction becomes exposed in trypsin-treated chloroplasts, so that the diuron inhibition site is by-passed (Steinback *et al.*, 1981). This seems not to occur with other electron acceptors such as dichlorophenolindophenol and paraquat (which will be discussed later) (Steinback *et al.*, 1981).

Is the relief of diuron inhibition that is caused by digestion with trypsin found with other inhibitors? Böger and Kunert (1979), measuring the Hill reaction with ferricyanide as an acceptor, found that pre-treatment with trypsin caused identical desensitisation of chloroplasts to diuron, fluometuron, atrazine, metribuzin and a pyridazinone, while Pallett and Dodge (1979) showed that tryptic digestion rendered the same reaction insensitive to monuron, bromacil, and metribuzin. Similarly Steinback *et al.* (1981) found that the dichlorophenolindophenol Hill reaction lost sensitivity to diuron, atrazine, bromacil and chloridazon (formerly pyrazon). Thus, the inhibition of electron transport caused by a wide range of structural types of herbicide is relieved by trypsin digestion.

On the other hand, inhibition by phenols and hydroxybenzonitriles has been found to be only slightly diminished (Pallett and Dodge, 1979; Oettmeier *et al.*, 1982) or increased by trypsin digestion. As an example of the latter, Steinback *et al.* (1981) found that the Hill reaction became more sensitive to bromonitrothymol and dinoseb following short trypsin treatments, although the sensitivity returned to nearer its original value after longer times of incubation.

Trypsin could also abolish the time-lag which occurs before hydroxybenzonitrile inhibitors exert their maximal inhibition (Reimer *et al.*, 1979, Böger and Kunert, 1979). [For instance, in the absence of trypsin, ioxynil caused approximately 5% inhibition after incubation with chloroplasts for half a minute at 0.5 μM but about 75% inhibition after 6 minutes (Reimer *et al.*, 1979).] The time lag was eliminated not only by trypsin but also by pre-illumination in the absence of trypsin (such pre-illumination has no effect on the inhibition caused by the main body of herbicides inhibiting photosynthesis; Böger and Kunert, 1979). It is not clear at present exactly how these results should be interpreted, but they clearly separate the hydroxybenzonitriles from the other herbicides.

Tryptic digestion of subchloroplast particles also offers an approach to the direct identification of the proteins which bind diuron and other herbicides. Croze et al. (1979) used a detergent to make small photosystem II particles which were able to carry out electron transport in the light and which retained a sensitivity to diuron. This sensitivity was destroyed by trypsin. These workers extracted control and trypsin-digested particles and examined the proteins present by electrophoresis on polyacrylamide gels. They found that the proteins in two bands, representing molecular masses of 27,000 and 32,000 daltons had been extensively degraded by the trypsin, to be replaced by a band of smaller molecular mass. Other small changes were also seen. Similar approaches by Mattoo et al. (1981) and Steinback et al. (1981) have supported the conclusion that a component of molecular mass 32,000 daltons was degraded by trypsin in parallel with the loss of other chloroplast activities.

2. *Direct measurement of the affinity of Hill inhibitor binding*

The interaction of photosynthetic herbicides with chloroplast membranes can be studied directly by the use of radioactive compounds, a technique introduced in the elegant experiments of Tischer and Strotmann (1977). The method is simple in principle. Chloroplasts are incubated with different concentrations of a radioactive Hill reaction inhibitor for a short period, and are then separated from the supernatant by centrifugation. Inhibitors become bound to the chloroplast membranes according to their affinity, and measurement of the amount of free (supernatant) and bound (pellet) radioactivity allows quantification of the binding constant. If a non-labelled herbicide lowers the amount of label bound to the membranes, the nature of the interaction between the two inhibitors can be analysed, and the binding constant of the non-radioactive ligand can be calculated. The analysis also allows calculation of the number of binding sites of an inhibitor, expressed on the basis of the chlorophyll content of the sample.

From such experiments with radioactive atrazine, metribuzin and phenmedipham, and a similar analysis using Hill reaction inhibition in place of bound radioactive ligand, Tischer and Strotmann were able to demonstrate that

(a) these compounds and diuron compete directly with one another for binding to chloroplast membranes and

(b) the dissociation constants calculated directly from studies with radioactive compound or indirectly from competition experiments, were the same as the inhibition constants computed from Hill reaction inhibition data.

Hence the site at which all the compounds compete is the site at which occupancy is required for inhibition to occur. The binding site was found to be present at the rate of one per 300–500 chlorophyll molecules. On the basis of the known concentrations of other electron carriers, this suggests that there is one binding site per electron transfer chain.

This technique has been employed by other workers to show that, with the possible exception of the phenol herbicides discussed below, all photosystem II herbicides displace each other in a competitive way from the chloroplast binding site. Van Assche and Carles (1982), for instance, added lenacil, propanil and chloridazon to the list of commercial Hill reaction inhibitors able to competitively inhibit the binding of radioactive atrazine.

In general it seems to have been assumed that interference of diuron type and phenolic inhibitors with each other's binding has been due to a competitive interaction (in this type of interaction diuron would decrease the affinity of, for example, ioxynil for the site but would not affect the total number of sites available to bind ioxynil). The question was investigated in a recent study exploring the relationship between the binding of diuron, ioxynil and a nitrophenol, 2-iodo-4-nitro-6-isobutylphenol (Oettmeier *et al.*, 1982). This last compound had an I_{50} of 0.03 μM and a

2-iodo-4-nitro-
6-isobutylphenol

dissociation constant of 0.1 μM (Oettmeier *et al.*, 1982). These authors concluded that, whereas the nitrophenol interfered with the binding of ioxynil in a competitive way, the interaction between diuron and each phenol was non-competitive. This would mean that diuron decreased the number of binding sites available for the phenols, while not affecting the affinity of the phenols for the remaining binding sites. However, an adjustment procedure used by Oettmeier and his colleagues to correct their data for low-affinity (non-specific) binding is considered by Laasch *et al.* (1982) to be inappropriate. From their own binding experiments these workers concluded that there was competitive displacement of 2-iodo-4-nitro-6-isobutylphenol by diuron, as well as of atrazine by dinoseb and ioxynil; further, diuron and ioxynil displaced each other in a competitive fashion. Additional work will presumably resolve these conflicting conclusions.

A competitive interaction of the sort discussed does not necessarily mean that the compounds bind to exactly the same site. The situation is

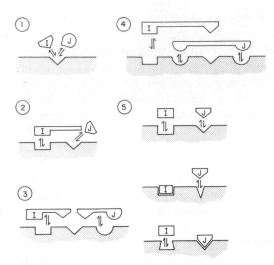

Fig. 2.4 Models for competitive interaction of two Hill reaction inhibitors I and J. (1) Classical model; I and J compete for the same binding site. I must resemble J structurally. (2) I and J are mutually exclusive because of steric hindrance. (3) I and J share a common binding group on the site. (4) The binding sites for I and J are distinct, but overlapping. (5) The binding of I to one inhibitor site causes a conformational change in the site that distorts or masks the J binding site (and vice versa). (Slightly modified from Siegel, 1975)

analogous to competitive inhibition in enzyme kinetics, and the correct conclusion to draw is that both inhibitors cannot bind simultaneously. Direct competition for the same site is the simplest case of this, but there are others, and some models are given in Fig. 2.4, adapted from Siegel (1975).

The use of direct binding studies in conjunction with trypsin digestion can also yield useful information about Hill inhibitors and their binding sites. For instance Tischer and Strotmann (1979) incubated chloroplasts with increasing concentrations of trypsin and with a single concentration of metribuzin for increasing periods of time, and found that the dissociation constant for the inhibitor rose only slightly while the total number of metribuzin binding sites fell away dramatically. In other words, with time or increasing trypsin concentration, some receptors were removed or destroyed completely by the enzyme while those remaining were only slightly modified in their affinity for metribuzin. As a second example Steinback *et al.* (1981) carried out a similar experiment using atrazine. They too found that the total number of binding sites fell away steadily, but was accompanied by substantial increases in the atrazine dissociation constant. They found further that the loss of binding sites paralleled the loss of ability to carry out electron transport, which supports the idea that the presence of the herbicide binding protein is essential for the function of components at the reducing side of photosystem II.

3. *Examination of chloroplasts from herbicide-resistant plants*

Although resistance (see Chapter 9) to herbicides is unusual, atrazine resistance had been recorded in six species up to 1979 (Pfister *et al.*, 1979), and it was shown that light-induced electron transport by chloroplasts isolated from three different weed biotypes was not inhibited by atrazine (Radosevich and De Villiers, 1976; Radosevich, 1977). Susceptibility to diuron was retained, however, reflecting the herbicide sensitivity of the intact plant. Detailed comparisons of chloroplasts from resistant and susceptible biotypes of these weeds (which include *Senecio vulgaris*, *Amaranthus retroflexus* and *Chenopodium album*) have now been made, and the implications for the herbicide binding site have been reviewed (Pfister and Arntzen, 1979).

In chloroplasts from susceptible *Senecio vulgaris* labelled diuron could be displaced by atrazine and *vice versa*. However, in chloroplasts from resistant plants diuron binding was only slightly diminished, while atrazine could neither bind to, nor displace radioactive diuron, from the membranes (Pfister *et al.*, 1979). As an extension of this work Pfister and Arntzen (1979) measured I_{50} values for various other Hill reaction inhibitors in resistant and susceptible chloroplasts, and expressed the results as the ratio of I_{50} values (resistant/susceptible). Some of their results are given in Table 2.3; due to the extreme insensitivity of resistant chloroplasts to some

Table 2.3   Susceptibility to Hill reaction inhibitors of chloroplasts from atrazine-susceptible and atrazine-resistant *Amaranthus retroflexus* plants[*]

Compound	I_{50} (µM) of chloroplasts from susceptible plants (S)	I_{50} (µM) of chloroplasts from resistant plants (R)	$\dfrac{R}{S}$	
Atrazine	0.36	*c.* 300	*c.* 1000	
Ametryne	0.043	20	46	
Metribuzin	0.21	54	260	
Diuron	0.06	0.081	1.4	
Chloridazon	6.1	*c.* 400		*c.* 65
Bromacil	0.25	5	20	
DNOC	35	5	0.14	
Ioxynil	0.7	0.45	0.64	
Bentazone	50	34	0.60	

[*]Data are taken from Pfister and Arntzen (1979).

compounds, values had to be estimated. As can be seen the degree of resistance to the triazines and, to a lesser extent, metribuzin was very large, and there was an intermediate resistance to bromacil and chloridazon. Atrazine-resistant chloroplasts were in fact more sensitive to phenols and this was also true for bentazone, though this was not a very good inhibitor in either case. On the basis of these results Pfister and Arntzen (1979) suggested that separate groups of herbicides bound to distinct but over-lapping binding sites; the area of overlap needed to be occupied for in-hibition to occur. The model therefore corresponds with number 3 in Fig. 2.4. A similar model was advocated by Trebst and Draber (1979). (Note that some evidence suggests that the phenol herbicide binding site is distinct from that to which other Hill inhibitors bind, as explained in other sections in this chapter, but the comments certainly apply to the sub-classes of Hill inhibitors binding at the main site.)

It has been of interest to examine in more detail the properties of chloro-plasts from atrazine-resistant plants. Photosystem II reactions were much less efficient in chloroplasts from atrazine-resistant weeds (Bowes et al., 1980; Holt et al., 1981) and this decreased fitness was reflected in the fact that resistant biotypes did not compete well with sensitive ones (Conard and Radosevich, 1979).

A number of differences have also been noted between the lipid com-positions of chloroplast membranes from sensitive and resistant biotypes of three weed species. Although the changes could be related to the triazine resistance they are not considered to be the direct cause of it (Pillai and St. John, 1981); resistance is thought to be due to changes in the binding protein (see below).

4. *Examination of the molecular nature of the binding site*

Recent studies have focused on the properties of the molecules which constitute the binding site, and photoaffinity labelling (p. 27) has been particularly useful. In this technique chloroplasts are incubated with radiolabelled photoreactive analogues of Hill reaction inhibitors, illumin-ated so that the analogue reacts with receptor groups which lie close to it, and then analysed to identify the component which carries the radioactivity.

Photoaffinity labelling has been carried out using ^{14}C-azido-atrazine (Gardner, 1981; Pfister et al., 1981b) with pea and *Amaranthus* chloro-plasts. The major labelled protein had a molecular mass of 32–34,000 daltons; it was synthesised by chloroplasts from resistant *Amaranthus*, but was not labelled by azido-atrazine (Steinback et al., 1982).

azido-atrazine

2-azido-4-nitro-6-isobutylphenol

Mullet and Arntzen (1981) found a similar protein to be labelled when they examined the effect of azido-atrazine on subchloroplast particles with photosystem II activity. In this same preparation these authors also found that the loss of sensitivity to atrazine paralleled the loss of the 32–34,000 dalton protein following either trypsin digestion or selective extraction with detergent (detergent is presumed not to cause breakage of covalent bonds). The detergent-extracted preparation retained some sensitivity to diuron and dinoseb, so inhibition by these herbicides does not have an absolute requirement for the 32–34,000 dalton protein and must presumably require the presence of other membrane components as well.

Results with *Chlamydomonas* mutants lacking various photosynthetic components also suggested that high-affinity diuron binding required the presence of components other than the single polypeptide which has received most attention so far (Shochat *et al.*, 1982).

The report by Gressel (1982) that *Spirodela* plants depleted of the 32,000 dalton protein are still sensitive to atrazine and diuron also suggests that the 32,000 dalton protein does not uniquely determine herbicide sensitivity. Gressel advocates caution in interpreting photolabelling experiments, pointing out that the photoaffinity reagent may become covalently attached to a protein adjacent to the one to which it is reversibly bound. To make this point more generally, a number of membrane components may contribute to the binding site, whereas only one will be so placed as to react with the photoaffinity label.

A study using tritiated 2-azido-4-nitro-6-isobutylphenol as the photo-affinity label led to the conclusion that the phenol herbicides bind to a different protein, of molecular mass 41,000 daltons (Oettmeier *et al.*, 1982). These authors proposed that this protein might be beneath the 32–34,000 dalton protein, close to Q_A and Q_B, so as to take account of the observations that trypsin readily abolishes diuron binding but not phenol binding, that photosystem II particles depleted of the 32–34,000 dalton protein lose most of their sensitivity to diuron while remaining relatively susceptible to phenolic inhibitors, that mutants losing the ability to bind atrazine became more sensitive to phenol herbicides, and that there is a time delay before phenol herbicide inhibition is maximal, consistent with impaired access to the binding site.

5. *Function of the site*

In proposing that a protein shield carried the herbicide binding site, Renger (1976) suggested that the protein was involved in regulating electron flow between Q_A and Q_B, and in preventing indiscriminate reaction between these electron carriers and other redox components. Some of the evidence described above supports this view. It is also known that, across a wide range of plant species, the protein is both rapidly synthesised and degraded, and of broadly similar structure (Hoffman-Falk *et al.*, 1982); this is again consistent with an important role.

It is known that bicarbonate, HCO_3^- (or CO_2) is required for photosynthetic electron transport (for reviews see Govindjee and van Rensen, 1978; Vermaas and Govindjee, 1981b), a requirement which is not shown by trypsin-treated chloroplasts (Khanna *et al.*, 1981). Recent evidence has indicated that this ion binds very close to the herbicide binding site and that bicarbonate-depleted chloroplasts show a decreased affinity for atrazine (Khanna *et al.*, 1981). Bicarbonate can be displaced from chloroplasts by silicomolybdate, and the fact that this can be antagonised by diuron is consistent with a model in which bicarbonate lies at the bottom of a herbicide binding site, overlaid by diuron (Stemler, 1977; Vermaas and van Rensen, 1981). This also fits with the observation that diuron and the triazine simeton interfered with bicarbonate binding (van Rensen and Vermaas, 1981).

Phenol herbicides also bind more tightly to chloroplasts carrying a bicarbonate ion though results obtained with ioxynil were easier to accommodate than those obtained with DNOC (Vermaas *et al.*, 1982).

Thus there is a site close to, or part of, the herbicide binding site which needs to be occupied by bicarbonate for electron transport to occur, perhaps as a result of a conformational change (Vermaas and Govindjee, 1982) (for the possible physiological significance of the bicarbonate binding site see Vermaas and Govindjee, 1981b). Bicarbonate binding could be the function of the Hill inhibitor binding site, but this would not provide any rationale for the existence of a site which is capable of binding structurally diverse Hill reaction inhibitors with high affinity.

An alternative possibility is that the natural ligand for the Hill inhibitor binding site is the special plastoquinone molecule Q_B, which, as mentioned above, would be displaced by the chemical so as to interrupt electron flow. Such a possibility is supported by recent results showing that plastoquinone analogues can displace bound herbicides from chloroplast membranes (Arntzen *et al.*, 1983).

6. *Summary of the properties of the binding site*

On the basis of a consensus of the work currently available we can build up a picture of the herbicide binding region as follows. Most of the photosystem II herbicides, but perhaps not the phenolic inhibitors, bind to a site which occurs at the rate of one per electron transfer chain, and which is probably located on a protein of molecular mass about 32,000 daltons. Triazine-resistance is associated with changes in this protein. The presence of other membrane components seems also to be required for herbicide-induced inhibition of electron transport. When the site is occupied, electron transfer between Q_A and Q_B is prevented. In many studies the phenolic herbicides have behaved differently from Hill inhibitors of other classes. These compounds may bind to a distinct but closely-associated protein, also located in the Q_A/Q_B region, having a molecular mass of about 41,000 daltons, and perhaps being less accessible from the stromal side of the membrane.

Trebst *et al.* (1983) summarise these ideas in a schematic diagram on which the following is based:

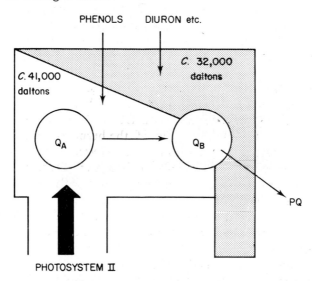

Since Laasch *et al.* (1982) found competitive interactions between the phenols and the other inhibitors (p. 71), they suppose that the two proteins interact to create a common binding site. There would, however, be specific sub-receptors for each type of herbicide.

Thus the exact relationships between the main binding site, the phenol binding site and the binding site for the special plastoquinone molecule Q_B still have to be clarified. However, this is a rapidly moving field and we can anticipate much progress in the near future.

D. Specific compounds and classes of compound

In the section which follows, particular points of interest will be mentioned relating to members of the various classes of Hill reaction inhibitor. General accounts of these herbicide classes are given by Ashton and Crafts (1981).

1. Ureas

Diuron, as the archetypal urea, has been widely used in studies aimed at increasing our understanding of the interaction of Hill reaction inhibitors with chloroplasts. Other ureas are included in Table 2.1 on the basis of structural analogy and/or the symptoms exhibited by treated plants. The comparative survey by Dicks (1978) serves to emphasise the fact that figures obtained in different laboratories must be compared only with great caution because of the different biological materials and assay techniques employed.

The urea herbicides have been reviewed by Geissbühler et al. (1975).

2. Cyclic ureas

Buthidazole, isocarbamid and methazole are urea derivatives in which the urea moiety is cyclised through a bridging group between the two nitrogen atoms (Table 2.1).

Methazole, the first member of this class to be introduced, is reported as about 100 times less active than diuron as a Hill reaction inhibitor (Zohner and Bayzer, 1977); in spinach chloroplasts it gave only about 20% inhibition at 500 μM (Verity et al., 1981a). Jones and Foy (1972) showed that cotton probably metabolised the compound by a route since confirmed in other plants, viz.

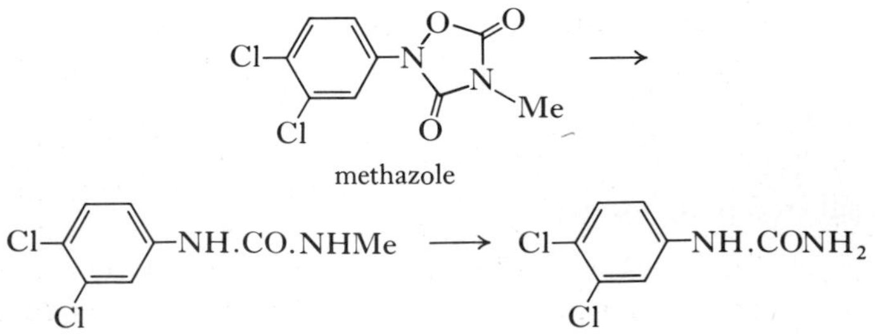

methazole

They found that the first metabolite, but not the second, was approximately as herbicidal as methazole itself, and that the symptoms of phytotoxicity of both compounds were typical of those caused by photosynthetic inhibitors. The first metabolite is closely related to diuron, and was shown by Good (1961) to cause 50% inhibition of the Hill reaction at 0.1 μM. It therefore seems quite probable that methazole is herbicidal because it acts as a precursor for a urea which inhibits the Hill reaction. A broad correlation was found between the extent of formation of this urea and the susceptibility of various plants to methazole (Verity et al., 1981a, b).

Buthidazole on the other hand is an active Hill reaction inhibitor in its own right, with an I_{50}, determined using several systems, in the range 0.05–0.5 μM (Hatzios et al., 1980).

There does not seem to be any specific information about the mode of action of the cyclic urea derivative isocarbamid (Table 2.1).

3. Triazines

Atrazine, like diuron, has been widely used in experiments to elucidate the herbicide binding site. Wider ranging aspects of the triazine herbicides have been reviewed by Esser et al. (1975), Knuesli (1976) and Ebert and Dumford (1976).

In a comparison of the effects of atrazine, cyanazine and the related compound procyazine on pea chloroplasts, Brewer et al. (1979) found that,

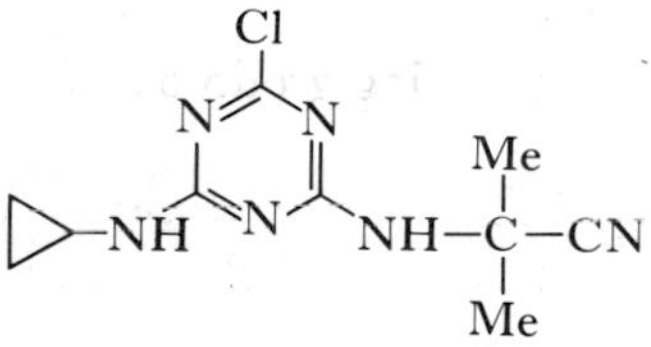

procyazine

while all three compounds were equally effective as inhibitors of photosynthetic electron transport, only the latter two showed a time delay before exhibiting their full inhibition. No inhibition at all could be detected after 30 minutes, though low I_{50} values were measured after 180 minutes. The value for cyanazine in Table 2.1 was obtained by extrapolating to infinite time. The authors attribute the time-dependence to slow partitioning of the compound into the chloroplast membranes from the aqueous medium, and this is supported by the fact that the initial inhibition level of cyanazine and procyazine was increased if the membranes were treated with detergent (Brewer et al., 1979). The observation of time-dependence over such a long period should be borne in mind when conducting or evaluating Hill reaction inhibition experiments with triazines, and could affect some of the values in the table.

Trietazine is only a poor inhibitor of the Hill reaction (Table 2.1) so it may act *in vivo* by the loss of an ethyl group to give simazine.

4. Triazinones

The triazinone herbicide isomethiozin seems not to be active in its own right, but owes its effect to conversion to metribuzin in the plant (Fedtke, 1982).

5. Bentazone

Bentazone is a contact herbicide used for selective post-emergence weed control in, e.g. soybeans (Worthing, 1979). It has been shown to inhibit photosynthetic carbon dioxide fixation (Retzlaff and Fischer, 1973) and photosynthetic electron transport (see Table 2.1 for references).

Bentazone is, however, a relatively ineffective inhibitor of the *in vitro* Hill reaction, with I_{50} values in the range 20–150 μM (Table 2.1). Also, with the phenolic herbicides, it falls within the class of Hill reaction inhibitors to which chloroplasts from atrazine-resistant plants were actually more sensitive (Pfister and Arntzen, 1979). Various other anomalies of the Hill reaction inhibition have been recorded and it is possible that a metabolite may be involved in the inhibition (Böger *et al.*, 1977). The phytotoxicity of bentazone is light-dependent, and ultrastructural changes consistent with an effect on photosynthetic electron transport occur within the chloroplast following bentazone treatment (Potter and Wergin, 1975; Meier *et al.*, 1980). Nevertheless the possibility that there is a more sensitive chloroplast target should not be overlooked (Potter and Wergin, 1975).

6. Chloridazon (formerly pyrazon)

This compound (Table 2.1) causes 50% inhibition of the Hill reaction at less than 10 μM, and the evidence suggests that this explains its herbicidal action. Related pyridazinones used as experimental herbicides will inhibit carotenoid synthesis (Chapter 6), but this is probably not significant in the herbicidal action of chloridazon itself (Hilton *et al.*, 1969).

7. Other miscellaneous compounds

The triazinedione hexazinone (Table 2.1) was found to depress CO_2 fixation by isolated soybean leaf cells, with an I_{50} of around 0.1 μM (Hatzios and Howe, 1982).

Oxadiazon (Table 2.1) is a weak inhibitor of the Hill reaction, having an I_{50} of approximately 20 μM against spinach chloroplasts (Trebst and Harth, 1974). Although this is a little high to be certain that inhibition of photosynthetic electron transport is its primary site of action, the fact that light is required for herbicidal activity (Matsunaka, 1970) is consistent with such a view.

The dose-response curve for inhibition of algal growth by the algicide quinonamide (Table 2.1) closely resembled that for inhibition of the Hill reaction in spinach chloroplasts, both having I_{50} values in the region of 2 μM (Bauer and Köcher, 1979). The site of action was not investigated in this study, but other halogenated naphthoquinones act at a site which is at least very close to that occupied by diuron (Pfister et al., 1981a).

8. Phenolic herbicides

Throughout this section on photosynthesis inhibitors we have pointed out that the hydroxybenzonitriles and nitrophenols (Table 2.2) have different properties from those of other inhibitors. This also emerged from a study by Trebst and Draber (1979), who made a series of alkyl-substituted nitrophenols of structure:

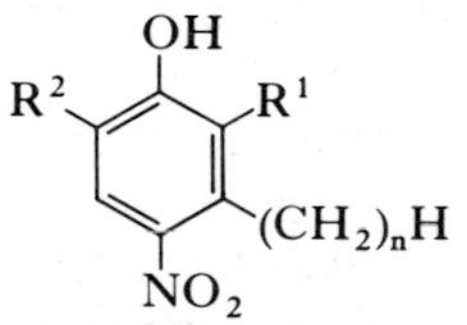

$$\text{structure with } R^2, \text{ OH, } R^1, (CH_2)_nH, NO_2$$

and carried out a detailed structure/activity correlation of their ability to inhibit photosynthetic electron transport. They concluded that it was the shape of R^1 and R^2 rather than their electronic and hydrophobic properties which determined the efficacy of the compounds as Hill reaction inhibitors. Steric factors were also found to be more important than electronic ones in a more limited study of structure/activity relationships in a series of benzonitriles related to bromoxynil (Szigeti et al., 1981). This had not been the case in earlier structure/activity correlations within other (non-phenolic) classes of inhibitor (reviewed by Trebst and Draber, 1979) and this is consistent with the suggestion that the nitrophenols may interact with a different receptor protein from the one involved in binding other classes of Hill reaction inhibitor.

Phenolic herbicides and other Hill inhibitors also behave differently in respect of the symptoms exhibited by treated plants. Bromoxynil and ioxynil are considered to be foliar-contact herbicides; injury appears as blistered or necrotic spots within 24 hours of treatment, and later extensive destruction of leaf tissue occurs, with chlorosis appearing around the necrotic areas. Dinoseb is also seen as a 'scorching' agent (Ashton and Crafts, 1981). On the other hand, most Hill reaction inhibitors do not give initial scorch, and chlorosis develops before necrosis (Ashton and Crafts, 1981).

On this basis, it seems unlikely that the hydroxybenzonitriles and phenols act solely through their effect on photosynthesis. The effects of these compounds on mitochondrial respiration were described in Chapter 1 and are relevant in the current discussion. To take the hydroxybenzonitriles first, although photosynthetic electron transport is the more sensitive target *in vitro*, the symptoms of treated plants suggest that the effects of uncoupling are first to show themselves *in vivo*; the scorch might result from rapid tissue death in the face of an inadequate ATP supply. Chlorosis develops subsequently, indicating that effects on photosynthesis may be important in the later stages of phytotoxicity. The evidence for this view is least convincing in the case of bromoxynil, which, *in vitro*, seems to be a considerably better inhibitor of photosynthesis (I_{50} in the range 1–18 μM) than it is an uncoupler of oxidative phosphorylation (half uncoupled at 50–60 μM in plant mitochondria). Photosynthesis inhibition could make a greater contribution to phytotoxicity in this case, though factors such as differential uptake, movement and metabolism could be operating *in vivo*.

Turning now to the nitrophenols, these compounds uncouple oxidative phosphorylation and inhibit photosynthetic electron transport at comparable micromolar concentrations, so the symptoms *in vivo* are again the best indicator that the uncoupling effect is expressed first. Nitrophenols are clearly good enough uncouplers for this to be the basis of their activity since they are used against non-photosynthetic organisms (p. 36).

Certain compounds which affect electron transport and phosphorylation in both mitochondria and chloroplasts have been described as 'inhibitory uncouplers' (Moreland, 1980). At relatively high concentrations the phenolic herbicides can additionally inhibit mitochondrial electron transport and uncouple photosynthetic phosphorylation and so they were included in this class. Mitochondrial and chloroplast functions depend crucially on the well-being of the organelle membranes, and it has been suggested that the compounds interact directly with membranes in a relatively non-specific way, leading to fluidity changes which are manifested as described above. Effects on other membranes may perhaps also occur (Moreland *et al.*, 1982). However, a compound which can be put in this class may, nevertheless, have a primary site of action in a particular part of the membrane since some membrane functions may be more sensitive than others, and some may be more essential to plant survival in the short term.

9. Structure-activity correlations

With the possible exception of the hydroxybenzonitriles, phenols and bentazone, we can conclude that the other Hill reaction inhibitors probably bind at the same site on a protein which has a functional role in electron transport. It does not necessarily follow however that compounds of different classes bind to the receptor through identical chemical interactions and, indeed, this is argued against by the changes in sensitivity of chloroplasts from atrazine-resistant plants to herbicides of other classes. However, it is certainly reasonable to look for common features within Hill reaction inhibitors, and this approach has been popular.

As was pointed out earlier the common structural requirements amongst Hill reaction inhibitors are a lipophilic moiety near the group –C(X)–N<, where X = O or NH (Draber et al., 1974; Büchel, 1972; Moreland, 1969; Trebst and Harth, 1974). This holds true for the ureas, acylanilides, phenyl carbamates, triazines, triazinones, uracils and the pyridazinone class of herbicides and also for the trifluoromethyl benzimidazole and pyrimidinone groups of Hill reaction inhibitors which do not boast a commercially successful member (Trebst and Draber, 1979).

Shipman (1981, 1982) carried out an analysis of the molecular properties shared by all active photosystem II herbicides. In terms of the requirements for binding to the site, a herbicide was considered to require a flat polar component about the size of a phenyl ring, to which component were attached hydrophobic substituents to enable the chemical to partition into lipid regions of the membrane. Shipman speculates that these properties may enable the inhibitor to combine with a hydrophobic region of the receptor while interacting through its polar region with an electric field generated by ionic links in the protein; these could have a natural function in stabilising Q_B^-.

Quantitative structure-activity correlations have been carried out within the various categories of compound (Trebst and Draber, 1979) with the aim of empirically determining an equation which describes the in vitro Hill reaction inhibition in terms of various parameters defining the properties of the molecule.* In general, hydrophobic parameters turned out to be important (with a certain steric constraint) and electronic parameters were useful in some cases (Trebst and Draber, 1979).

The sugar beet herbicide metamitron was discovered following this approach. A retrospective structure/activity analysis of analogues of metribuzin (Table 2.1) led to the view that groups in the position of the

---

A fuller account of such a procedure, applied to inhibitors of acetylcholinesterase, is given on p. 128.

–SMe substituent contributed to inhibition mainly by steric interaction with the binding site. Consequently alkyl groups were tried in this position, and metamitron was amongst the compounds synthesised (Draber et al., 1974; Schmidt et al., 1975).

E. Why do treated plants die?

At first sight inhibition of ATP and NADPH formation and consequent reduction of CO_2 fixation and depletion of sucrose might appear to be sufficient explanation of the herbicidal action of Hill reaction inhibitors. However, this starvation hypothesis does not receive current support, principally because the symptoms of plants treated and kept in the light are not the same as those of untreated plants maintained in darkness (Ashton, 1965; Sweetser and Todd, 1961).

Instead it is thought that plant death is due to destructive reactions which follow a failure of the mechanism that normally protects the photosynthetic apparatus against excessive illumination (Ridley, 1977). Stanger and Appleby (1972) initially proposed that, by preventing the formation of NADPH, diuron stopped the reduction and regeneration of carotenoid pigments which had been epoxidised as part of the chlorophyll protection mechanism. Thus the carotenoids could no longer carry out their protective function, and pigment breakdown would follow. In support of such a mechanism, Giannopolitis and Ayers (1978) reported that NADPH or NADH could prevent chlorophyll loss from illuminated chloroplasts.

However, Ridley (1977) pointed out a number of defects in this hypothesis. One objection is that the de-epoxidase system is now known to be on the inner face of the thylakoid membrane, whereas $NADP^+$ is reduced on the stromal side (Fig. 2.2), and a second is that NADPH has since been shown to be involved in the epoxidation step, rather than de-epoxidation, as envisaged in Stanger and Appleby's hypothesis (see Ridley, 1977, for original references). Ridley went on to make a detailed study of the kinetics of loss of photosynthetic pigments following diuron treatment of pea chloroplasts, as a result of which he suggested that when diuron blocks electron transport, conformational changes in the membrane that allow spillover of excitation energy from the pigment systems serving photosystem II to those serving photosystem I are prevented. Protection via the normal mechanism involving carotenoids is then no longer adequate and pigment breakdown ensues. When electron transport between the photosystems resumes following the addition of (for instance) a donor between the photosystems, the conformational change which allows spillover can occur again and protection is restored.

The damaging species is likely to be an excited form of molecular oxygen known as singlet oxygen which is generated when excess light energy absorbed by chlorophyll is dissipated by reaction not with a carotenoid but with O_2. A contribution may also be made by the direct reaction of excited chlorophyll molecules (triplet chlorophyll) with susceptible materials (e.g. unsaturated fatty acids) to give reactive radical species which participate in further destructive reactions (Dodge, 1982). The involvement of oxygen is supported by the observation of Pallett and Dodge (1980) that loss of chlorophyll by flax cotyledons incubated in the light with diuron was greatly diminished when the experiment was conducted under argon rather than oxygen. Further, a quencher of singlet oxygen diminished damage to chloroplasts incubated *in vitro* without an electron acceptor, whereas the inclusion of a singlet oxygen generator in such a system led to promotion of damage (see Dodge, 1982).

Damage occurs both to photosynthetic pigments and to membrane lipids, which undergo peroxidation by reaction with singlet oxygen (Pallett and Dodge, 1976). This has been confirmed recently in a study of the effect of diuron on *Euglena* cells, in which ethane formation was used as a measure of peroxidation of linolenic acid, the principal chloroplast fatty acid (Elstner and Oswald, 1980 and references therein).

It has been suggested (see Churchill and Klepper, 1979) that nitrite (NO_2^-), which can accumulate in the absence of reduced ferredoxin from photosynthetic electron transport, could be the cause (or partial cause) of plant death. This might be a contributing factor under some nitrogen regimes, but cannot be obligatory since plants are damaged by photosynthesis inhibitors under conditions in which no nitrite accumulation is seen (Churchill and Klepper, 1979; Fedtke, 1977, 1982).

Descriptions of the various changes in chloroplast ultrastructure and biochemistry that can be observed when plants are grown in the presence of diuron and other Hill reaction inhibitors, are given by Lichtenthaler *et al.* (1980), Pallett and Dodge (1980) and Wergin and Potter (1975).

III. SITE OF ACTION OF DIBROMO-METHYL-ISOPROPYLBENZO-QUINONE

DBMIB (2,5-dibromo-3-methyl-6-isopropyl-1,4-benzoquinone), which is not used as a herbicide, was introduced by Trebst *et al.* (1970). It is probably active only after prior reduction to the hydroquinone form by the chloroplast system (Reimer *et al.*, 1979). At low concentrations it acts specifically after plastoquinone in the electron transfer sequence (Fig. 2.1) and prevents reoxidation of the reduced form (plastoquinol) by cytochrome f and plastocyanin (Izawa, 1977; Trebst, 1980). At higher concentrations,

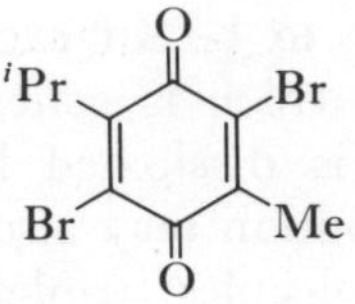

DBMIB

however, an additional inhibitory effect on the reduction of plastoquinone is seen (Trebst, 1980) probably as a result of competition by DBMIB for the site to which the special plastoquinone molecule Q_B binds (Velthuys, 1981).

An attempt was made to discover herbicidal DBMIB analogues in a study which included alkyl-substituted halo- and nitrophenols (Trebst et al., 1979; Trebst and Draber, 1979). However, these compounds inhibited photosynthetic electron transport at the same site as diuron (though not in exactly the same way). Interestingly, diphenylether derivatives of the phenols were found to act at the same site as DBMIB and it was confirmed in some cases that radioactive metribuzin could not be displaced from chloroplast membranes by such compounds (Trebst et al., 1978; Trebst, 1979; Trebst and Draber, 1979). However, no new herbicide was discovered from any of these studies.

The dinitroaniline trifluralin has been shown to inhibit the Hill reaction at 11–30 μM (Moreland et al., 1972; Robinson et al., 1977) with effects which are similar to, but not identical with, those of DBMIB. However, inhibition of cell division has been recorded at concentrations of 1 nM (see Chapter 5) so the effect on photosynthetic electron transport is probably slight in practice.

Table 2.4 Bipyridinium herbicides

Name	Structure
Diquat (as the dibromide)	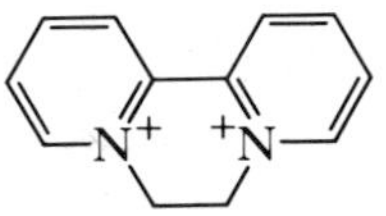 2 Br⁻
Paraquat (as the dichloride)	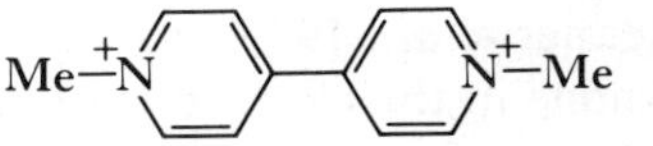 2 Cl⁻

IV. HERBICIDES INTERCEPTING ELECTRONS FROM COMPONENTS OF PHOTOSYNTHESIS

A. Bipyridinium herbicides

1. *Introduction*

The mode of action of the bipyridinium herbicides (Table 2.4) has been well understood for some time now (Calderbank, 1968; Dodge; 1971). A comprehensive account of all aspects of the bipyridinium herbicides is available (Summers, 1980).

The herbicidal cation is reduced by the light reactions of photosynthesis to form a relatively stable free radical. In the presence of oxygen the bipyridinium free radical becomes oxidised to form the original ion, which is then free to react again, and an activated oxygen species which destroys the plant tissue (Fig. 2.5).

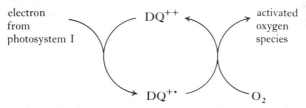

Fig. 2.5 Formation of activated oxygen species in the presence of diquaternary bipyridinium compounds (DQ)

2. *Action on intact plant tissue*

The key discoveries on the mode of action of bipyridinium herbicides in whole plants are described by Mees (1960) and Homer *et al.* (1960). Subsequent studies have largely confirmed and extended this work.

Bipyridinium herbicides kill plants very quickly, but only if light is present. Mees (1960) exposed broad bean plants to diquat by dipping one leaflet in 3 mM herbicide for 24 hours in the dark. The leaflet was then cut off so that no more herbicide could enter the plants, which were left in the dark or exposed to light. Illuminated plants showed signs of injury within 10 minutes, whereas those maintained in darkness looked normal for several hours, though, if the treated leaflet was left on the plants that were kept completely in the dark, they died after 5 days, showing that diquat will kill in the absence of light. This latter effect is unlikely to be important in the field.

Besides light, oxygen is important for the action of paraquat. Mees (1960) treated mustard and bean leaf discs in the dark with diquat for 18 hours, and then exposed them to varying levels of oxygen, diluted with nitrogen. Damage was assessed by measuring the inhibition of oxygen uptake compared with discs similarly treated with diquat but subsequently exposed to air in the dark. It was clear that oxygen was necessary to achieve the maximum toxic effect.

Mees (1960) reasoned that, if a photosynthetic reaction was required for the rapid herbicidal action of diquat, pre-treatment of the plant with the herbicidal Hill reaction inhibitor monuron might delay the effect of the bipyridinium compound. He was able to demonstrate that when leaves of various intact plants were dipped in 100 μM monuron for 18 hours, and then transferred to an illuminated solution containing 100 μM monuron and 1 mM diquat, the plants died up to 24 hours after plants treated solely with diquat. As we have seen, phytotoxicity of Hill reaction inhibitors is also attributed to damaging effects caused by activated oxygen species (though this occurs more slowly than in the case of paraquat), so that observation of the alleviating effect of, for example, monuron is likely to depend on delivering exactly the correct dose on the site of action.

3. *Action on isolated chloroplasts*

Although their herbicidal activity was not discovered until 1955, bipyridinium compounds were used as oxidation-reduction indicators before this. Michaelis and Hill (1933) showed that methyl viologen (paraquat), with a redox potential of -0.446 volts, accepted one electron in a non-biological system to form a stable free radical. The redox potential for diquat is -0.349 volts (Calderbank, 1968). Reference to the reaction scheme of photosynthesis drawn against a redox potential scale (Fig. 2.1) shows that light reaction I of photosynthesis generates a reducing potential capable of adding an electron to diquat or paraquat.

The bipyridinium herbicides accept electrons from the reducing side of photosystem I, in the region of ferredoxin (Dodge, 1982) (Fig. 2.1). As an example of the sort of experimental evidence which supports this interpretation Zweig *et al.* (1965) showed that although diquat did not inhibit the reduction of various Hill acceptors (including ferricyanide) it did inhibit $NADP^+$ reduction, with 50% inhibition occurring at 3.3 μM. Although the inhibition of $NADP^+$ reduction by diuron may be overcome by the addition of DCPIP/ascorbate, that produced by diquat was unaffected, and diquat inhibited the $NADP^+$ reduction that occurred in the presence of diuron and DCPIP/ascorbate (Zweig *et al.*, 1965).

Although one effect of the interception of electrons is to prevent $NADP^+$ being reduced to NADPH, the damage to the plant is caused by the activated oxygen species formed during the bipyridinium radical autoxidation.

As with Hill reaction inhibitors, pigments and lipids become degraded (Harris and Dodge, 1972). Elstner *et al.* (1980) have pointed out that, while bipyridinium herbicides are dependent upon an electron donor to photosystem I for their activity, the natural donors (i.e. photosystem II and the electron transfer components) become inhibited relatively early during the course of paraquat poisoning, though pigment loss and membrane destruction continue for considerably longer (Harris and Dodge, 1972; Böger and Kunert, 1978). Following a study of the kinetics of chlorophyll loss and lipid peroxidation (measured as ethane formation (p. 85)) in paraquat-treated spinach sub-chloroplast particles Elstner *et al.* (1980) suggest that paraquat phytotoxicity develops in two phases. Firstly, they envisage the production of reactive oxygen species dependent upon photosynthetic electron flow leading to pigment and lipid damage and inhibition of photosystem II and components of the electron transport chain. Secondly, they propose a further phase when the electrons cannot come from the now inactivated photosynthetic electron transport. However, the chloroplasts are already damaged at this stage and Elstner *et al.* (1980) suggest that this enables electrons to pass to photosystem I directly from chlorophyll or unsaturated fatty acids.

4. *Nature of the damaging species*

Davenport (1963) illuminated chloroplasts with diquat plus ferrimyoglobin in a vacuum, and observed spectral changes consistent with the formation of the complex between ferrimyoglobin and hydrogen peroxide; the oxygen necessary for the formation of the peroxide is, of course, available from the splitting of water by photosystem II. The production of hydrogen peroxide may also be demonstrated in chloroplasts in which oxygen evolution is avoided by using reduced 2,6-dichlorophenolindophenol as an electron source, thus by-passing light reaction II (Fig. 2.1) (Dodge, 1971). If paraquat is added to such a system the uptake of oxygen from air may be demonstrated with an oxygen electrode. The presence of hydrogen peroxide in the system can be inferred by switching off the light and adding catalase, an enzyme which converts hydrogen peroxide to water and oxygen, and observing the liberation of oxygen.

Once the formation of hydrogen peroxide was recognised, a good deal of work was carried out on the nature of the activated oxygen species formed in the presence of paraquat (Summers, 1980). After the diquaternary salt (DQ^{++}) has been photoreduced to the radical cation $(DQ^{+\cdot})$, candidate toxicants could be formed by the reactions:

$$DQ^{+\cdot} + O_2 \longrightarrow DQ^{++} + O_2^{-\cdot}$$
$$\text{superoxide}$$

$$DQ^{+\cdot} + O_2^{-\cdot} \longrightarrow DQ^{++} + O_2^{--} \xrightarrow{2H^+} H_2O_2$$
$$\text{peroxide}$$

$$DQ^{+\cdot} + H_2O_2 \longrightarrow DQ^{++} + OH^- + HO^\cdot$$
$$\text{hydroxy radical}$$

Based on a consideration of the kinetics of formation and reaction of candidate species and on their reactivity, the superoxide ion ($O_2^{-\cdot}$) was favoured over hydrogen peroxide as the most important toxicant (Farrington et al., 1973; Farrington, 1976). It was thought that the hydroxy radical has so short a life that, even if it were formed, it would not be a likely toxicant in vivo (Farrington, 1976). The role of superoxide as primary toxicant is supported by a demonstration that paraquat damage can be alleviated by the presence of a copper chelate of D-penicillamine having superoxide dismutating activity (Youngman et al., 1979; Youngman and Dodge, 1979). Nevertheless it is still possible that the importance of superoxide may lie in its ability to generate other reactive oxygen species (Youngman and Dodge, 1979; Dodge, 1982).

5. Structure-activity correlations

Although no commercial herbicide in this class has been introduced since 1964, further active compounds have been reported. The factors which are important in determining whether a compound has a bipyridinium-type action are as follows (Summers, 1979, 1980):

(a) stability in aqueous solution at physiological pH values;

(b) ability to be reduced to a stable radical cation in aqueous solution by a one-electron transfer which is reversed by oxygen;

(c) the potential at which the one-electron reduction occurs;

(d) the dimensions of the molecule – only compact molecules are active.

B. Nitrodiphenylethers

The nitrodiphenylethers (Table 2.5) are broad-spectrum herbicides which require light to exert their herbicidal effect (Matsunaka, 1975; Yih and Swithenbank, 1975) and thus probably interfere with chloroplast function.

Nitrodiphenylethers can inhibit the Hill reaction. For instance fluorodifen and nitrofen have I_{50} values of 4.5 μM and 22 μM respectively

(Moreland *et al.*, 1970) and nitrofluorfen inhibited photosynthetic electron transport, apparently between plastoquinone and cytochrome f, with an

Table 2.5 Nitrodiphenylether herbicides

Name	Structure
Acifluorfen	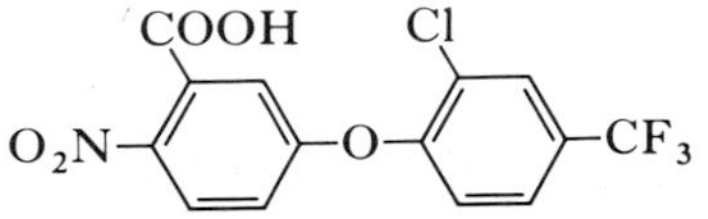
Bifenox	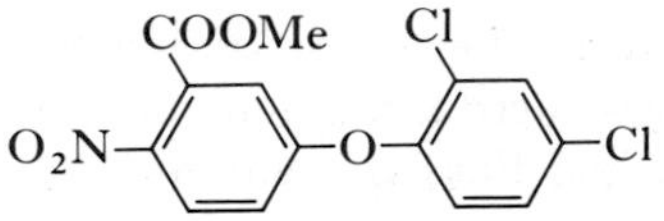
2,4-Dichlorophenyl 3-methoxy-4-nitrophenyl ether	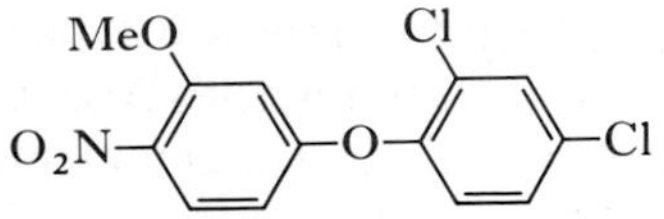
Fluorodifen	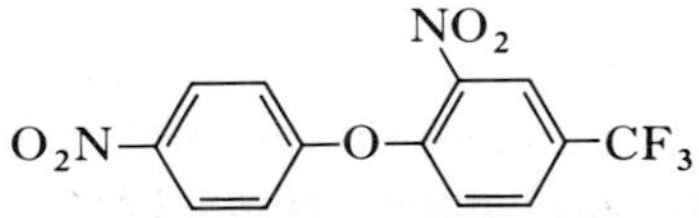
Nitrofen	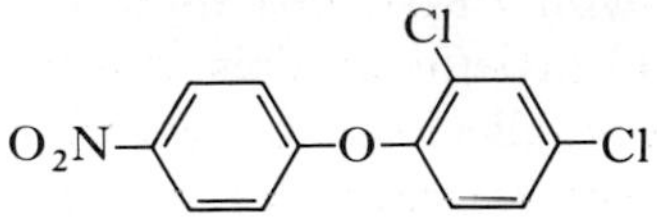
Nitrofluorfen	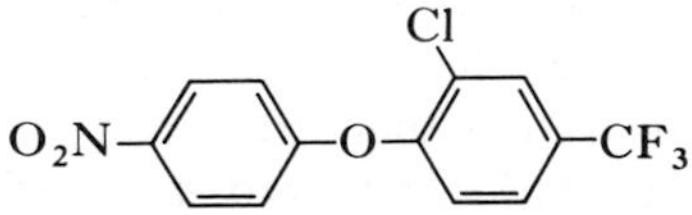
Oxyfluorfen	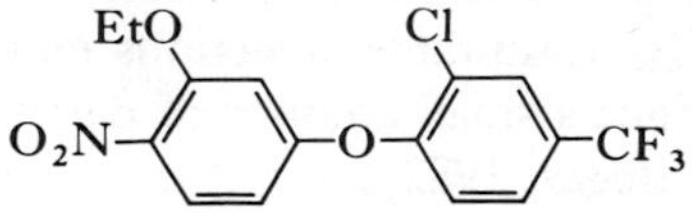

I_{50} of about 5 μM (Bugg et al., 1980). Nevertheless the Hill reaction I_{50} for nitrofen is still rather high for this to be the primary site of action. In addition the compounds are regarded as contact herbicides, unlike most Hill reaction inhibitors (Ashton and Crafts, 1981). Re-examining the situation, Lambert et al. (1979) found that an inhibitory effect of nitrofen on photophosphorylation correlated with growth inhibition in the alga Bumilleriopsis. Photophosphorylation was also inhibited in spinach chloroplasts, though here inhibition of electron flow and uncoupling were also evident (Lambert et al., 1979).

The solution to these anomalies has perhaps been found in a series of recent experiments (reviewed by Orr and Hess, 1982, and Sandmann and Böger, 1982). A consensus of results obtained indicates that the nitro-diphenylethers may act by becoming reduced in the plant to radical species which initiate destructive reactions in membrane lipids, leading to cell leakage. The principal lines of evidence for the view are indicated below.

Kunert and Böger (1981) studied the strong and rapid bleaching effect caused by oxyfluorfen (1 μM) in cultures of the alga Scenedesmus and found that carotenoids and subsequently chlorophyll were destroyed concurrently with the inhibition of photosynthetic oxygen evolution and the formation of ethane, used as a measure of lipid peroxidation (p. 85). Ethane was also formed by paraquat-treated algae. This has since been made the basis of a quantitative assay to measure the concentration of nitrodiphenylether required to give half-maximal ethane production in a 15 hour period. The most active compound found so far is the methyl ester of acifluorfen, which gave this effect at 0.09 μM (Lambert et al., 1983).

In a second approach to the investigation, light-irradiated cucumber cotyledons pre-loaded with [86]Rb[+] ions rapidly became leaky following treatment with nitrodiphenylethers. This was confirmed for other ionic species, and found to be alleviated by α-tocopherol, a known in vivo scavenger of lipophilic free radicals (Orr and Hess, 1982).

The source of the electrons which reduce the herbicides to radical species is not certain. Although it has been suggested that the electrons could come directly from carotenoids (see Orr and Hess, 1982) this would not explain the demonstration that ureas can alleviate damaging effects caused by nitrodiphenylethers (Pritchard et al., 1980; Kunert and Böger, 1981; Lambert et al., 1983). These results suggest instead that the electrons could come from the photosynthetic electron transport chain at a point beyond the quinone Q_B. Respiratory reactions may also be able to activate nitrodiphenylethers since a small amount of damage can be caused in the dark (Sandmann and Böger, 1982).

Evidence obtained so far suggests that active oxygen species are not involved in mediating the toxic effect of nitrodiphenylethers (as they are in the case of paraquat and, indeed, diuron-type herbicides). Instead it is thought that the herbicide radical abstracts a proton from an unsaturated fatty acid, giving a radical which can subsequently undergo peroxidation by reaction with molecular oxygen (Orr and Hess, 1982).

Lambert *et al.* (1983) examined the structural features required to cause good peroxidative activity (assayed as hydrocarbon formation) in the alga *Bumilleriopsis*. The *para*-nitro group was an important feature though compounds in which this was replaced by p-NO or p-NH.OH still retained good activity. An oxygen bridge was preferable to a sulphur one and activity was considerably enhanced by appropriate substituents adjacent to the nitro group; the groups -OEt, -OMe, -CO.NH.Me and -COOMe made positive contributions to activity.

REFERENCES

Anderson, J. M. (1981). *FEBS Lett.* **124**, 1–10

Arnon, D. I., Tsujimoto, H. Y. and Tang, G. M.-S. (1981). *Proc. Natl. Acad. Sci. U.S.A.* **78**, 2942–2946.

Arntzen, C. J., Steinback, K. E., Vermaas, W. and Ohad, I. (1983). In "Pesticide Chemistry: Human Welfare and the Environment" (J. Miyamoto and P. C. Kearney, eds), Vol. 3, pp. 51–58. Pergamon Press, Oxford

Ashton, F. M. (1965). *Weeds* **13**, 164–168

Ashton, F. M. and Crafts, A. S. (1981) "Mode of Action of Herbicides" John Wiley and Sons, New York

Bauer, K. and Köcher, H. (1979). *Z. Naturforsch.* **34**C, 961–963

Belbachir, O., Matringe, M., Chevallier, D. and Tissut, M. (1980). *Pestic. Biochem. Physiol.* **14**, 309–313

Bishop, N. I. (1958). *Biochim. Biophys. Acta.* **27**, 205–206

Böger, P. (1982). *Physiol. Plant.* **54**, 221–224

Böger, P. and Kunert, K. J. (1978). *Z. Naturforsch.* **33**C, 688–694

Böger, P. and Kunert, K. J. (1979). *Z. Naturforsch.* **34**C, 1015–1025

Böger, P., Beese, B. and Miller, R. (1977). *Weed Res.* **17**, 61–67

Bowes, J., Crofts, A. R. and Arntzen, C. J. (1980). *Arch. Biochem. Biophys.* **200**, 303–308

Brewer, P. E., Arntzen, C. J. and Slife, F. W. (1979). *Weed Sci.* **27**, 300–308

Bucha, H. C. and Todd, C. W. (1951). *Science* **114**, 493–494

Büchel, K. H. (1972). *Pestic. Sci.* **3**, 89–110

Bugg, M. W., Whitmarsh, J., Rieck, C. E. and Cohen, W. S. (1980). *Plant Physiol.* **65**, 47–50

Calderbank, A. (1968). *Adv. Pest Control Res.* **8**, 127 235

Churchill, K. and Klepper, L. (1979). *Pestic. Biochem. Physiol.* **12**, 156–162

Conard, S. G. and Radosevich, S. R. (1979). *J. Appl. Ecol.* **16**, 171–177

Croze, E., Kelly, M. and Horton, P. (1979). *FEBS Lett.* **103**, 22–26

Davenport, H. E. (1963). *Proc. Roy. Soc. Ser. B* **157**, 332–345

De Mur, A. R., Swader, J. A. and Youngner, V. B. (1972). *Pestic. Biochem. Physiol.* **2**, 337–341

Desmoras, J. and Jacquet, P. (1964). Proc. 16th Int. Symp. Crop Protection, pp. 633–642

De Villiers, O. T., Van der Merwe, M. J. and Koch, H. M. (1979). *S. Afr. J. Sci.* **75**, 315–316

Dicks, J. W. (1974). *Biochem. Educ.* **2**, 69–71

Dicks, J. W. (1978). *Pestic. Sci.* **9**, 59–62

Dodge, A. D. (1971). *Endeavour* **30**, 130–135

Dodge, A. D. (1982). In "Biochemical Responses Induced by Herbicides" (D. E. Moreland, J. B. St. John and F. D. Hess, eds) pp. 57–77. American Chemical Society, Washington, D. C.

Draber, W., Büchel, K. H., Dickoré, K., Trebst, A. and Pistorius, E. (1969). In "Progress in Photosynthesis Research" (H. Metzner, ed.) Vol. III, pp. 1789–1795. Tübingen

Draber, W., Büchel, K. H., Timmler, H. and Trebst, A. (1974). In "Mechanism of Pesticide Action" (G. K. Kohn, ed.) pp. 100–116. American Chemical Society, Washington, D. C.

Ebert, E. and Dumford, S. W. (1976). *Residue Rev.* **65**, 1–103

Elstner, E. F. and Oswald, W. (1980). *Z. Naturforsch.* **35 C**, 129–135

Elstner, E. F., Lengfelder, E. and Kwiatkowski, G. (1980). *Z. Naturforsch.* **35 C**, 303–307

Esser, H. O., Dupuis, G., Ebert, E., Marco, G. and Vogel, C. (1975). In "Herbicides: Chemistry, Degradation and Mode of Action" (P. C. Kearney and D. D. Kaufman, eds) Vol. 1, pp. 129–208. Dekker, New York

Farrington, J. A. (1976). Proc. British Crop Protection Conf. "Weeds", Vol. 1, pp. 225–228

Farrington, J. A., Ebert, M., Land, E. J. and Fletcher, K. (1973). *Biochim. Biophys. Acta* **314**, 372–381

Fedtke, C. (1973). *Pestic. Sci.* **4**, 653–664

Fedtke, C. (1977). *Pestic. Sci.* **8**, 152–156

Fedtke, C. (1982). "Biochemistry and Physiology of Herbicide Action". Springer-Verlag, Berlin

Frank, R. and Switzer, C. M. (1969). *Weed Sci.* **17**, 344–348

Gardner, G. (1981). *Science* **211**, 937–940

Geissbühler, H., Martin, H. and Voss, G. (1975). In "Herbicides: Chemistry, Degradation and Mode of Action" (P. C. Kearney and D. D. Kaufman, eds) Vol. 1, pp. 209–291. Dekker, New York

Giannopolitis, G. N. and Ayers, G. S. (1978). *Weed Sci.* **26**, 440–443

Gimmler, H. (1977). In "Photosynthesis I" (A. Trebst and M. Avron, eds), Encycl. Plant Physiol., New Series, Vol. 5, pp. 448–472. Springer-Verlag, Berlin

Good, N. E. (1961). *Plant Physiol.* **36**, 788–803

Govindjee and van Rensen, J. J. S. (1978). *Biochim. Biophys. Acta* **505**, 183–213

Green, D. H., Schuler, J. and Ebner, L. (1966). Proc. 8th Br. Weed Control Conf., Vol. 2, pp. 363–371

Gregory, R. P. F. (1977). "Biochemistry of Photosynthesis", 2nd edn. Wiley Interscience, London

Gressel, J. (1982). *Plant Sci. Lett.* **25**, 99–106

Gromet-Elhanan, Z. (1968). *Biochem. Biophys. Res. Commun.* **30**, 28–31

Gysin, H. and Knuesli, E. (1960). *Adv. Pest Control. Res.* **III**, 289–358

Harris, N. and Dodge, A. D. (1972). *Planta* **104**, 210–219

Hatzios, K. K. and Howe, C. M. (1982). *Pestic. Biochem. Physiol.* **17**, 207–214

Hatzios, K. K., Penner, D. and Bell, D. (1980). *Plant Physiol.* **65**, 319–321

Hill, R. (1965). In "Essays in Biochemistry" (P. N. Campbell and G. D. Greville, eds) Vol. 1, pp. 121–151. Academic Press, London and New York

Hill, R. and Bendall, F. (1960). *Nature (Lond.)* **186**, 136–137

Hilton, J. L., Monaco, T. J., Moreland, D. E. and Gentner, W. A. (1964). *Weeds* **12**, 129–131

Hilton, J. L., Scharen, A. L., St. John, J. B., Moreland, D. E. and Norris, K. H. (1969). *Weed Sci.* **17**, 541–547

Hoffman-Falk, H., Mattoo, A. K., Marder, J. B., Edelman, M. and Ellis, R. J. (1982). *Proc. Natl. Acad. Sci. U.S.A.* **257**, 4583–4587

Hoffmann, C. E. (1972). In "Herbicides, Fungicides, Formulation Chemistry" (A. S. Tahori, ed.) pp. 65–86. Gordon and Breach, New York

Hoffmann, C. E., McGahen, J. W. and Sweetser, P. B. (1964). *Nature (Lond.)* **202**, 577–578

Holt, J. S., Stemler, A. J. and Radosevich, S. R. (1981). *Plant Physiol.* **67**, 744–748

Homer, R. F., Mees, G. C. and Tomlinson, T. E. (1960). *J. Sci. Food Agric.* **11**, 309–315

Izawa, S. (1977). In "Photosynthesis I" (A. Trebst and M. Avron, eds) Encycl. Plant Physiol., New Series, Vol. 5, pp. 266–282, Springer-Verlag, Berlin

Izawa, S. (1980). *Methods in Enzymology* **69**, 413–434

Jones, D. W. and Foy, C. L. (1972). *Pestic. Biochem. Physiol.* **2**, 8–26

Kaplan, S. and Arntzen, C. J. (1982). In "Photosynthesis: Energy Conversion by Plants and Bacteria" (Govindjee, ed.) Vol. 1, pp. 65–150. Academic Press, London and New York

Katoh, S. (1972). *Plant Cell Physiol.* **13**, 273–286

Kerr, M. W. and Wain, R. L. (1964). *Ann. Appl. Biol.* **54**, 447–450

Khanna, R., Pfister, K., Keresztes, A., van Rensen, J. J. S. and Govindjee (1981). *Biochim. Biophys. Acta* **634**, 105–116

Knuesli, E. (1976). In "Pesticide Chemistry in the 20th Century" (J. R. Plimmer, ed.) pp. 76–92. American Chemical Society, Washington, D.C.

Kunert, K. J. and Böger, P. (1981). *Weed Sci.* **28**, 169–173

Laasch, H., Pfister, K. and Urbach, W. (1982). *Z. Naturforsch.* **37**C, 620–631

Lambert, R., Kunert, K. J. and Böger, P. (1979). *Pestic. Biochem. Physiol.* **11**, 267–274

Lambert, R., Sandmann, G. and Böger, P. (1983). In "Pesticide Chemistry: Human Welfare and the Environment" (J. Miyamoto and P. C. Kearney, eds) Vol. 3, pp. 97–102. Pergamon Press, Oxford

Lichtenthaler, H. K., Burkard, G., Grumbach, K. H. and Meier, D. (1980). *Photosynth. Res.* **1**, 29–43

Mantai, K. E. (1970). *Plant Physiol.* **45**, 563–566

Matsunaka, S. (1970). *Zasso Kenkyo* **10**, 40–43: [*Chemical Abstracts* (1971) **75**, 34331]

Matsunaka, S. (1975). In "Herbicides. Chemistry, Degradation and Mode of Action" (P. C. Kearney and D. D. Kaufman, eds) Vol. 2, pp. 709–739. Dekker, New York

Mattoo, A. K., Dick, U., Hoffman-Falk, H. and Edelman, M. (1981). *Proc. Natl. Acad. Sci. U.S.A.* **78**, 1572–1576

Mees, G. C. (1960). *Ann. Appl. Biol.* **48**, 601–612

Meier, D., Lichtenthaler, H. K. and Burkard, G. (1980). *Z. Naturforsch.* **35**C, 656–664

Michaelis, L. and Hill, E. S. (1933). *J. Gen. Physiol.* **16**, 859–873

Mine, A. and Matsunaka, S. (1975). *Pestic. Biochem. Physiol.* **5**, 444–450

Moreland, D. E. (1967). *Ann. Rev. Plant Physiol.* **18**, 365–386

Moreland, D. E. (1969). In "Progress in Photosynthesis Research" (H. Metzner, ed.) Vol. III, pp. 1693–1714. Tübingen

Moreland, D. E. (1980). *Ann. Rev. Plant Physiol.* **31**, 597–638

Moreland, D. E. and Blackmon, W. J. (1970). *Weed Sci.* **18**, 419–426

Moreland, D. E. and Hill, K. L. (1962). *Weeds* **10**, 229–236

Moreland, D. E., Blackmon, W. J., Todd, H. G. and Farmer, F. S. (1970). *Weed Sci.* **18**, 636–642

Moreland, D. E., Farmer, F. S. and Hussey, G. G., (1972). *Pestic. Biochem. Physiol.* **2**, 342–353

Moreland, D. E., Huber, S. C. and Novitky, W. P. (1982). In "Biochemical Responses Induced by Herbicides" (D. E. Moreland, J. B. St. John and F. D. Hess, eds) pp. 79–96. American Chemical Society, Washington, D.C.

Mullet, J. E. and Arntzen, C. J. (1981). *Biochim. Biophys. Acta* **635**, 236–248

Muschinek, G., Garab, G. I., Mustardy, L. A. and Faludi-Daniel, A. (1979). *Weed Res.* **19**, 101–107

Oettmeier, W., Masson, K. and Johanningmeier, U. (1982). *Biochim. Biophys. Acta* **679**, 376–383

Orr, G. L. and Hess, F. D. (1982). In "Biochemical Responses Induced by Herbicides" (D. E. Moreland, J. B. St. John and F. D. Hess, eds) pp. 131–152. American Chemical Society, Washington, D.C.

Pallett, K. E. and Dodge, A. D. (1976). Proc. British Crop Protection Conf. "Weeds", Vol. 1, pp. 235–240

Pallett, K. E. and Dodge, A. D. (1979). *Pestic. Sci.* **10**, 216–220

Pallett, K. E. and Dodge, A. D. (1980). *J. Exp. Bot.* **31**, 1051–1066

Pfister, K. and Arntzen, C. J. (1979). *Z. Naturforsch.* **34C**, 996–1009

Pfister, K., Radosevich, S. R. and Arntzen, C. J. (1979). *Plant Physiol.* **64**, 995–999

Pfister, K. Lichtenthaler, H. K., Burger, G., Musso, H. and Zahn, M. (1981a). *Z. Naturforsch.* **36C**, 645–655

Pfister, K., Steinback, K. E., Gardner, G. and Arntzen, C. J. (1981b). *Proc. Natl. Acad. Sci. U.S.A.* **78**, 981–985

Pillai, P. and St. John, J. B. (1981). *Plant Physiol.* **68**, 585–587

Potter, J. R. and Wergin, W. P. (1975). *Pestic. Biochem. Physiol.* **5**, 458–470

Pritchard, M. K., Warren, G. F. and Dilley, R. A., (1980). *Weed Sci.* **28**, 640–645

Radosevich, S. R. (1977). *Weed Sci.* **25**, 316–318

Radosevich, S. R. and De Villiers, O. T. (1976). *Weed Sci.* **24**, 229–232

Regitz, G. and Ohad, I. (1976). *J. Biol. Chem.* **251**, 247–252

Reimer, S., Link, K. and Trebst, A. (1979). *Z. Naturforsch.* **34C**, 419–426

Renger, G. (1976). *Biochim. Biophys. Acta.* **440**, 287–300

Renger, G. (1979). *Z. Naturforsch.* **34C**, 1010–1014

Renger, G., Erixon, K., Döring, G. and Wolff, Ch. (1976). *Biochim. Biophys. Acta* **440**, 279–286

Retzlaff, G. and Fischer, A. (1973). *Mitt. Biol. Bundesanst. Land Forstwirt.* **151**, 179–180

Ridley, S. M. (1977). *Plant Physiol.* **59**, 724–732

Robinson, S. J., Yocum, C. F., Ikuma, H. and Hayashi, F. (1977). *Plant Physiol.* **60**, 840–844

2. HERBICIDES INTERFERING WITH PHOTOSYNTHESIS

Sandmann, G. and Böger, P. (1982). In "Biochemical Responses Induced by Herbicides" (D. E. Moreland, J. B. St. John and F. D. Hess, eds) pp. 111–130. American Chemical Society, Washington, D.C.

Schmidt, R. R. and Fedtke, C. (1977). *Pestic. Sci.* **8**, 611–617

Schmidt, R. R., Draber, W., Eue, L. and Timmler, H. (1975). *Pestic. Sci.* **6**, 239–244

Schulz, H. (1969). In "Progress in Photosynthesis Research" (H. Metzner, ed.), Vol. III, pp. 1752–1760. Tübingen

Shahak, Y., Crowther, D. and Hind, G. (1981). *Biochim. Biophys. Acta* **636**, 234–243

Shipman, L. L. (1981). *J. Theoret. Biol.* **90**, 123–148

Shipman, L. L. (1982). In "Biochemical Responses Induced by Herbicides" (D. E. Moreland, J. B. St. John and F. D. Hess, eds), pp. 23–35, American Chemical Society, Washington, D.C.

Shochat, S., Owens, G. C., Hubert, P. and Ohad, I. (1982). *Biochim. Biophys. Acta* **681**, 21–31

Siegel, I. H. (1975). "Enzyme Kinetics". John Wiley and Sons, New York

Stanger, C. E. and Appleby, A. P. (1972). *Weed Sci.* **20**, 357–363

Steinback, K. E., Pfister, K. and Arntzen, C. J. (1981). *Z. Naturforsch.* **36 C**, 98–108

Steinback, K. E., Pfister, K. and Arntzen, C. J. (1982). In "Biochemical Responses Induced by Herbicides" (D. E. Moreland, J. B. St. John and F. D. Hess, eds) pp. 37–55. American Chemical Society, Washington, D.C.

Stemler, A. (1977). *Biochim. Biophys. Acta* **460**, 511–522.

Summers, L. A. (1979). In "Advances in Pesticide Science" (H. Geissbühler, ed.) Pt 2, pp. 244–247. Pergamon Press, Oxford

Summers, L. A. (1980). "The Bipyridinium Herbicides". Academic Press, London and New York

Sweetser, P. B. and Todd, C. W. (1961). *Biochim. Biophys. Acta* **51**, 504–508

Szigeti, Z., Sárvári, E. and Bujtás, C. (1981). *Weed Res.* **21**, 37–41

Tischer, W. and Strotmann, H. (1977). *Biochim. Biophys. Acta* **460**, 113–125

Tischer, W. and Strotmann, H. (1979). *Z. Naturforsch.* **34 C**, 992–995

Trebst, A. (1979). *Z. Naturforsch.* **34 C**, 986–991

Trebst, A. (1980). *Methods in Enzymology* **69**, 675–715

Trebst, A. and Avron, M. (eds) (1977). "Photosynthesis I". Encycl. Plant Physiol., New Series, Vol. 5. Springer-Verlag, Berlin

Trebst, A. and Draber, W. (1979). In "Advances in Pesticide Science" (H. Geissbühler, ed.) Pt 2, pp. 223–234. Pergamon Press, Oxford

Trebst, A. and Harth, E. (1974). *Z. Naturforsch.* **29 C**, 232–235

Trebst, A. and Wietoska, H. (1975). *Z. Naturforsch.* **30 C**, 499–504

Trebst, A., Harth, E. and Draber, W. (1970). *Z. Naturforsch.* **25 B**, 1157–1159

Trebst, A., Wietoska, H., Draber, W. and Knops, H. J. (1978). *Z. Naturforsch.* **33 C**, 919–927

Trebst, A., Draber, W. and Donner, W. T. (1983). In "Pesticide Chemistry: Human Welfare and the Environment" (J. Miyamoto and P. C. Kearney, eds) Vol. 3, pp. 85–90. Pergamon Press, Oxford
Chem. (IUPAC), Kyoto. Pergamon Press, Oxford

Van Assche, C. J. and Carles, P. M. (1982). In "Biochemical Responses Induced by Herbicides" (D. E. Moreland, J. B. St. John and F. D. Hess, eds) pp. 1–21 American Chemical Society, Washington, D.C.

van Rensen, J. J. S. (1982). *Physiol. Plant.* **54**, 515–521
van Rensen, J. J. S. and Kramer, H. J. M. (1979). *Plant Sci. Lett.* **17**, 21–27
van Rensen, J. J. S. and Vermaas, W. F. J., (1981). *Physiol. Plant.* **51**, 106–110
van Rensen, J. J. S., Van der Vet, W., van Vliet, W. P. A. (1977). *Photochem. Photobiol.* **25**, 579–583
Velthuys, B. R. (1981). *FEBS Lett.* **126**, 277–281
Verity, J., Walker, A. and Drennan, D. S. H. (1981a). *Weed Res.* **21**, 207–316.
Verity, J., Walker, A. and Drennan, D. S. H. (1981b). *Weed Res.* **21**, 317–324
Vermaas, W. F. J. and Govindjee (1981a). *Photochem. Photobiol.* **34**, 775–793
Vermaas, W. F. J. and Govindjee (1981b). *Proc. Ind. Natl. Sci. Acad.* **B47**, 581–605
Vermaas, W. F. J. and Govindjee (1982). *Biochim. Biophys. Acta* **680**, 202–209
Vermaas, W. F. J. and van Rensen, J. J. S. (1981). *Biochim. Biophys. Acta* **636**, 168–174
Vermaas, W. F. J., van Rensen, J. J. S. and Govindjee (1982). *Biochim. Biophys. Acta* **681**, 242–247
Vernon, L. P. and Avron, M. (1965). *Ann. Rev. Biochem.* **34**, 269–296
Vernon, L. P. and Zaugg, W. S. (1960). *J. Biol. Chem.* **235**, 2728–2733
Wergin, W. P. and Potter, J. R. (1975). *Pestic. Biochem. Physiol.* **5**, 265–279
Wessels, J. S. C. and van der Veen, R. (1956). *Biochim. Biophys. Acta* **19**, 548–549
Worthing, C. R. (ed.) (1979). "The Pesticide Manual", 6th edn. British Crop Protection Council, London
Yih, R. Y. and Swithenbank, C. (1975). *J. Ag. Food Chem.* **23**, 592–593
Youngman, R. J. and Dodge, A. D. (1979). *Z. Naturforsch.* **34C**, 1032–1035
Youngman, R. J., Dodge, A. D., Lengfelder, E. and Elstner, E. F. (1979). *Experientia* **35**, 1295–1296
Younis, H. M. and Mohanty, P. (1980). *Chem.–Biol. Interact.* **32**, 179–186.
Zohner, A. and Bayzer, H. (1977). Proc. 11th Int. Velsicol. Symp., paper No. 4, 3 pp.
Zweig, G., Shavit, N. and Avron, M. (1965). *Biochim. Biophys. Acta* **109**, 332–346

3 | Insecticides Inhibiting Acetylcholinesterase

I. BIOCHEMICAL BACKGROUND

Most of the insecticides in current use act by interfering with the passage of impulses in the insect nervous system. Therefore a brief account of nervous transmission will first be given. A very clear introductory review of neurophysiology in general has been written by Dowson (1977) while a more detailed treatment has been provided by Shankland (1976).

Insects depend, like mammals, on an integrated nervous system which enables external stimuli to be translated into effective action. The nervous system contains specialised cells, called neurones, which consist of a cell body and a long process called an axon (Fig. 3.1). The axon terminates near a receiving cell of some sort (e.g. another nerve cell or a muscle cell) and the gap between the cells is called a synapse. Studies have shown that axonal transmission in insects is mediated by ionic mechanisms (these are more fully described on p. 141) and that in the case of most synapses, when the impulse reaches the end of the axon it triggers the release of a transmitter substance into the synapse. The exact mechanism of transmitter release is not fully understood. It is known, however, that arrival of the impulse at the nerve ending leads to a flow of calcium into the cell, and that, in some way, this brings about release of the transmitter, perhaps from vesicles in which it may normally be stored. The transmitter diffuses across the gap and combines with its receptor on the post-synaptic mem-

99

brane. The binding alters the ionic permeability of the membrane and this results in the post-synaptic cell passing an impulse if it is a nerve cell or in muscle contraction if it is a muscle cell. Secondary messengers in the target cell may also be activated.

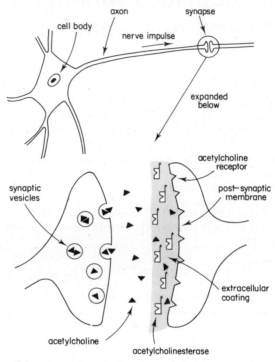

Fig. 3.1 Diagram of a nerve cell and a cholinergic synapse

A number of different chemicals have been implicated in transmission at various insect synapses but we shall be concerned only with transmission mediated by acetylcholine. A substantial body of evidence, reviewed by Leake and Walker (1980), supports the view that acetylcholine is the transmitter at central nervous system synapses in insects. In order for the nervous system to operate properly it is necessary that, once the appropriate message has been passed, excess acetylcholine should be removed from the synapse, both to prevent repetitive firing and to allow a succeeding message to be transmitted. This removal is effected by the enzyme acetylcholinesterase (EC 3.1.1.7), which catalyses hydrolysis of the ester bond, as follows:

$$(CH_3)_3\overset{+}{N}CH_2CH_2OCO \cdot CH_3 \longrightarrow (CH_3)_3\overset{+}{N}CH_2CH_2OH + CH_3CO_2H$$

acetylcholine   choline   acetic acid

This chapter discusses the inhibition of the esterase by organophosphorus and carbamate insecticides. The result of this inhibition is that acetylcholine accumulates in the synapses so that proper nerve function is impaired. This leads ultimately to the death of the insect.

II. ORGANOPHOSPHORUS AND CARBAMATE INSECTICIDES THAT INHIBIT ACETYLCHOLINESTERASE

The great economic significance of organophosphorus and carbamate insecticides (Table 3.1) has resulted in an extensive literature. In recent years monographs have appeared on organophosphorus compounds (Eto, 1974; Fest and Schmidt, 1973) and carbamates (Kuhr and Dorough, 1976) and both subjects have been reviewed as part of more general texts (e.g. Main, 1976; O'Brien, 1976; Fukuto, 1979). In addition a review on the biochemistry of acetylcholinesterase has been published (Rosenberry, 1975).

Table 3.1   Insecticides which inhibit acetylcholinesterase*

Name	Structure
A. Organophosphorus compounds Acephate	$MeCO.NH$ MeO $>PO.SMe$
Azamethiphos	$(MeO)_2PO.SCH_2-N$ ... (with fused oxazolo-pyridine ring bearing Cl)
Azinphos-ethyl	$(EtO)_2PS.SCH_2-N$ ... (fused benzotriazinone ring)
Azinphos-methyl	$(MeO)_2PS.SCH_2-N$ ... (fused benzotriazinone ring)

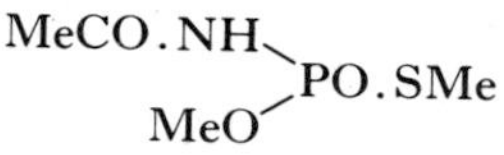

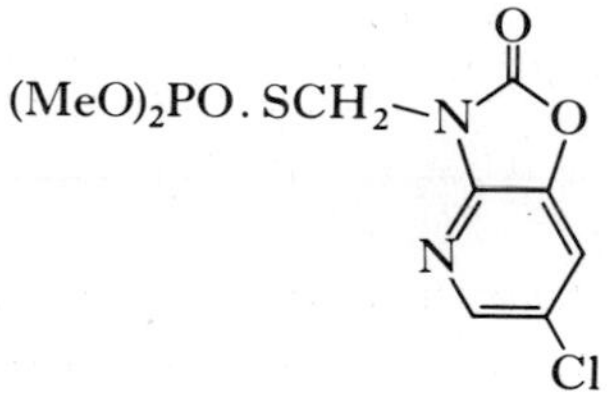

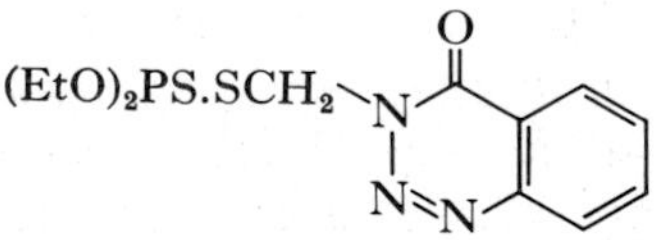

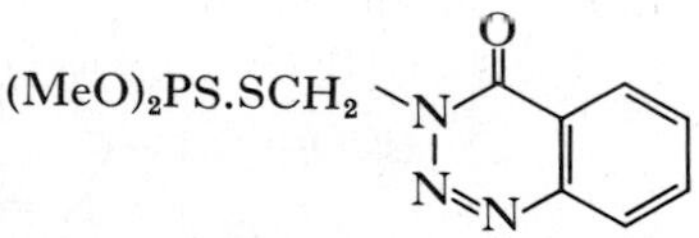

*Stereochemistry at chiral centres and at double bonds is not defined in the table; for details of this see Worthing (1979).

Name	Structure
Bromophos	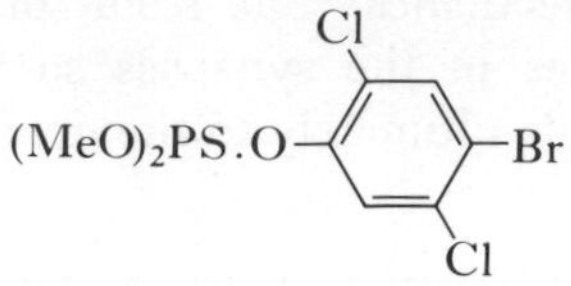
Bromophos-ethyl	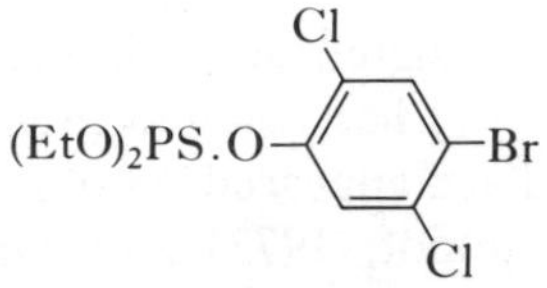
Carbophenothion	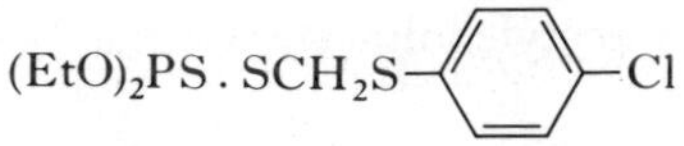
Chlorfenvinphos	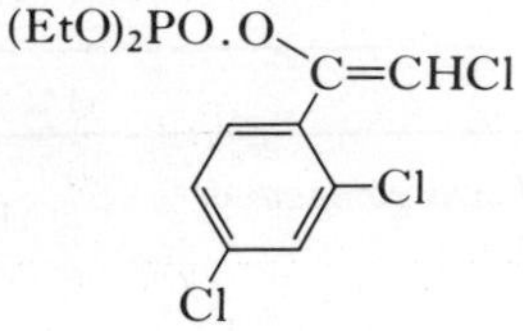
Chlormephos	$(EtO)_2PS.SCH_2Cl$
S-4-Chlorophenylthiomethyl *O*,*O*-dimethyl phosphorodithioate	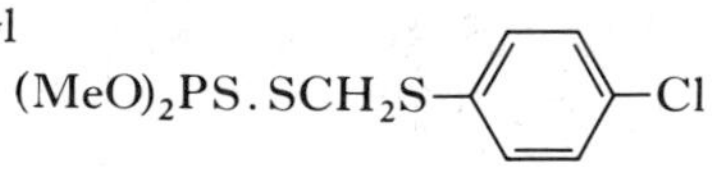
Chlorphoxim	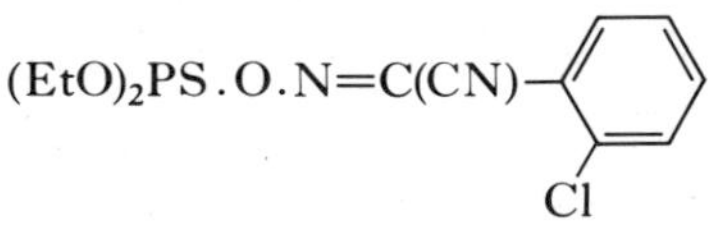
Chlorpyrifos	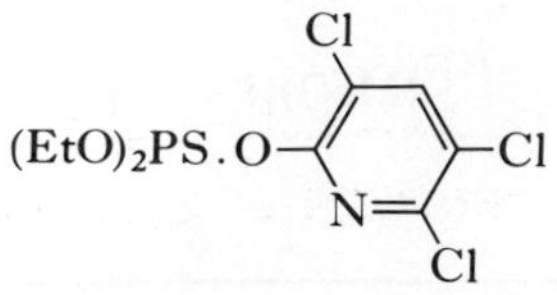

Name	Structure
Chlorpyrifos-methyl	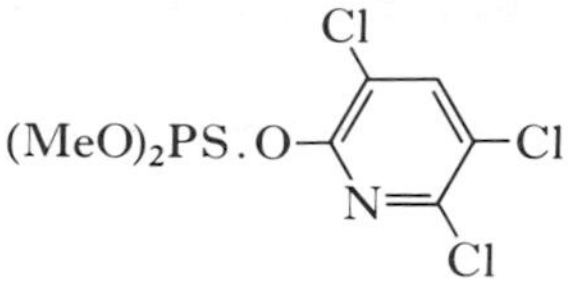

Chlorthiophos

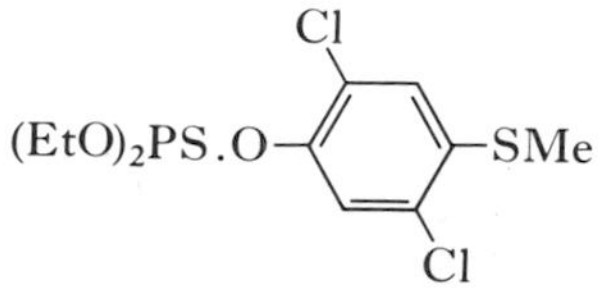

$(EtO)_2PS.O$ —〈ring, Cl〉— SMe   and

$(EtO)_2PS.O$ —〈ring, Cl, MeS〉— Cl   and $(EtO)_2PS.O$ —〈ring, SMe, Cl〉— Cl

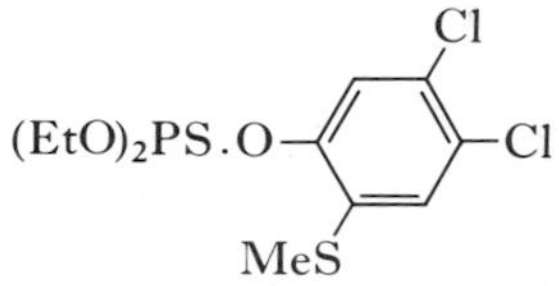

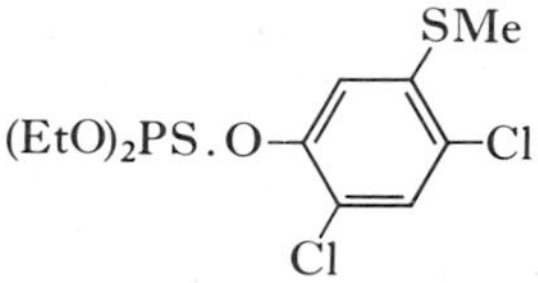

Crotoxyphos

$(MeO)_2PO.OC(Me)=CH.CO.OCH$—〈ring〉
                                                 |
                                                Me

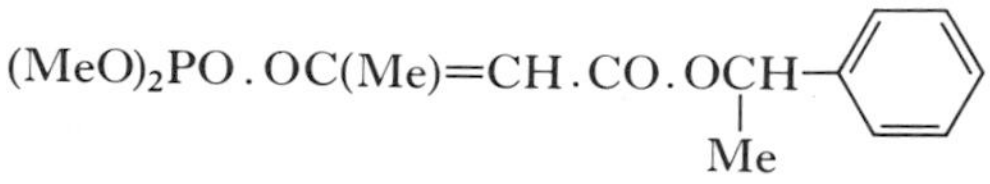

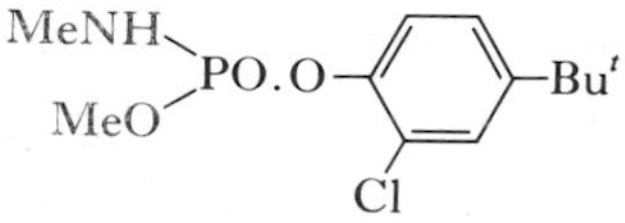

Crufomate

MeNH
        $PO.O$—〈ring, Cl〉—Bu^t
MeO

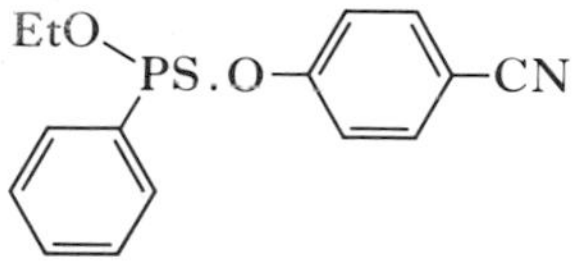

Cyanofenphos

EtO
      $PS.O$—〈ring〉—CN
〈ring〉

Cyanophos

$(MeO)_2PS.O$—〈ring〉—CN

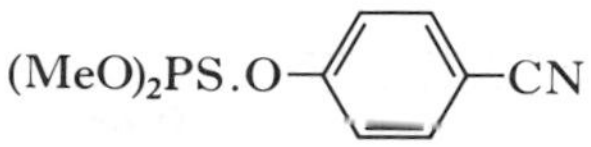

Demephion $(MeO)_2PS.OCH_2CH_2SMe$ and $(MeO)_2PO.SCH_2CH_2SMe$

Name	Structure
Demeton	$(EtO)_2PS.OCH_2CH_2SEt$ and $(EtO)_2PO.SCH_2CH_2SEt$
Demeton-S-methyl	$(MeO)_2PO.SCH_2CH_2SEt$
Demeton-S-methyl sulphone	$(MeO)_2PO.SCH_2CH_2SO_2Et$
Dialifos	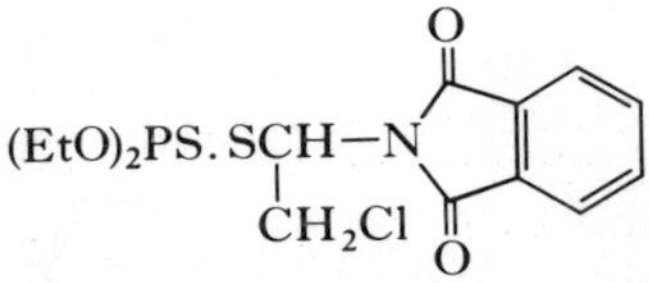
Diazinon	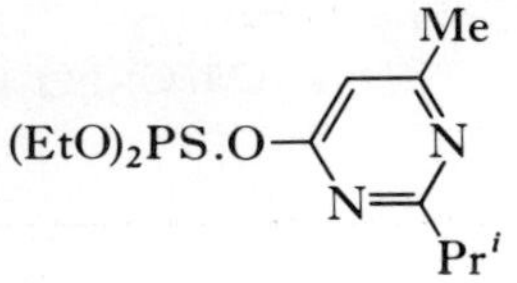
Dichlofenthion	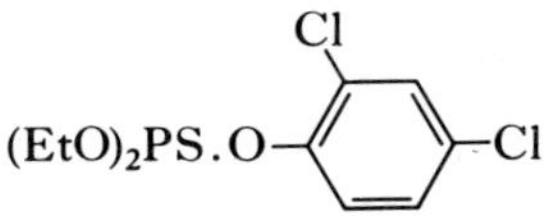
O-2,4-Dichlorophenyl O-ethyl phenylphosphonothioate	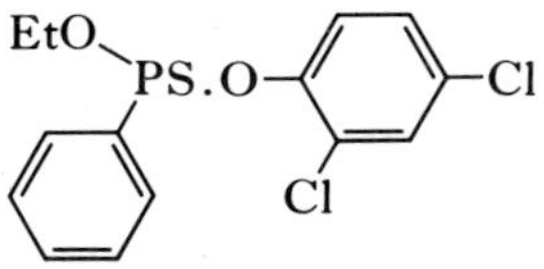
Dichlorvos	$(MeO)_2PO.OCH=CCl_2$
Dicrotophos	$(MeO)_2PO.OC(Me)=CH.CONMe_2$

Name	Structure
O,O-Diethyl O-5-phenyl-isoxazol-3-yl phosphorothioate	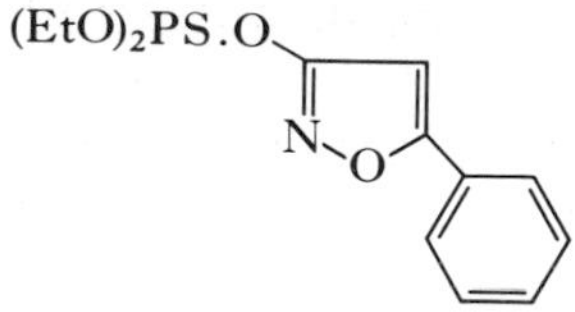
Dimefox	$(Me_2N)_2PO.F$
Dimethoate	$(MeO)_2PS.SCH_2CONHMe$
1,3-Di(methoxycarbonyl) prop-1-en-2-yl dimethyl phosphate	$(MeO)_2PO.OC=CHCOOMe$ $\quad\quad\quad\quad\quad \mid$ $\quad\quad\quad\quad CH_2COOMe$
Dimethyl 4-(methylthio)phenyl phosphate	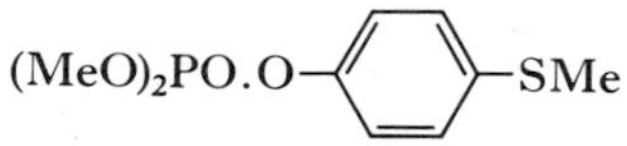
Dioxathion	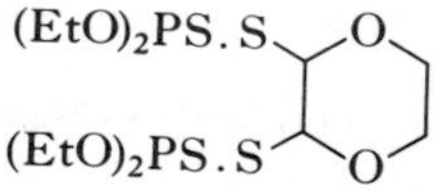
Disulfoton	$(EtO)_2PS.SCH_2CH_2SEt$
EPN	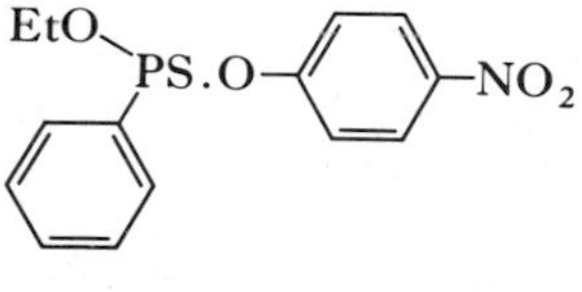
Ethion	$(EtO)_2PS.SCH_2SPS(OEt)_2$
Ethoprophos	$EtOPO(SPr)_2$

Name	Structure
S-Ethylsulphinylmethyl *O,O*-di-isopropyl phosphorodithioate	$(^{i}PrO)_2PS.SCH_2SOEt$
S-(2-Ethylsulphinyl-1-methylethyl) *O,O*-dimethyl phosphorothioate	$(MeO)_2PO.SCH(Me)CH_2SOEt$
Etrimfos	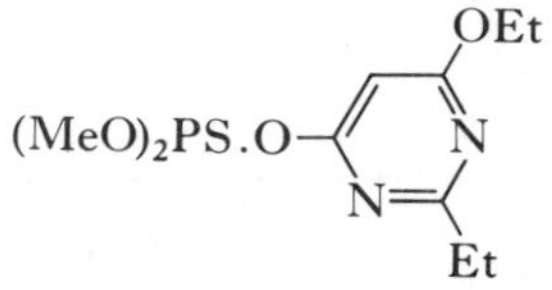
Fenamiphos	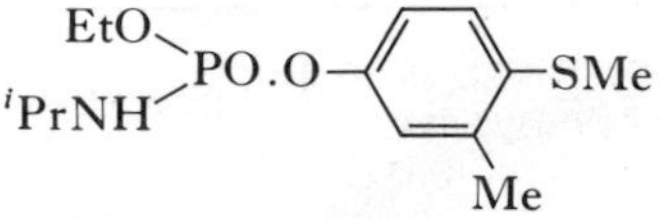
Fenchlorphos	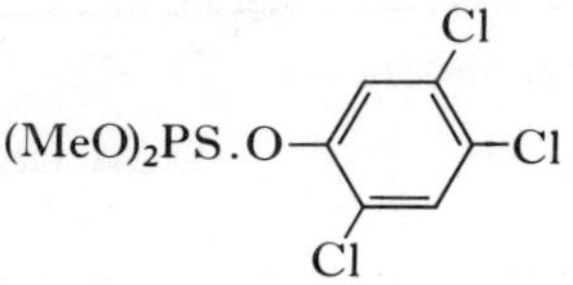
Fenitrothion	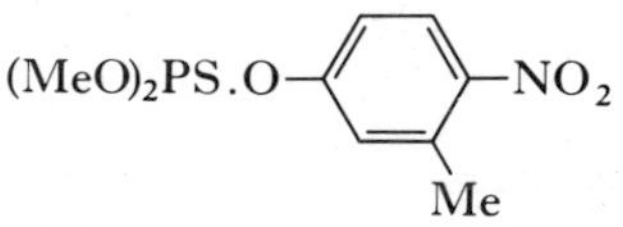
Fensulfothion	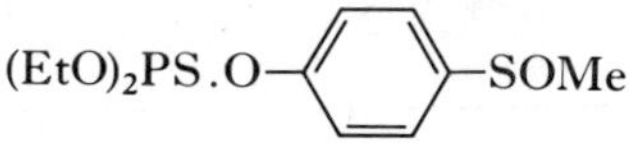
Fenthion	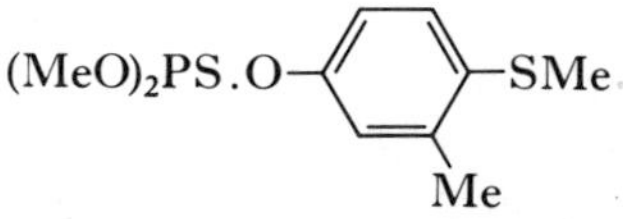
Fonofos	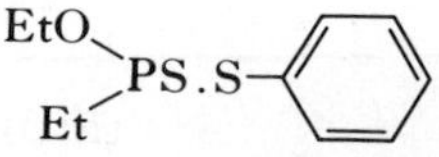
Formothion	$(MeO)_2PS.SCH_2CO.N.CHO$ Me

Name	Structure

Fosthietan

$(EtO)_2PO.N$

Heptenophos

$(MeO)_2PO.O$

Cl

Iodofenphos

$(MeO)_2PS.O$

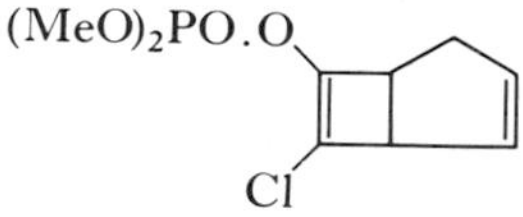

Isofenphos

EtO
^iPrNH $PS.O$

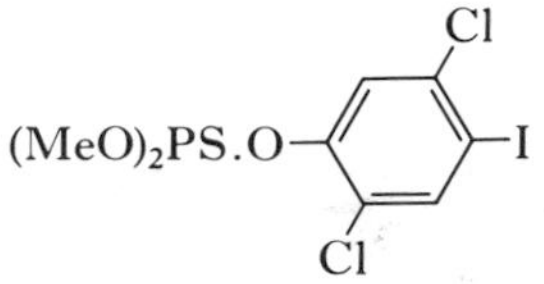

$COOPr^i$

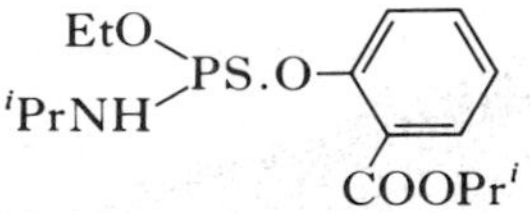

Isothioate

$(MeO)_2PS.SCH_2CH_2SPr^i$

Leptophos

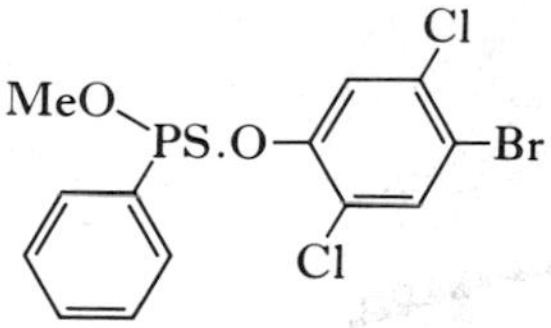

MeO
$PS.O$—Br

Cl   Cl

Malathion

$(MeO)_2PS.SCHCOOEt$
$\qquad\quad CH_2COOEt$

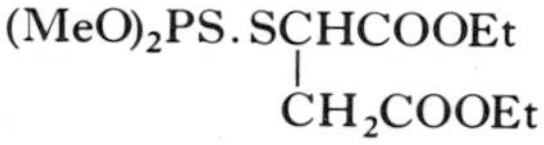

Mecarbam

$(EtO)_2PS.SCH_2CO.N.COOEt$
$\qquad\qquad\qquad\; Me$

Name	Structure

Menazon

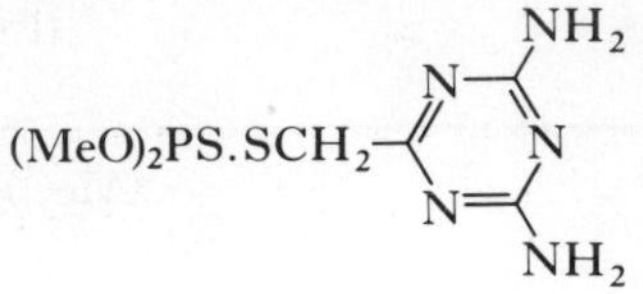

$(MeO)_2PS.SCH_2-$ (triazine ring with two NH_2 groups)

Mephosfolan

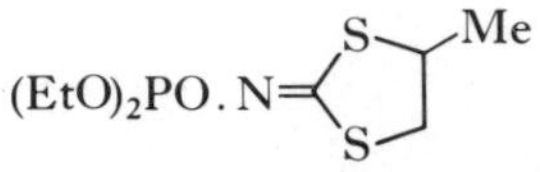

$(EtO)_2PO.N=$ (dithiolane ring with Me)

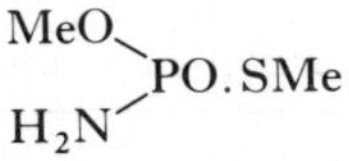

Methamidophos

MeO
      $\backslash$
       $PO.SMe$
      $/$
H_2N

Methidathion

$(MeO)_2PS.SCH_2$ (thiadiazolidinone ring with OMe)

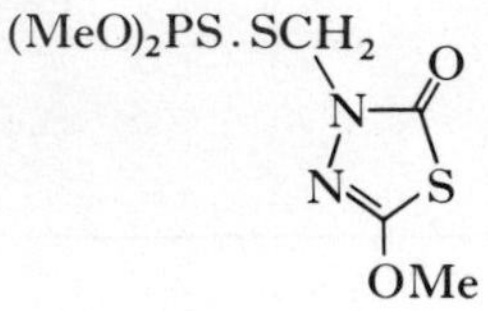

2-Methoxy-4*H*-benzo-1,3,2-
dioxaphosphorin 2-sulphide

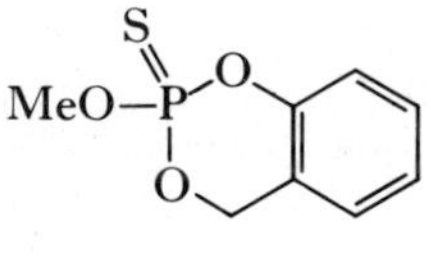

$MeO-P$ (benzo-dioxaphosphorin ring, S double bond)

4-(Methylthio)phenyl
dipropyl phosphate

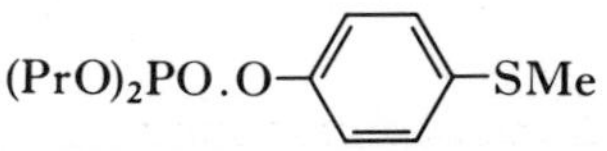

$(PrO)_2PO.O-$ (phenyl) $-SMe$

Mevinphos

$(MeO)_2PO.OC(Me)=CH.COOMe$

Monocrotophos

$(MeO)_2PO.OC(Me)=CH.CONHMe$

Name	Structure
Naled	$(MeO)_2PO.OCH(Br)CBrCl_2$
Omethoate	$(MeO)_2PO.SCH_2.CONHMe$
Oxydemeton-methyl	$(MeO)_2PO.SCH_2CH_2SOEt$
Parathion	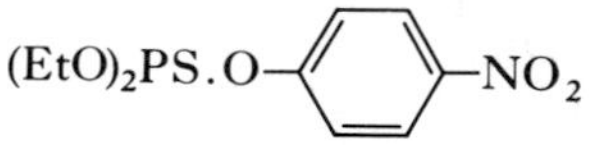 $(EtO)_2PS.O-\!\!\!\!\bigcirc\!\!\!\!-NO_2$
Parathion-methyl	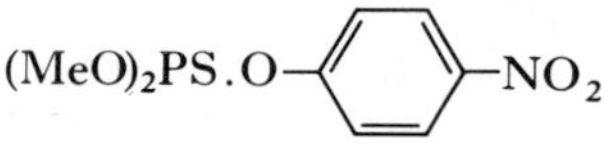 $(MeO)_2PS.O-\!\!\!\!\bigcirc\!\!\!\!-NO_2$
Phenthoate	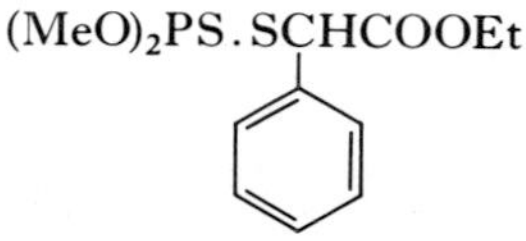 $(MeO)_2PS.SCHCOOEt$
Phorate	$(EtO)_2PS.SCH_2SEt$
Phosalone	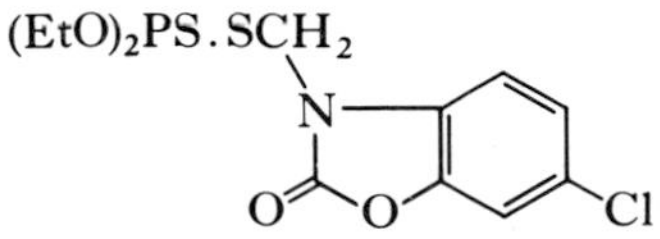 $(EtO)_2PS.SCH_2$
Phosfolan	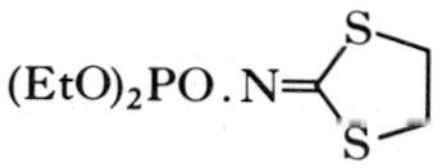 $(EtO)_2PO.N\!\!=\!\!\langle$
Phosmet	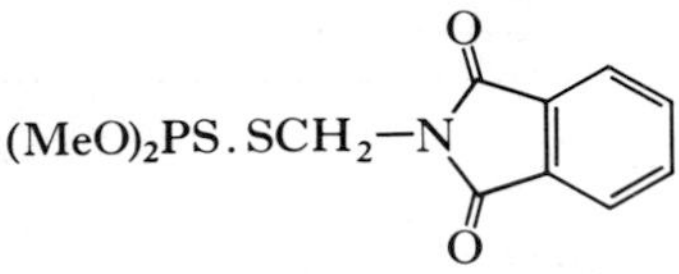 $(MeO)_2PS.SCH_2-N$

Name	Structure
Phosphamidon	$(MeO)_2PO.OC(Me)=C(Cl)CO.NEt_2$
Phoxim	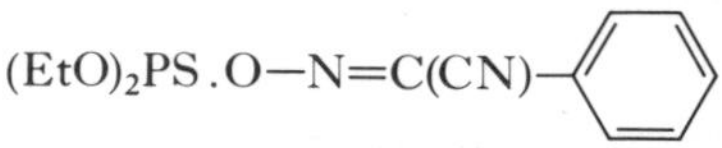 $(EtO)_2PS.O-N=C(CN)-$
Pirimiphos-ethyl	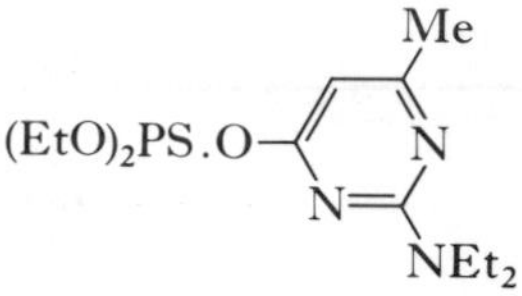 $(EtO)_2PS.O-$
Pirimiphos-methyl	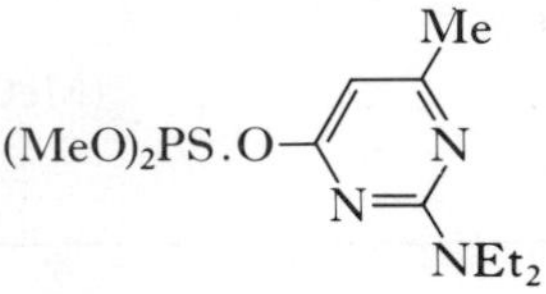 $(MeO)_2PS.O-$
Profenofos	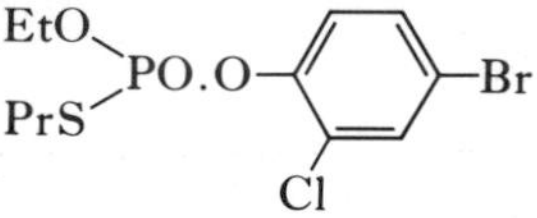
Propetamphos	
Prothiofos	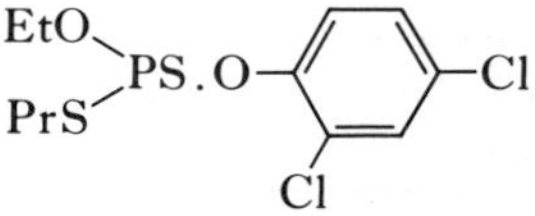
Prothoate	$(EtO)_2PS.SCH_2CO.NHPr^i$
Quinalphos	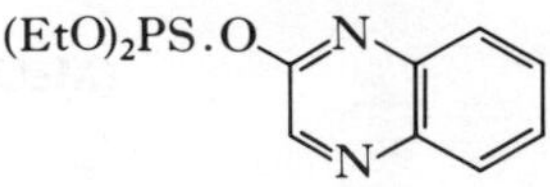 $(EtO)_2PS.O-$

Name	Structure
Schradan	$(Me_2N)_2PO.OPO(NMe_2)_2$
Sulfotep	$(EtO)_2PS.OPS(OEt)_2$
Sulprofos	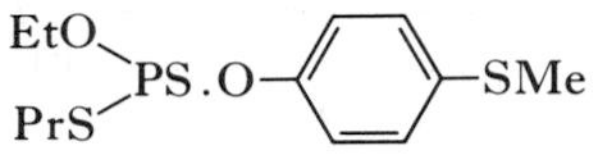
Temephos	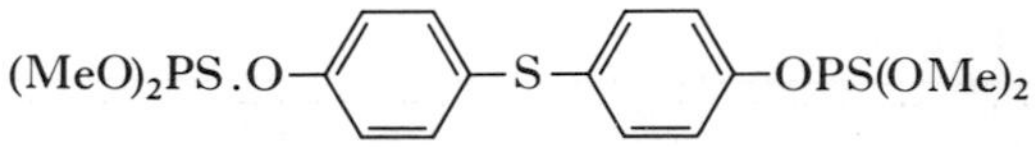
TEPP	$(EtO)_2PO.OPO(OEt)_2$
Terbufos	$(EtO)_2PS.SCH_2SBu^t$
Tetrachlorvinphos	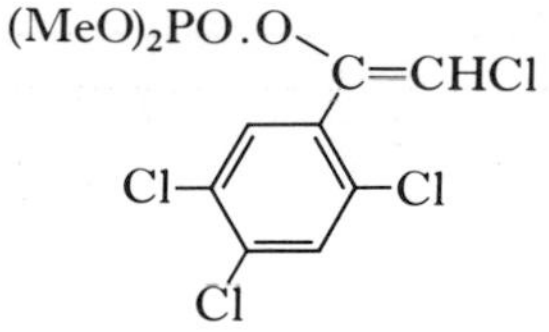
O,O,O',O'-Tetrapropyl dithiopyrophosphate	$(PrO)_2PS.OPS(OPr)_2$
Thiometon	$(MeO)_2PS.SCH_2CH_2SEt$
Thionazin	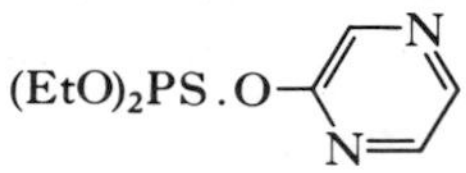

Name	Structure

Triazophos

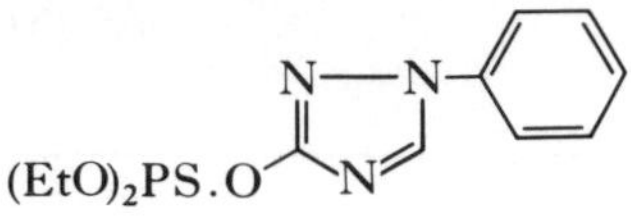

$(EtO)_2PS.O$

Trichloronate

EtO
$PS.O$
Et
Cl
Cl
Cl

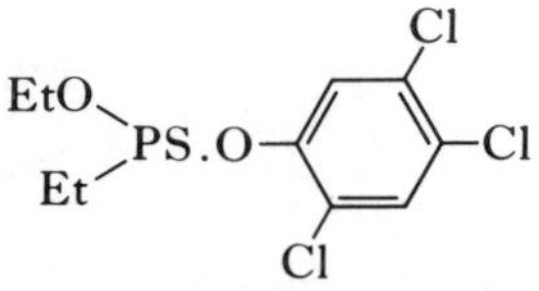

Trichlorphon

$(MeO)_2PO.CH(OH)CCl_3$

Vamidothion

$(MeO)_2PO.SCH_2CH_2SCH(Me)CONHMe$

B. Carbamates
Aldicarb

$MeSC(Me)_2.CH=N–OCO.NHMe$

Aldoxycarb

$MeSO_2.C(Me)_2.CH=N–OCO.NHMe$

Aminocarb

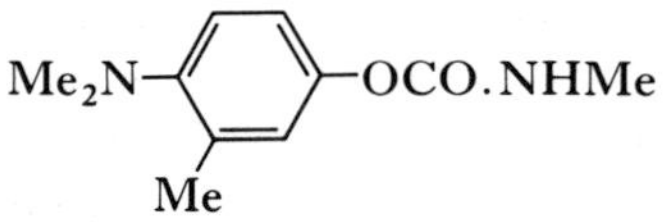

$Me_2N–$ $–OCO.NHMe$
Me

Bendiocarb

$–OCO.NHMe$
O O
Me Me

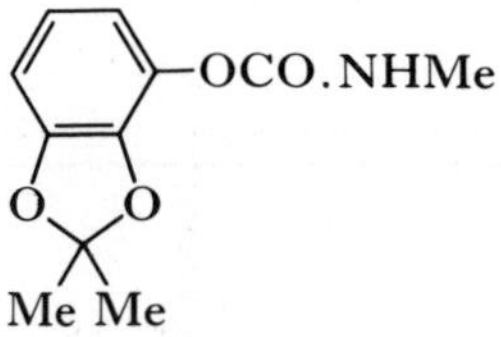

Name	Structure

Bufencarb

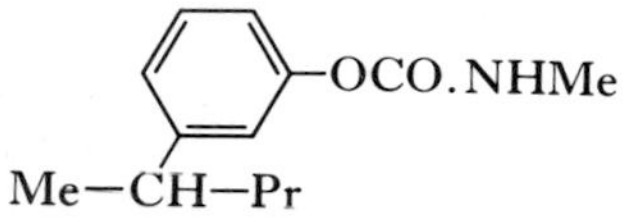

Me—CH—Pr

and

CHEt$_2$

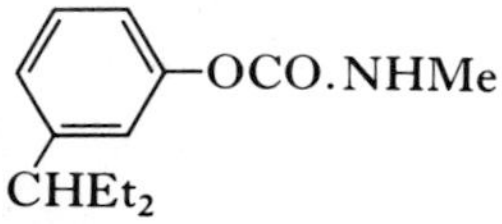

Butocarboxim

MeSCH(Me).C(Me)=N–OCO.NHMe

Butoxycarboxim

MeSO$_2$.CH(Me).C(Me)=N–OCO.NHMe

2-sec-Butylphenyl methylcarbamate

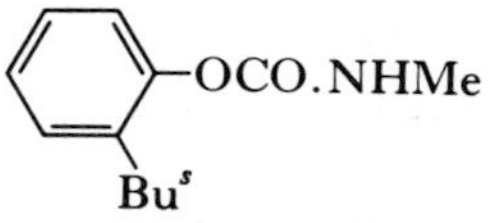

Bu^s

Carbaryl

Carbofuran

Me Me

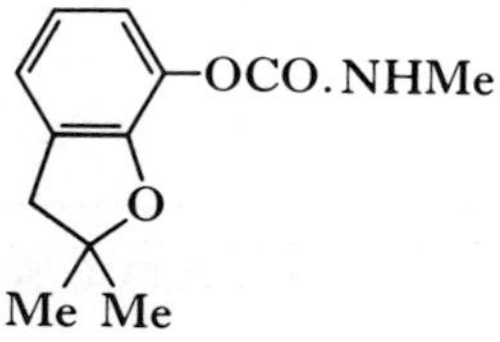

Dimetilan

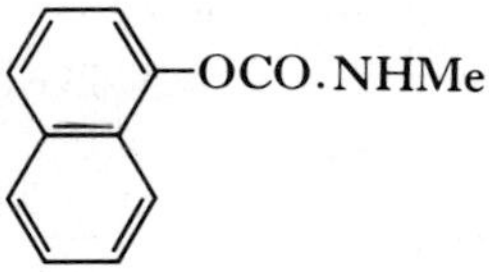

Me

Me$_2$NCO—N$_N$—OCO.NMe$_2$

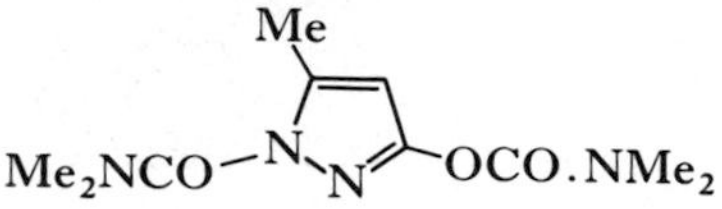

Name	Structure
Dioxacarb	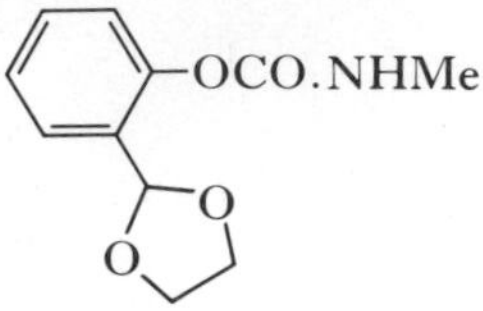
Ethiofencarb	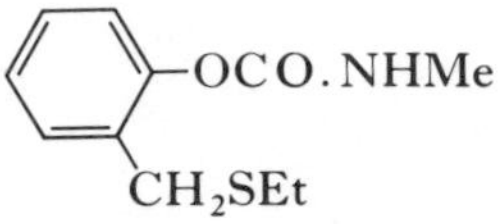
Formetanate	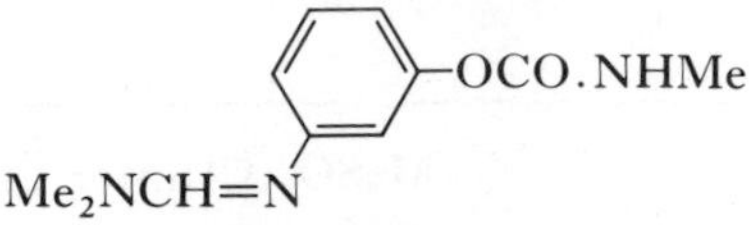
Isoprocarb	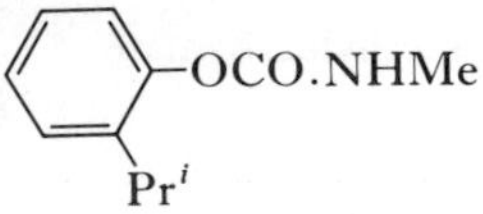
Methiocarb	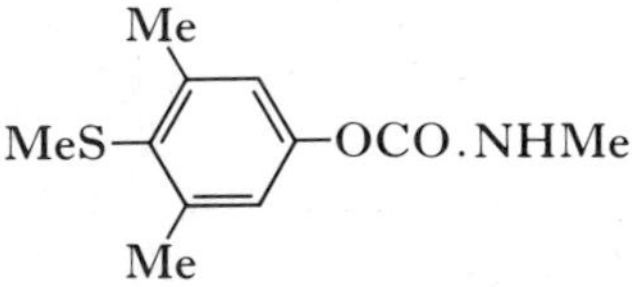
Methomyl	MeSC(Me)=N–OCO.NHMe
Nitrilacarb	NCCH$_2$CH$_2$C(Me)$_2$.CH=N–OCO.NHMe
Oxamyl	Me$_2$NCO.C(SMe)=N–OCO.NHMe
Pirimicarb	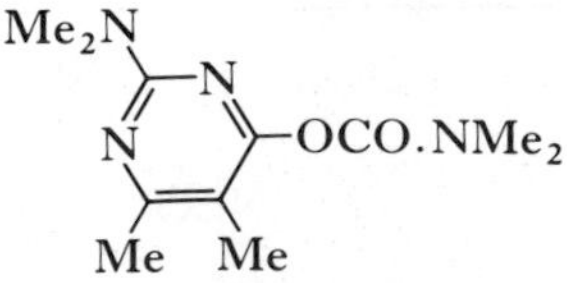

Name	Structure

Promecarb

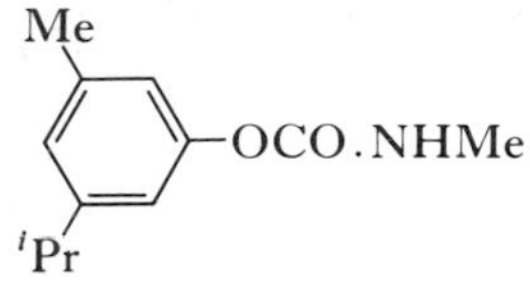

Me / ring / ^{i}Pr — OCO.NHMe

Propoxur

ring — OCO.NHMe, OPr^i

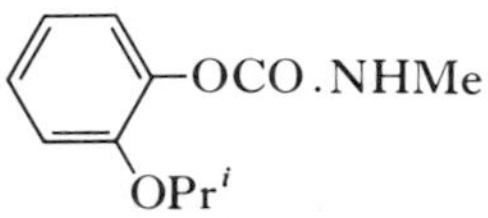

Thiofanox

$MeSCH_2C(Bu^t)=N-OCO.NHMe$

m-Tolyl methylcarbamate

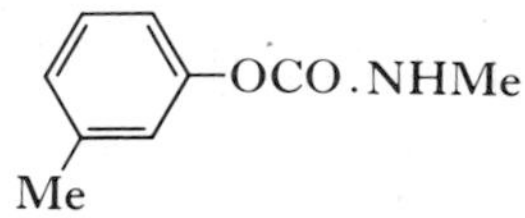

ring — OCO.NHMe, Me

3,4-Xylyl methylcarbamate

Me — ring — OCO.NHMe, Me

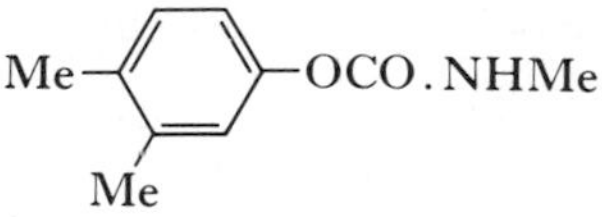

3,5-Xylyl methylcarbamate

Me / ring / Me — OCO.NHMe

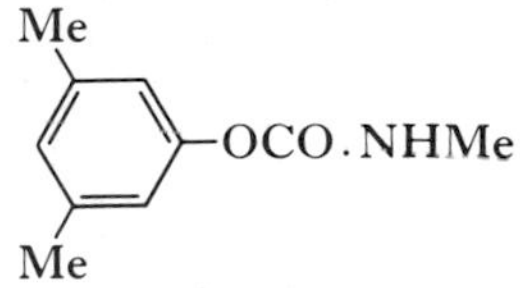

A. Nomenclature of organophosphorus insecticides

Some knowledge of the nomenclature of organophosphorus compounds is required to allow discussion of their biochemistry. The names used in Table 3.1 are those given by Worthing (1979) and follow the scheme:

phosphate

phosphonate

phosphorothioate

S-alkylphosphorothioate (also thiolate)

S-alkylphosphorodithioate

phosphoramidate

phosphonothioate

phosphonodithioate

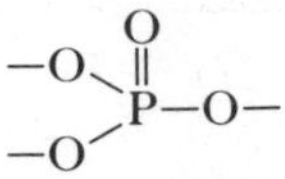

Biochemical studies on organophosphorus compounds are often hampered by the presence in the compounds of trace impurities (see, for example, Hart and O'Brien, 1976) usually produced during manufacture. The effect is often to contaminate a compound of the type $(RO)_2.PS.OR^1$ with one of the type $(RO)_2.PO.SR^1$. Difficulties then arise because the latter is a very much more powerful inhibitor of acetylcholinesterase than the former. Nevertheless mixtures such as these, for example demeton, are used commercially.

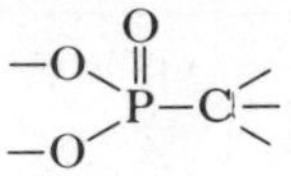

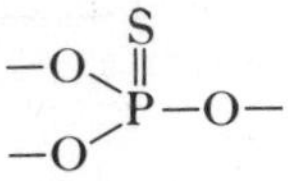

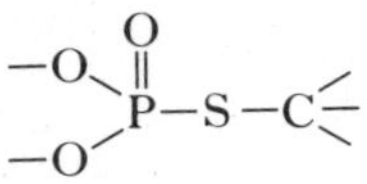

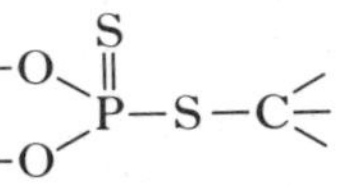

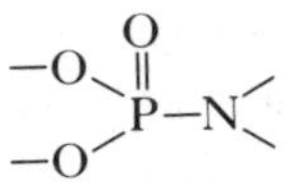

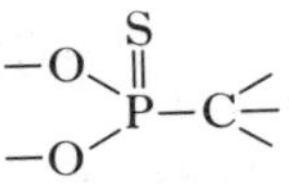

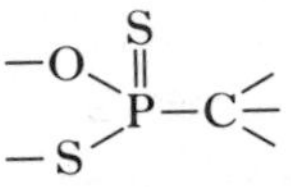

B. Properties of acetylcholinesterase

The enzyme has been isolated from a variety of insect sources, notably heads of housefly (*Musca domestica*). Purification has generally been effected by conventional biochemical methods based on the physico-chemical properties of the enzyme (e.g. Huang and Dauterman, 1973) but better purification has been achieved by recently developed affinity chromatographic procedures (Steele and Smallman, 1976a; Tripathi and O'Brien, 1977). In this technique the crude protein is passed through a column packed with an inert polymeric matrix material to which a reversible competitive inhibitor of acetylcholinesterase has been linked. Acetylcholinesterase is retained on the column because it complexes with the immobilised inhibitor, and it is then displaced from the column by a high-affinity inhibitor which can subsequently be separated from the enzyme e.g. by further gel filtration or dialysis.

It has been reported that the fundamental monomeric unit of fly head acetylcholinesterase is a polypeptide of molecular mass *c.* 80,000 daltons

(Steele and Smallman, 1976b; O'Brien, 1979) but the enzyme can also be extracted in a variety of aggregated forms (Steele and Smallman, 1976a,b; Tripathi *et al.*, 1978) some of which have an associated collagenous component (Tripathi *et al.*, 1978). In the electric fish (*Torpedo californica*), which is a rich source of the enzyme, acetylcholinesterase has been shown to be associated both with a collagenous component and with tissue phospholipids (Lwebuga-Mukasa *et al.*, 1976; Viratelle and Bernhard, 1980) and it has been suggested that the collagenous component anchors the enzyme to the extracellular coating of the post-synaptic cell membrane (Lwebuga-Mukasa *et al.*, 1976). It is the current view (O'Brien, 1979) that a similar situation exists in insects (Fig. 3.1) so that a portion of the acetylcholine released into the synapse will be hydrolysed by the esterase before the remainder reaches its receptor.

It has been suggested that the aggregated molecular forms referred to above may vary in their susceptibility to organophosphorus compounds. Thus Tripathi and O'Brien (1973) reported that, in houseflies which had been exposed to insecticides, some forms of acetylcholinesterase isolated from the housefly thorax were more sensitive than enzyme forms from housefly head, with one thoracic enzyme being particularly sensitive. The various forms also differed in their reaction with acetylthiocholine as substrate. Steele and Maneckjee (1979), on the other hand, found that all forms of the thoracic and head enzymes were similar in their reaction with acetylthiocholine while Steele and Smallman (1976c) found that the organophosphorus compound, paraoxon, inhibited head and thoracic enzymes to an identical extent. The role of aggregated molecular forms in toxicity is therefore in doubt. It should also be noted that the existence of these aggregated forms could be an artefact arising from the purification process, and it is not certain whether or not they occur *in vivo* (Devonshire, 1980).

Although insecticide activity is often better correlated with inhibition of the thoracic enzyme, Steele and Maneckjee (1979) attribute this to physiological factors rather than to biochemical differences in the enzymes.

C. Nature of the active centre of acetylcholinesterase
1. *General considerations*

The interaction of acetylcholine with the active centre of acetylcholinesterase is usually represented as shown in Fig. 3.2. The active centre contains two subsites. One of these (often referred to as the anionic site) interacts with the trimethylammonium group and the other (the esteratic site) contains a serine ($HOCH_2.CH(NH_2).CO_2H$) residue which is involved in the cleavage of the ester group; these sites are located at a specific distance from each other. The modern view (O'Brien, 1976) takes account of the fact that the reactive serine is surrounded, like every amino acid in a protein, by a variety of other amino acids, any one or group of which, by virtue of the three-dimensional folding of the protein chain, might be situated in the region of the active centre and could act as a binding site for a matching substrate or inhibitor. Two things follow from this. Firstly, molecules which are more complex than acetylcholine could interact with binding areas on the enzyme which are not used to bind the substrate. (This does not, of course, exclude the reasonable assumption that inhibitors with some resemblance to acetylcholine interact with the same region of the enzyme as does the natural substrate.) Secondly, it is probable that some of the binding areas will differ in enzymes from different species, so that one might expect inhibitors to interact differently with these different enzymes.

With these considerations in mind the various binding areas which have been proposed will now be briefly described, with particular emphasis on insect enzyme. The importance of these areas in binding inhibitors will be described later.

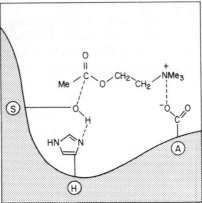

Fig. 3.2 Diagram of the active site of acetylcholinesterase. S, H and A represent serine, histidine and an acidic amino acid respectively

2. *Anionic or trimethyl-binding site* (*trimethylammonium site*)

This site has usually been referred to as the anionic site because of the indication that it contains an ionised carboxylic acid (Krupka, 1966) and it is widely believed that this helps to bind the positively-charged trimethyl-ammonium group of the substrate. The enzyme is known to be negatively charged at physiological pH and Nolte *et al.* (1980) have concluded that there are 6–9 negative charges distributed near the active centre and that electrostatic factors do govern the interaction of the active site of acetyl-cholinesterase with positively-charged species.

On the other hand Hasan *et al.* (1980, 1981) concluded that a positive charge was not necessary for good substrate binding and that the anionic site should be regarded as a trimethyl binding site. This is consistent with previous proposals that the *N*-methyl groups contribute more to binding, via hydrophobic interactions, than does the positive charge (Wilson, 1971). A unifying hypothesis is that cationic inhibitors are initially attracted to the (negative) enzyme by non-specific electrostatic forces and that binding to the active site is then more specific (see Nolte *et al.*, 1980). Since the exact contributions of ionic and hydrophobic bonding to this site are not certain, we will refer to it as the trimethylammonium site throughout this chapter, since it obviously binds the trimethylammonium (Me_3N^+) group of acetylcholine.

In the case of the housefly head enzyme Hellenbrand and Krupka (1975) have concluded that there is a broad hydrophobic area in the region of this site. They proposed that appropriate groups on organophosphorus and carbamate inhibitors could bind to this region, whereas such interactions would be detrimental to the binding of true substrates.

3. *Esteratic site*

It appears that the active centre serine is also surrounded by hydrophobic areas which can interact with appropriate groups on an inhibitor (Hellenbrand and Krupka, 1975; Fukuto, 1979). Examples of this will be mentioned later (p. 133).

4. *Phenyl-binding site*

A glance at Table 3.1 shows that many organophosphorus and carbamate insecticides contain an aromatic ring and it therefore follows that the enzyme must be able to bind this group. Hellenbrand and Krupka (1975) have postulated that the binding site for the ring lies between the trimethylammonium site and the esteratic site. Although they noted that this binding site could play an important part in the interaction of organophosphorus and carbamate inhibitors with the enzyme, they considered that the mode of interaction might be subtly different in each class of compound.

The forces involved in binding the aromatic ring are unknown. It is possible that hydrophobic interactions could occur but charge-transfer complex formation may also play a part (Hetnarski and O'Brien, 1973, 1975a,b). Such complexes are formed when an electron-rich species, typically an aromatic ring, transfers an electron to an electron-deficient acceptor. The model acceptor normally used is the electron-deficient alkene, tetracyanoethylene, and the extent to which a donor transfers an electron to this compound is used as the basis for a comparative assay. Hetnarski and O'Brien have shown that, for several series of inhibitors containing an aromatic ring, a relationship exists between the ability of a compound to bind to acetylcholinesterase and its ability to act as an electron donor to tetracyanoethylene. However, O'Brien (1976) has himself pointed out that there is no direct evidence for the formation of such complexes in the enzyme reaction and Su and Lien (1980) have argued that charge-transfer phenomena involve electronic and steric factors which are, in themselves, important for binding. Indeed the fact that 4-nitrobenzyl N-methylcarbamate binds well to the enzyme but does not form a charge-transfer complex with tetracyanoethylene (Hetnarski and O'Brien, 1973) argues against the involvement of such complexes.

5. *Allosteric site*

An allosteric site is a region of the enzyme distinct from the catalytic site, to which molecules bind and thus modify the catalytic activity of the enzyme by inducing conformational changes. It is likely, on kinetic grounds, that acetylcholinesterase contains such sites (Rosenberry, 1975) but pesticides are not known to bind to them.

6. *Indophenyl acetate site*

It has been proposed that acetylcholinesterase contains a site which binds indophenyl acetate (see O'Brien, 1976) but since this is not the natural substrate and the group does not appear in insecticidal structures we will not consider this site further.

D. Mechanism of the acetylcholinesterase reaction

It is generally accepted that the enzyme reaction is as follows:

$$\text{E–OH} + \text{AcCh} \rightleftharpoons \text{E–OH} . \text{AcCh} \underset{\text{ChH}}{\xrightarrow{\hspace{0.5cm}}} \text{E–OAc} \xrightarrow{\text{H}_2\text{O}} \text{E–OH} + \text{AcOH} \qquad (1)$$

In equation (1) E–OH represents the enzyme, AcCh is acetylcholine, E–OH.AcCh is the intermediate reversible complex between enzyme and substrate, E–OAc is acetylated enzyme, ChH is choline and AcOH is acetic acid. The enzyme (E–OH) contains a serine residue in its active centre. The hydroxy group of the amino acid is not normally a good nucleophile (a *nucleophile* is a reagent which attacks an electron-deficient centre such as the carbonyl carbon atom of an ester). However, it is thought that bonding between the serine and the imidazole group of a histidine residue acts to remove the hydrogen atom from the OH group (Fig. 3.2). The resulting $-O^-$ ion is much more reactive than the OH group, so that, following the initial binding of acetylcholine to the enzyme, the serine attacks the carbonyl group of the ester. Choline is liberated and the serine becomes acetylated. The acetylserine is finally hydrolysed to acetate with regeneration of the free hydroxy group. All of the steps are rapid. In the case of the housefly head enzyme acetylation is the slowest, and therefore the rate-determining, step (Hellenbrand and Krupka, 1970).

E. Mechanism of the inhibition of acetylcholinesterase by organo-phosphorus and carbamate insecticides

It is now generally agreed that organophosphorus compounds and carbamates react with acetylcholinesterase by a mechanism which is analogous to that of the physiological reaction, i.e.:

$$E\text{-OH} + AB \underset{k_{-1}}{\overset{k_1}{\rightleftharpoons}} E\text{-OH}.AB \overset{k_2}{\underset{BH}{\searrow}} E\text{-OA} \overset{k_3}{\longrightarrow} E\text{-OH} + AOH \qquad (2)$$

The enzyme is represented by E–OH and AB is the inhibitor (A is the phosphorylating or carbamylating group and B is the leaving group, equivalent to choline in the physiological reaction). The meaning of the rate constants (k_{+1}, etc.) will be explained below.

Thus an organophosphorus compound forms a reversible complex with the enzyme. Elimination of the leaving groups follows, to give a derivative of the enzyme in which serine is phosphorylated. A carbamate likewise leads to a carbamylated enzyme. In general, formation of enzyme-inhibitor and enzyme-substrate complexes is considered to involve non-covalent interactions, but Kamoshita et al. (1979a) have proposed that the complex between acetylcholinesterase and phenyl N-methylcarbamates involves formation of a covalent bond between the serine hydroxy group and the inhibitor (see p. 130).

Although the phosphorylated or carbamylated enzyme is itself usually hydrolysed, reaction is very much slower in these cases (Reiner, 1971), so that the enzyme is no longer able to deal efficiently with bursts of acetylcholine liberated into the synapse as a result of the insect's normal nervous activity. The transmitter builds up and proper nerve function is disrupted.

Considerable effort has gone into the elucidation of eqn (2) and the evidence for each step will now be discussed.

1. *Formation of the reversible complex, EOH.AB*

(*a*) *Reduction of inhibitory effect by the substrate*

The presence of acetylcholine reduces the inhibitory effect of organo-phosphorus and carbamate compounds on acetylcholinesterase (Augustinsson and Nachmansohn, 1949), suggesting that competition for the same binding site occurs.

(*b*) *Structural resemblance of some inhibitors to acetylcholine*

The relationship between the structure of inhibitors of acetylcholinesterase and their activity on the enzyme will be considered below in a separate section, but both in the organophosphorus and carbamate series there is evidence that a structural resemblance of the inhibitor to the natural substrate appears to improve the inhibitory effect; this is consistent with complex formation.

(*c*) *Kinetic evidence*

In eqn (2) and following the usual biochemical nomenclature the extent to which the EOH.AB complex forms is governed by its dissociation constant K_i (p. 15) where:

$$K_i = \frac{[\text{EOH}][\text{AB}]}{[\text{EOH.AB}]}$$

(It should be noted that the literature is confusing in this area since other symbols, notably K_d and K_a are used as well as K_i.)

The rate of phosphorylation or carbamylation is given by the rate constant k_2; the overall potency of an inhibitor depends on both K_i and k_2 and is represented by the so-called bimolecular reaction constant $k_i = k_2/K_i$ (Main, 1964). It will be seen that, for a compound which binds well (low K_i) and phosphorylates or carbamylates rapidly (high k_2) k_i will be large.

The rate of dephosphorylation and decarbamylation is described by k_3. For organophosphorus compounds k_3 is usually very small and can be neglected in kinetic treatments; for carbamates k_3, although small, is finite and should ideally be considered.

Because k_2 is usually large the concentration of EOH.AB is invariably very small, and cannot be demonstrated directly. Its involvement has, however, been established by Main and his colleagues who developed a kinetic treatment for the scheme and experimental methods which allowed this treatment to be tested. We shall look at these studies in some detail since they form the basis for much of the subsequent work on acetyl-cholinesterase inhibitors.

Based on the assumption that [AB] does not change during the course of the reaction, Main (1964) developed an equation for the rate of inhibition of acetylcholinesterase by organophosphorus compounds. This equation, on integration, gave the following:

$$\frac{1}{[AB]} = \frac{\Delta t}{2.303\,\Delta \log v} \cdot \frac{k_2}{K_i} - \frac{1}{K_i} \tag{3}$$

where v is the rate of the substrate reaction catalysed by the enzyme remaining unphosphorylated after reaction with inhibitor for time, t. Experimental verification of eqn (3) proved difficult. The requirement that [AB] does not change is met, experimentally, by using high concentrations of AB so that it is not significantly depleted by formation of EOH.AB and EOA. However, under these circumstances, the inhibition is so rapid as to be complete in a few minutes. Main and his colleagues overcame this problem by designing special reaction vessels so that the (organophosphorus) inhibitor could be reacted with the enzyme for a very short time (Main and Iverson, 1966) and then diluted with a large excess of substrate. Dilution effectively reduces [AB] to zero and ensures that all the EOH.AB complex dissociates to EOH and AB. The rate of the substrate reaction catalysed by the enzyme remaining unphosphorylated after reacting for time t could then be measured by its reaction with substrate. A plot of the logarithm of this velocity (v) against t at several inhibitor concentrations was constructed (Fig. 3.3a) and the slope of each of these lines represents the rate of change of the logarithm of the velocity with time, $\Delta \log v/\Delta t$, from which $\Delta t/2.303\,\Delta \log v$ could be calculated.

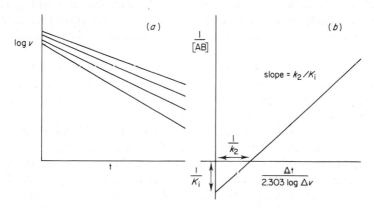

Fig. 3.3 Kinetic plots used in the graphical determination of K_i and k_2 for acetylcholinesterase inhibitors

The parameter $1/[AB]$ was then plotted against this quantity (Fig. 3.3b). Main (1964) showed that, as demanded by eqn (3), this plot gave a straight line. As can be seen from eqn (3), when $\Delta t/2.303\ \Delta\log v$ is zero $1/[AB] = -1/K_i$, i.e. the intercept on the $1/[AB]$ axis gives $1/K_i$. Similarly when $1/[AB]$ is zero $\Delta t/2.303\ \Delta\log v = 1/k_2$.

Thus the linear plot confirmed the formation of the EOH.AB complex and allowed determination of the kinetic constants.

These procedures were then applied successfully to the inhibition of acetylcholinesterase by a series of carbamates (O'Brien et al., 1966) and by the carbamate eserine (Main and Hastings, 1966). This and other work on carbamate inhibitors (Hastings et al., 1970), confirmed the validity of eqn (3) and the presence of an intermediate EOH.AB complex.

It should be noted that, because the plot is double-reciprocal in nature and the intercepts are near the origin, a slight difference in the slope of the line could have a large effect on measured K_i and k_2 values. Since the slope gives $k_2/K_i = k_i$ this means that measured values of K_i and k_2 may vary considerably while k_i does not markedly alter.

Although Main's procedure has proved extremely useful it does have some drawbacks. In particular it is not very easy to measure exactly the small time intervals between addition of the inhibitor to the enzyme and the dilution with substrate; the time taken for adequate mixing also presents a problem. In addition the procedure requires large amounts of enzyme, so that the activity remaining after dilution is large enough to measure. Several groups have attempted to overcome these difficulties using an assay procedure in which the enzyme is incubated with inhibitor and a chromogenic substrate, i.e. a substrate which, on reaction, gives a product with a diagnostic absorption spectrum. The reaction can be monitored continuously and, since the substrate competes with the inhibitor for the active site of the enzyme, the rate of inhibition is reduced (which is an advantage experimentally). However, the presence of substrate gives different kinetics and the analysis must take account of this fact.

The method was used by Post (1971) who determined K_i, k_2 and k_3 for the inhibition of spider mite enzyme by a carbamate. Although determinations need to be made at several inhibitor concentrations the method is sparing of enzyme, is claimed to be convenient and the kinetic treatment (which is complex) does not assume $k_3 = 0$.

A similar technique was then applied to organophosphorus inhibitors by O'Brien and co-workers (Hart and O'Brien, 1973). The kinetic analysis gives an integrated rate-equation identical in form to the Main eqn (3) and its graphical solution is identical.

Hart and O'Brien (1974, 1976) went on to apply their method to the study of very powerful organophosphorus inhibitors using stopped-flow techniques, an extension also recognised by Post (1971). In this procedure the enzyme is rapidly mixed with substrate and inhibitor and the course of the reaction is monitored using an oscilloscope, during a period of a few seconds. Most recently stopped-flow instrumentation has been coupled to automated data processing (Horton et al., 1977).

In spite of these technical advances, differences in the values of the kinetic constants are still found when results from different laboratories, or even the same laboratory (Hart and O'Brien, 1974), are compared. An explanation for this has recently been proposed by Nishioka et al. (1976), who studied the inhibition of bovine erythrocyte acetylcholinesterase by a large number of carbamates. Using complex integrated rate equations they determined K_i and k_2 values using both the Main and O'Brien methods under various conditions. They concluded that measured K_i and k_2 values depend on the concentration of inhibitor used and that reproducible values can be obtained by either method when this is rather low. It was suggested that, at the high inhibitor concentrations normally employed to ensure that [AB] remains constant throughout the experiment, further inhibitor molecules bind, primarily to the enzyme inhibitor complex, but also to the carabamylated enzyme. Nishioka et al. (1976) suggested that a similar situation might exist in the case of organophosphorus inhibitors.

2. *Formation of a phosphorylated or carbamylated enzyme*

(a) *Labelling studies*

The kinetic evidence discussed above indicates that, after the formation of the reversible complex EOH.AB, an essentially irreversible step occurs. Labelling studies have provided direct evidence on the chemical nature of this reaction.

[32]P-labelled diisopropyl fluorophosphate forms a stable adduct with acetylcholinesterase at the low concentrations that cause inhibition. The inhibitor is thought to combine at the active centre since the derivative is scarcely formed in the presence of acetylcholine (Michel and Krop, 1951). When the enzyme-inhibitor adduct was hydrolysed, radiolabelled phosphate and serine phosphate were recovered (Sanger, 1963), giving direct chemical proof of the formation of phosphorylated enzyme, and locating serine as the site of phosphorylation.

In a similar study Schaffer et al. (1973) treated acetylcholinesterase with a tritium-labelled organophosphorus inhibitor and degraded the treated esterase with proteolytic enzymes. When the resulting peptides were analysed it was found, as expected, that the inhibitor had reacted with a serine residue. Peptide analysis also provided information on the sequence of amino acids around the serine. It should be noted that the validity of such experiments depends on the use of pure acetylcholinesterase, which is difficult to obtain.

(b) *Release of the leaving group from the inhibitor*

Several groups of workers have shown that the leaving group of an inhibitor is released as inhibition proceeds; for example 'O'Brien *et al.* (1966) showed that release of radiolabelled 3,5-diisopropylphenol from the corresponding N-methylcarbamate paralleled the rate of inhibition and Rosenberry and Bernhard (1971) performed similar experiments using a carbamate containing a leaving group that formed a fluorescent product.

(c) *Equivalent kinetic behaviour of acetylcholinesterase inhibited by AB and by AB'*

According to eqn (2), acetylcholinesterase samples treated with organophosphorus (or carbamate) inhibitors having the same phosphoryl (or carbamyl) group, but with different leaving groups, should form identical phosphorylated (or carbamylated) enzymes. This has been demonstrated by several groups of workers.

Aldridge and Davison (1953) measured the rate of spontaneous reactivation of acetylcholinesterase treated with four organophosphorus inhibitors of the type $(CH_3O)_2PO-X$, and found identical rates of recovery, and Reiner and Simeon-Rudolf (1966) measured the spontaneous reactivation of acetylcholinesterase inhibited by the insecticides carbaryl and propoxur, and by phenyl N-methylcarbamate. All three compounds are N-monomethyl carbamates, and they formed inhibited enzyme samples that underwent reactivation at essentially similar rates.

3. *Reactivation of phosphorylated or carbamylated acetylcholinesterase*

(a) *Rate of spontaneous reactivation*

The overall rate of hydrolysis of organophosphorus and carbamate compounds by acetylcholinesterase is governed by the rate of dephosphorylation or decarbamylation, these being the slowest steps. The rate of deacetylation of acetylcholinesterase when it reacts with its normal substrate, acetylcholine, is at least 295,000 molecules of acetylcholine per active centre per minute at 37°C, and it may be higher. The corresponding rates of spontaneous reactivation for phosphorylated or carbamylated enzymes are lower by a factor of at least 10^5 or 10^6 (Aldridge, 1971).

Reiner (1971) has tabulated information on the spontaneous reactivation of acetylcholinesterase from various species treated with a variety of organophosphorus and carbamate compounds. The half-life of reactivation varies not only with the nature of the group bound to the enzyme, but also with the origin of the enzyme. It is therefore difficult to make valid generalisations correlating the structure of this group with the rate of reactivation. All commercial carbamate insecticides have either an N-methyl or an N,N-dimethylcarbamyl group, and, in the cases that have been examined so far, the half-lives of spontaneous reactivation of acetyl-cholinesterase inhibited by N-methyl- or N,N-dimethylcarbamates are measured in minutes, while diethylphosphorylacetylcholinesterases, which are formed by a large number of organophosphorus insecticides, have a half-life of hours, or, in most species, days.

(b) Reactivation by reagents

Many phosphorylacetylcholinesterases can be reactivated by reagents such as hydroxylamine and some oximes, which displace the phosphoryl group on the serine of the enzyme by nucleophilic attack. Since such reactivation occurs more quickly than the spontaneous reactivation with water, these compounds may be used as antidotes in organophosphorus poisoning. Some of them may utilise the catalytic machinery of the enzyme to form a complex that aids their action (Wilson, 1967; Englehard et al., 1967). It appears that not all these reactivating agents are effective on acetylcholin-esterases that have been inhibited by carbamates (Wilson et al., 1960; O'Brien, 1968).

Loss of an alkyl group from an alkoxy group on a phosphoryl residue attached to the active-site serine greatly slows the process of dephosphoryl-ation, and makes the enzyme insensitive to reactivating agents; this is referred to as the 'aging phenomenon'.

F. Relationship between the structure of organophosphorus and carbamate compounds and their inhibitory effects on acetylcholinesterase

As Table 3.1 indicates the range of compounds which inhibit acetyl-cholinesterase is very large; in fact for such a vital enzyme it is remarkably non-specific. In discussing structure-activity correlations we shall, for convenience, deal separately with carbamates and organophosphorus compounds but will, when appropriate, illustrate similarities and differences between the two classes of compound. Carbamates will be considered first. In the course of the review further information on the nature of the enzyme active centre will emerge.

As previously discussed, the rate of formation of phosphorylated or carbamylated enzyme is determined by the strength of binding between the organophosphorus or carbamate inhibitor, measured by K_i, and by the subsequent rate of phosphorylation or carbamylation of the serine residue at the enzyme active site, measured by k_2, so a proper understanding of structure-activity relationships at the level of the isolated enzyme ideally requires a knowledge of the value of these two parameters for each inhibitor. The physical meaning of K_i was outlined on p. 15. A value of k_2 of 10 min^{-1} means that x moles of EOH.AB would be converted at an initial rate of $10x$ moles min^{-1}, but, since the rate of the reaction is proportional to the concentration of EOH.AB, it will rapidly fall below this value as EOH.AB is used up.

1. Carbamates

Most commercial carbamates are phenyl N-methylcarbamates, i.e. carbamate esters derived from substituted phenols. Early work on the inhibition of a mammalian enzyme by a series of these led to the conclusion that virtually all of the differences in anticholinesterase activity were due to differences in complexing ability (K_i) (O'Brien et al., 1966). Recent detailed studies by Fujita and his co-workers both on mammalian enzyme (Nishioka et al., 1977) and insect enzymes (Kamoshita et al., 1979a,b) confirm this. Their study of the housefly enzyme (Kamoshita et al., 1979a) illustrates their approach.

The unsubstituted phenyl N-methylcarbamate is taken to be the reference point and K_i and k_2 are measured for this compound. Incorporation of a substituent into the phenyl ring will change these parameters. The nature of such a change depends on the properties of the substituent, such as its size, hydrophobicity, electron-withdrawing or electron-donating ability, ability to form hydrogen bonds and so on. These properties can be obtained for many substituents from standard tables or can be determined if required (for details see Nishioka et al., 1977).

For each substituted compound therefore, there is an equation which relates, for example, K_i to these properties, although at this stage the contribution of each property to binding is not known. However, solution of the entire series of simultaneous equations, usually by computer, does give the contribution of each property and provides an equation describing the series, such that the best fit is obtained between the observed K_i values and the K_i values calculated from the equation.

Kamoshita et al. (1979a) tested over seventy mono- and di-substituted phenyl N-methylcarbamates (some of their data are shown in Table 3.2) and found that, whereas k_2 showed only a three-fold variation, K_i varied as much as five hundred and fifty fold. This wide variation was shown to be due to a combination of hydrophobic, electronic and hydrogen-bonding factors.

Table 3.2   Kinetic constants[a] for the inhibition of fly head acetylcholinesterase by aryl N-methylcarbamates of structure

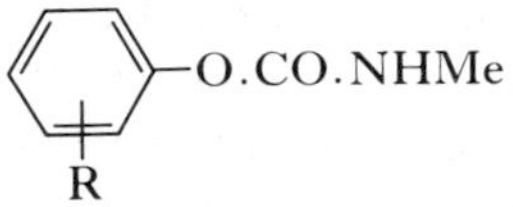

$$\text{(ring)}—\text{O.CO.NHMe}$$

R	K_i (μM)[b]	k_2 (min^{-1})	Common name
H	121	1.13	
2-Pr^i	1.59	1.15	Isoprocarb
2-OPr^i	0.7	1.43	Propoxur
2-NO$_2$	2.94	1.22	
3-Me	23.9	1.14	m-Tolyl methylcarbamate
3-Pr^i	1.2	1.55	
3-Bu^t	0.734	1.24	
3-OPr^i	5.72	0.85	
4-NO$_2$	6.1	1.11	
3-Me-5-Pr^i	0.0114	0.95	Promecarb
3,4-Me$_2$	8.1	0.96	3,4-Xylyl methylcarbamate
3,5-Me$_2$	3.73	0.87	3,5-Xylyl methylcarbamate

[a]From Kamoshita et al. (1979a).
[b]K_m (for acetylthiocholine) $= 17$ μM (Steele and Maneckjee, 1979)

Hydrophobic bonding was important in the binding of carbamates having lipophilic groups in the 2- and 3-positions. This presumably accounts for the tight binding of compounds with branched alkyl or alkoxy groups in those positions (Table 3.2). It is noteworthy that many of these compounds have a more or less obvious structural similarity to acetylcholine, suggesting that the alkyl or alkoxy groups bind at or near the trimethylammonium site. This view is supported by the fact that the distance from the carbonyl carbon of the carbamate to the centre of the alkyl group in, for example, 2-isopropoxy or 3-isopropyl compounds, is close to that between the carbonyl carbon and the nitrogen in acetylcholine (reviewed by Metcalf, 1971).

Electronic effects of substituents were rather complex. Both electron-withdrawing and electron-donating 4-substituents improved complex formation and two possible reaction pathways were proposed to account for this (Fig. 3.4). Compounds with electron-withdrawing 4-substituents were considered to follow route a and those with electron-donating groups take route b. The reasoning behind this is as follows.

Electron-withdrawing groups take electrons away from the carbamate group and thus make the carbonyl carbon more positive. It was therefore suggested that nucleophilic attack of the serine on the carbonyl function is the critical step in this group of compounds. Conversely electron-donating substituents push electrons towards the carbamate group and

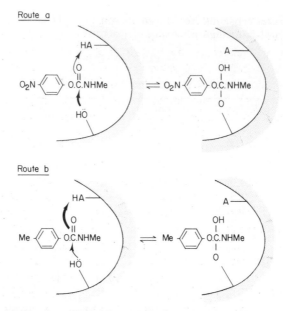

Fig. 3.4 Proposed routes of formation of the reversible complex between acetyl-cholinesterase and phenyl N-methylcarbamates containing (a) an electron-withdrawing and (b) an electron-donating group in the 4-position of the phenyl ring. OH represents serine and HA represents a hydrogen donor

thus make the carbonyl oxygen relatively negative. It was therefore proposed that protonation of this oxygen is the critical step in the case of the second group of inhibitors.

In the case of 2-substituted compounds Kamoshita et al. (1979a) concluded that route a of Fig. 3.4 was followed, possibly because the 2-substituent sterically hinders the protonation of the carboxyl oxygen required in route b.

On the other hand 3-substituted compounds were considered to form the reversible complex via route b.

Coming finally to the effects of hydrogen bonding on K_i, Kamoshita et al. (1979a) concluded that hydrogen bonding aided complex formation and that the enzyme contained a hydrogen bond donor. They proposed that the donor is situated so that it can interact with acceptors located at the 2- and 3-positions although interaction with 2-alkoxy groups is easier than with 3-alkoxy groups.

It is of interest to note that these authors regard the reversible complex as being the tetrahedral intermediate in Fig. 3.4, i.e. one in which there is a covalent bond between the carbamate and the enzyme, rather than the non-covalent interactions usually envisaged. The rate of conversion of this intermediate to the carbamylated enzyme is given by k_2. As we have seen the magnitude of this rate constant does not vary much throughout the series. It is relevant to note here that, whereas most commercial carbamates have one methyl substituent on the nitrogen atom of the carbamyl group, some have two. Hastings *et al.* (1970) measured k_2 values for three pairs of monomethyl and dimethyl analogues and found that carbamylation was faster with only one methyl substituent.

Before leaving this section it should be noted that Kamoshita *et al.* (1979a) have stated that they were able to use their derived equation to successfully design novel carbamates which had low K_i values.

The second major group of commercial carbamates (the oxime carbamates, Table 3.1) has the general structure $R^1(R^2)C=N-O.CO.NHMe$. One of the most important of these, aldicarb, emerged from an attempt to make a structural analogue of acetylcholine (Payne *et al.*, 1966) but it seems that the analogy is more apparent than real since Hastings *et al.* (1970) found that the compound did not complex well to acetylcholinesterase, having a K_i of 10.3 mM (*cf.* K_m for acetylthiocholine of 17 μM; Steele and Maneckjee, 1979). However, this is compensated for by a k_2 value of 146 min^{-1} which is very much larger than those of the aryl N-methylcarbamates. In plants, aldicarb is converted to its sulphoxide which is a more active inhibitor than aldicarb itself (Payne *et al.*, 1966) and the effectiveness of the parent compound depends on the hydrophilic nature of the derived sulphoxide which makes the compound systemic in plants (Weiden, 1971).

Other oxime carbamates such as methomyl and, particularly, oxamyl have less resemblance to acetylcholine but, for example, the former compound is more active as an inhibitor of housefly head enzyme than is aldicarb (Payne *et al.*, 1966). A detailed analysis of K_i and k_2 values for oxime carbamates has apparently not been carried out but I_{50} values for a number of compounds related to aldicarb have been reported (Payne *et al.*, 1966).

In reviewing these and other data, Kuhr and Dorough (1976) concluded that oxime carbamates differ from the aromatic compounds in that k_2 rather than K_i is important in determining overall inhibition.

2. *Organophosphorus compounds*

A very wide variety of organophosphorus compounds has found use as insecticides so it is probably not surprising that no extensive, internally-consistent analysis of the kind described above for carbamates has yet been carried out for organophosphorus esters. It is also unfortunate that kinetic studies on organophosphorus compounds are often hampered by the presence of trace impurities (Hart and O'Brien, 1976). Furthermore the results may vary with the experimental method; for example k_2 values for paraoxon have been measured as 14.0 min^{-1} (Hart and O'Brien, 1973) and 215 min^{-1} (Hart and O'Brien, 1974) using different methods. Thus, although several groups of workers have determined K_i and k_2 values for various types of inhibitor (Hart and O'Brien, 1973, 1974, 1976; Hastings and Dauterman, 1976; Huang *et al.*, 1974; Chiu *et al.*, 1969) general conclusions based on enzyme data obtained in different laboratories with enzyme from varying sources and under different experimental conditions must be treated with caution.

Nevertheless some trends can be discerned. It appears that steric and hydrophobic effects operate in much the same way as they do in carbamates. Many of the compounds studied have an aliphatic leaving group which bears only slight resemblance to choline. As a result the compounds do not bind very tightly; for example the omethoate analogue, $(EtO)_2PO.SCH_2$ $CO.NH.Me$, has a K_i on the housefly enzyme of 2 mM (*cf.* the K_m value for acetylthiocholine of 17 μM noted earlier. However, the corresponding prothoate analogue has NHMe replaced by NHPr^i, a group which is both more lipophilic and more resembles the trimethylammonium group of acetylcholine, and the K_i reduces to 1.09 mM; the NPr$_2$ analogue has a K_i of 0.03 mM (Huang *et al.*, 1974). Similar results were reported for bovine erythrocyte enzyme by Hart and O'Brien (1973) for compounds of the general structure $(EtO)_2PO.O(CH_2)_nBu^t$. K_i fell as n was increased and binding was maximal where n was 3 or 4 (rather than n = 2 which would represent a compound isosteric with acetylcholine). It seems reasonable to conclude that, in both cases, the more lipophilic groups were binding to the hydrophobic area near the trimethylammonium site.

Related results in the aryl series were obtained by Hollingworth *et al.* (1967), who studied analogues of fenitrothion in which the sulphur atom was replaced by oxygen, and the substituent *ortho* to the nitro group on the aromatic ring was varied. As this substituent was changed from H to methyl to isopropyl, K_i (on housefly head enzyme) fell from 37 μM to 11 μM to 3.3 μM respectively and this was attributed to increased binding to the trimethylammonium site.

Steric and hydrophobic factors are also involved in the interaction of chiral organophosphorus inhibitors with acetylcholinesterase. Chiral compounds are those in which an atom is linked to four different substituents, so that two isomeric compounds [termed enantiomers and denoted by *R*-(rectus) and *S*-(sinister)] are possible depending on the configuration of the substituents around this atom. Chiral pesticides include EPN and fonofos; the enantiomers of the latter are:

(S)   (R)

Because each enantiomer must interact differently with the enzyme surface it is not surprising that the enantiomers have different kinetic constants. This has been demonstrated for the P=O analogues of both fonofos (Lee et al., 1978) and EPN (Nomeir and Dauterman, 1979). In the former study bimolecular constants (k_i) were recorded rather than K_i but, in the case of EPN, one enantiomer bound about twice as well as the other. Particularly striking differences in K_i have been reported by Wustner and Fukuto (1974; see also Fukuto, 1979) for isomers of EtCH(Me).O.PO.(Et).SCH$_2$CH$_2$NMe$_2$, which has a bulky group attached to the chiral phosphorus atom and an additional chiral centre in its O-alkyl group. The configuration at the phosphorus atom was particularly important. Thus, when the configuration at the chiral centre in the alkoxy group was kept constant and the effect of the configuration at phosphorus was examined, the K_i was about 10 μM for one isomer and around 2 mM for the other, a two hundred fold difference. Changing the configuration at the chiral centre in the O-alkyl group gave only slight differences. These authors also found that the isomers with a very low K_i also had a significantly larger k_2 than those isomers with a high K_i. It should be noted that, since the S-alkyl group probably binds near the trimethylammonium site, then the alkyl and alkoxy groups can be visualised as interacting with hydrophobic areas near the serine group (Fukuto, 1979).

Let us now turn to the reaction of organophosphorus compounds with the active-site serine. It will be recalled that, in the case of carbamates, reaction with serine has been depicted (Kamoshita et al., 1979a) as being reversible and leading to the tetrahedral enzyme-inhibitor complex (Fig. 3.4). However, the situation with organophosphorus compounds is different, since attack of a nucleophile at a tetrahedral phosphorus atom does not lead to a reaction intermediate analogous to that formed by attack at a carbonyl group. In the following discussion we will adopt the generally-held view that the reaction of organophosphorus compounds with serine leads irreversibly to phosphorylated enzyme and that the rate of this reaction is given by k_2.

In the case of organophosphorus compounds, k_2 is generally very large and therefore plays a major role in determining overall inhibitory potency. Much early work (summarised by O'Brien, 1976) led to the conclusion that the inhibitory activity of organophosphorus compounds was improved if electron-withdrawing groups were attached to the phosphorus atom. Evidence for this includes the finding that inhibition of acetylcholinesterase by phosphates of the type $(EtO)_2.PO.OAr$ closely paralleled the sensitivity of the esters to base-catalysed hydrolysis (Aldridge and Davison, 1952). The simplest interpretation of the data is that electron-withdrawing substituents make the phosphorus atom relatively positive and therefore more susceptible to attack by the serine in the enzyme active site, i.e. electronic substituent effects act primarily on k_2 (O'Brien, 1976) rather than on K_i.

However, this is not always the case; thus Hart and O'Brien (1976) found that for substituted compounds such as $EtO(PrS).PO.OAr$ (cf. profenofos, prothiophos) k_2 could not be correlated with the electron-withdrawing capacity of the aryl moiety. It should be noted that the compounds in question are chiral (the enantiomers of profenofos have recently been separated by Leader and Casida, 1982), and Hart and O'Brien (1976) concluded that special steric or other constraints are involved in the enzyme-inhibitor reaction. This is no doubt true for the interaction with all carbamates and organophosphorus compounds and the nature of these constraints will only become clear when the structure of the active centre, preferably with bound inhibitor present, is known.

3. Comparison of organophosphorus compounds and carbamates

Although, as we have seen, steric and hydrophobic effects seem to operate in the same sort of way in each class of compound, workers in the area have long been puzzled as to why closely related organophosphorus

Table 3.3   Comparison of the kinetic constants for inhibition of bovine erythrocyte acetylcholinesterase by selected pairs of organophosphorus compounds[a] and carbamates[b]

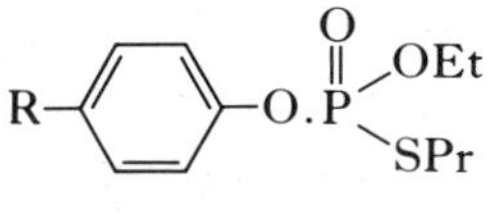

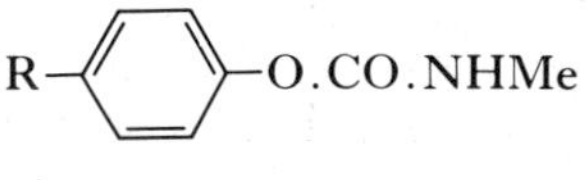

$$R-\!\!\!\left\langle \bigcirc \right\rangle\!\!\!-O.P\!\!\begin{array}{c} O \\ \| \\ \end{array}\!\!\begin{array}{c} OEt \\ SPr \end{array} \qquad R-\!\!\!\left\langle \bigcirc \right\rangle\!\!\!-O.CO.NHMe$$

R	K_i (mM)	k_2 (min^{-1})	R	K_i (mM)	k_2 (min^{-1})
H	1.93	4.83	H	3.02	0.86
Cl	0.62	6.27	Cl	3.16	0.74
Me	1.80	3.79	Me	1.61	0.98
NO$_2$	1.81	263	NO$_2$	0.103	0.81

[a]From Hart and O'Brien (1976); [b]From Nishioka et al. (1977).

compounds and carbamates behave differently in their reactions with acetylcholinesterase. As an example the data in Table 3.3 show that introduction of the nitro group dramatically increases k_2 of the phosphorus compound but not the carbamate. On the other hand K_i is substantially lowered in the carbamate but not in the phosphate.

Some years ago Hastings et al. (1970) concluded that the substrate and carbamate reactions were more closely related than were those of the substrate and organophosphorus compounds, and they suggested that phosphorylation might occur at a different region of the enzyme active centre from the site of acetylation. Hellenbrand and Krupka (1975) have also discussed the possibility that the two classes may interact differently with the phenyl binding site, particularly in view of the obvious chemical and structural differences between them.

Evidence that the compounds do indeed interact with different regions of the active site has recently been obtained by Järv et al. (1977). Their argument is based on the fact that the geometry of nucleophilic attack at carbonyl carbon (present in the substrate and in carbamates) differs from that of nucleophilic attack at tetrahedral phosphorus. To be precise, the planar carbonyl group is attacked from either side of the plane in which its atoms lie whereas, in the case of a phosphorus compound, the nucleophile attacks along the line of the bond from phosphorus to the leaving group as follows:

$$(\text{Ser} \quad \overset{..}{\text{O}}\text{H} \;\rightarrow\; \overset{\text{O}}{\overset{\|}{\underset{}{\text{P}}}}{-}\text{X})$$

This led Järv et al. (1977) to the view that, if the nucleophile (i.e. the serine OH) is relatively fixed in the enzyme, it is not possible for the leaving group in an organophosphorus compound to occupy the same site on the enzyme as the leaving group in acetylcholine and in a carbamate. They then compared data on the acetylation of the enzyme by a series of substrates such as $CH_3 COOR$ (R represents an alkyl group) and phosphorylation by an analogous series of inhibitors of structure RO.(Me).PO.SBu. They found that each reaction responded in the same way to changes in R, indicating that the R group binds to the same region of the enzyme in each case. However, in the case of the substrate, OR is the leaving group while in the phosphate it is not (SBu leaves). In other words the leaving group in phosphorus esters binds to a region of the enzyme which is different from that which binds the leaving group of acetylcholine and, by implication, of carbamates.

G. Effects of organophosphorus and carbamate insecticides on insects

We have already seen that phosphorylated and carbamylated acetyl-cholinesterases have appreciable life times, so that, if organophosphorus and carbamate insecticides really do kill insects by inhibiting the enzyme, it should be possible to isolate inhibited acetylcholinesterase from poisoned insects. O'Brien (1961) treated houseflies with four commercial organophosphorus insecticides and showed that the acetylcholinesterase of dead flies was 74 to 99% inhibited 320 minutes after application of the LD_{50}. Kolbezen et al. (1954) have extracted inhibited acetylcholinesterase from flies rendered prostrate by carbamates. Booth and Metcalf (1970) demonstrated by histochemical techniques the inhibition of acetylcholinesterase in insects treated with organophosphorus compounds and carbamates.

However, we do not know the actual level and duration of acetylcholinesterase inhibition that is required to obtain a lethal effect, nor whether inhibition in a particular area of the nervous system is critical (Devonshire, 1980). There are reports that only relatively low levels of inhibition are necessary to cause death (Dahm, 1971). However, Smissaert et al. (1975) have calculated that in susceptible spider mites 98% of the critical acetylcholinesterase would have to be inactivated to cause harmful effects.

A likely consequence of acetylcholinesterase inhibition is the accumulation of acetylcholine, and Smallman and Fisher (1958) have shown that acetylcholine levels may rise to 260% of normal when insects are treated with organophosphorus insecticides. However, it must be noted that accumulation of acetylcholine can also occur as a result of stress or DDT poisoning (Lewis et al., 1960) so that it is not necessarily a specific response to acetylcholinesterase inhibition.

What effect is this accumulation of the transmitter substance likely to have on nervous transmission? The post-synaptic response caused by acetylcholine is referred to as the 'excitatory post-synaptic potential', and changes in this and other parameters of nervous conduction may be determined from electrophysiological experiments. The application of various organophosphorus compounds and the carbamate eserine to the cockroach nerve cord prolonged this potential, presumably due to the persistence of acetylcholine in the synapse (Narahashi, 1971). Such compounds, as well as insecticidal carbamates (Metcalf, 1971) eventually cause a block in nervous transmission.

A further correlation between activity on the isolated enzyme and effects on the nerve cord was provided by the demonstration that the rates of ganglionic response, as measured on the cockroach nerve cord, for a series of insecticidal carbamates, were directly proportional to the inhibitory activity on acetylcholinesterase in vitro (Metcalf, 1971). More recently Uchida et al. (1975) reported that the ability of members of a series of carbamates to inhibit the esterase correlated well with their ability to induce excitation in insect nerve cords and that this in turn was directly related to insecticidal activity. However, there are often complications involved in establishing these correlations, since the insecticide must penetrate into the insect and then run the gauntlet of the various detoxification processes before reaching its target. In such instances it has been shown that, when oxidative metabolism is suppressed by a suitable synergist (see p. 330) then a reasonable relationship exists between enzyme inhibition and toxicity to insects (Metcalf, 1971; see also Kamoshita et al., 1979a,b). Similarly differences in the insecticidal activity of isomeric organophosphorus compounds usually parallel the differences referred to above in their anticholinesterase activity.

It is relevant to mention here that examples are known where resistance to carbamate or organophosphorus pesticides has turned out to be due to an alteration in acetylcholinesterase, rendering it less susceptible to inhibition; this has been demonstrated in mites (Smissaert, 1964), in the housefly (Plapp and Tripathi, 1978; Devonshire, 1975) and in the cattle tick (Nolan and Schnitzerling, 1976); other cases are reviewed by Devonshire (1980). In the housefly, decreased inhibition is due to decreased binding, i.e. a higher K_i (Plapp and Tripathi, 1978). These data provide additional evidence that acetylcholinesterase is the site of action of the compounds.

Insects treated with organophosphorus and carbamate insecticides show initial hyperactivity followed by convulsions then paralysis. However, insects treated with organophosphorus compounds apparently show subtly different symptoms of poisoning from those treated with carbamates (Miller, 1976), while carbamates also produce effects much earlier after treatment (Miller et al., 1973).

The problems of trying to establish the precise chain of events between inhibition of the enzyme and insect death have proved to be very severe. Obviously excess acetylcholine would have widespread harmful effects, amongst which may be the uncontrolled hormone release known to accompany insect poisoning (Maddrell, 1980). It has been suggested (Gerolt, 1976) that the ultimate cause of death is probably degeneration of vital tissues following loss of water from the insect integument, a response induced by organophosphorus compounds and carbamates as well as other insecticides. Gerolt (1976) considered that the sensitive element could be of nervous origin and located in the integument itself. Dehydration of the tissues may cause the appearance of vacuoles in the insect ganglia. This effect, first described by Kruger (1931), has recently been demonstrated by Collins et al. (1979) in flies treated with the carbamate insecticide carbofuran.

Organophosphorus compounds have been used commercially as ovicides to control insects in the egg stage. In this case the biochemical site of action may not always be acetylcholinesterase since at least some insect embryos are killed before it is possible to detect any acetylcholinesterase activity. Krysan and Guss (1971) showed that paraoxon was a powerful inhibitor of insect egg lipase (EC 3.1.1.3), an enzyme which hydrolyses triglycerides, and they suggest this as a possible site of action for some organophosphorus ovicides.

REFERENCES

Aldridge, W. N. (1971). *Bull. Wld. Hlth. Org.* **44**, 25–30

Aldridge, W. N. and Davison, A. N. (1952). *Biochem. J.* **51**, 62–70

Aldridge, W. N. and Davison, A. N. (1953). *Biochem. J.* **55**, 763–766

Augustinsson, K. B. and Nachmansohn, D. (1949). *J. Biol. Chem.* **179**, 543–559

Booth, G. M. and Metcalf, R. L. (1970). *Ann. Entomol. Soc. Am.* **63**, 197–204

Chiu, Y. C., Main, A. R. and Dauterman, W. C. (1969). *Biochem. Pharmacol.* **18**, 2171–2177

Collins, C., Kennedy, J. M. and Miller, T. (1979). *Pestic. Biochem. Physiol.* **11**, 135–158

Dahm, P. A. (1971). *Bull. Wld. Hlth. Org.* **44**, 215–219

Devonshire, A. L. (1975). *Biochem. J.* **149**, 463–469

Devonshire, A. L. (1980). In "Insect Neurobiology and Pesticide Action", pp. 473–480. Society of Chemical Industry, London

Dowson, R. J. (1977). *Pestic. Sci.* **8**, 651–660

Englehard, N., Prchal, K. and Nenner, M. (1967). *Angew. Chem. Int. Ed. Engl.* **6**, 615–626

Eto, M. (1974). "Organophosphorus Pesticides: Organic and Biological Chemistry" CRC Press, Cleveland

Fest, C. and Schmidt, K.-J. (1973). "The Chemistry of Organophosphorus Pesticides". Springer-Verlag, Berlin

Fukuto, T. R. (1979). In "Neurotoxicology of Insecticides and Pheromones" (T. Narahashi, ed.) pp. 277–295. Plenum Press, New York

Gerolt, P. (1976). *Pestic. Sci.* **7**, 604–620

Hart, G. J. and O'Brien, R. D. (1973). *Biochemistry* **12**, 2940–2945

Hart, G. J. and O'Brien, R. D. (1974). *Pestic. Biochem. Physiol.* **4**, 239–244

Hart, G. J. and O'Brien, R. D. (1976). *Pestic. Biochem. Physiol.* **6**, 85–90

Hasan, F. B., Cohen, S. G. and Cohen, J. B. (1980). *J. Biol. Chem.* **255**, 3898–3904

Hasan, F. B., Elkind, J. L., Cohen, S. G. and Cohen, J. B. (1981). *J. Biol. Chem.* **256**, 7781–7785

Hastings, F. L. and Dauterman, W. C. (1976). *J. Econ. Entomol.* **69**, 69–72

Hastings, F. L., Main, A. R. and Iverson, F. (1970). *J. Agric. Food Chem.* **18**, 497–502

Hellenbrand, K. and Krupka, R. M. (1970). *Biochemistry* **9**, 4665–4672

Hellenbrand, K. and Krupka, R. M. (1975). *Croat. Chem. Acta* **47**, 345–353

Hetnarski, B. and O'Brien, R. D. (1973). *Biochemistry* **12**, 3883–3887

Hetnarski, B. and O'Brien, R. D. (1975a). *J. Med. Chem.* **18**, 29–33

Hetnarski, B. and O'Brien, R. D. (1975b). *J. Agric. Food Chem.* **23**, 709–713

Hollingworth, R. M., Fukuto, T. R. and Metcalf, R. L. (1967). *J. Agric. Food Chem.* **15**, 235–241

Horton, G. L., Lowe, J. R. and Lieske, C. N. (1977). *Analyt. Biochem.* **78**, 213–228

Huang, C. T. and Dauterman, W. C. (1973). *J. Insect Biochem.* **3**, 325–334

Huang, C. T., Dauterman, W. C. and Hastings, F. L. (1974). *Pestic. Biochem. Physiol.* **4**, 249–253

Järv, J., Aaviksaar, A., Godovikov, N. and Lobanov, D. (1977). *Biochem. J.* **167**, 823–825

Kamoshita, K., Ohno, I., Fujita, T., Nishioka, T. and Nakajima, M. (1979a). *Pestic. Biochem. Physiol.* **11**, 83–103

Kamoshita, K., Ohno, I., Kasamatsu, K., Fujita, T. and Nakajima, M. (1979b). *Pestic. Biochem. Physiol.* **11**, 104–116

Kolbezen, M. J., Metcalf, R. L. and Fukuto, T. R. (1954). *J. Agric. Food Chem.* **2**, 864–870

Kruger, F. (1931). *Z. Angew. Entomol.* **18**, 344–353

Krupka, R. M. (1966). *Biochemistry* **5**, 1988–1998

Krysan, J. L. and Guss, P. L. (1971). *Biochim. Biophys. Acta* **239**, 349–352

Kuhr, R. J. and Dorough, H. W. (1976). "Carbamate Insecticides; Chemistry, Biochemistry and Toxicology". CRC Press, Cleveland

Leader, H. and Casida, J. E. (1982). *J. Agric. Food Chem.* **30**, 546–551

Leake, L. D. and Walker, R. J. (1980). "Invertebrate Neuropharmacology". Blackie and Son, Glasgow

Lee, P. W., Allahyari, R. and Fukuto, T. R (1978). *Pestic. Biochem. Physiol.* **8**, 146–157

Lewis, S. E., Waller, J. B. and Fowler, K. S. (1960). *J. Insect Physiol.* **4**, 128–137

Lwebuga-Mukasa, J. S., Lappi, S. and Taylor, P. (1976). *Biochemistry* **15**, 1425–1434

Maddrell, S. H. P. (1980). In "Insect Neurobiology and Pesticide Action", pp. 329–334. Society of Chemical Industry, London

Main, A. R. (1964). *Science* **144**, 992–993

Main, A. R. (1976). In "Biology of Cholinergic Function" (A. M. Goldberg and I. Hanin, eds) pp. 269–353 Raven Press, New York

Main, A. R. and Hastings, F. L. (1966). *Science* **154**, 400–402

Main, A. R. and Iverson, F. (1966). *Biochem. J.* **100**, 525–531

Metcalf, R. L. (1971). *Bull. Wld. Hlth. Org.* **44**, 43–78

Michel, H. O. and Krop, S. (1951). *J. Biol. Chem.* **190**, 119–125

Miller, T. (1976). *Pestic. Biochem. Physiol.* **6**, 307–319

Miller, T., Kennedy, J. M., Collins, C. and Fukuto, T. R. (1973). *Pestic. Biochem. Physiol.* **3**, 447–455

Narahashi, T. (1971). In "Advances in Insect Physiology" (J. W. L. Beament, J. E. Treherne and V. B. Wigglesworth, eds) Vol. 8, pp. 1–93. Academic Press, London and New York

Nishioka, T., Kitamura, K., Fujita, T. and Nakajima, M. (1976). *Pestic. Biochem. Physiol.* **6**, 320–337

Nishioka, T., Fujita, T., Kamoshita, K. and Nakajima, M. (1977). *Pestic. Biochem. Physiol.* **7**, 107–121

Nolan, J. and Schnitzerling, H. J. (1976). *Pestic. Biochem. Physiol.* **6**, 142–147

Nolte, H.-J., Rosenberry, T. L. and Neumann, E. (1980). *Biochemistry* **19**, 3705–3711

Nomeir, A. A. and Dauterman, W. C. (1979). *Pestic. Biochem. Physiol.* **10**, 121–127

O'Brien, R. D. (1961). *J. Econ. Entomol.* **54**, 1161–1164

O'Brien, R. D. (1968). *Molec. Pharmacol.* **4**, 121–130

O'Brien, R. D. (1976). In "Insecticide Biochemistry and Physiology" (C. F. Wilkinson, ed.) pp. 271–296. Plenum Press, New York

O'Brien, R. D. (1979). In "Advances in Pesticide Science" (H. Geissbühler, ed.) Pt 3, pp. 449–457. Pergamon Press, Oxford

O'Brien, R. D., Hilton, D. B. and Gilmour, L. (1966). *Molec. Pharmacol.* **2**, 593–605

Payne, L. K., Stansbury, H. A. and Weiden, M. H. J. (1966). *J. Agric. Food Chem.* **14**, 356–365

Plapp, F. W. and Tripathi, R. K. (1978). *Biochem. Genet.* **16**, 1–11

Post, L. C. (1971). *Biochim. Biophys. Acta* **250**, 121–130

Reiner, E. (1971). *Bull. Wld. Hlth. Org.* **44**, 109–112

Reiner, E. and Simeon-Rudolf, V. (1966). *Biochem. J.* **98**, 501–505

Rosenberry, T. L. (1975). *Adv. Enzymol.* **43**, 103–218

Rosenberry, T. L. and Bernhard, S. A. (1971). *Biochemistry* **10**, 4114–4120

Sanger, F. (1963). *Proc. Chem. Soc.* 73–83

Schaffer, M. K., Michel, H. O. and Bridges, A. F. (1973). *Biochemistry* **12**, 2946–2950

Shankland, D. L. (1976). In "Insecticide Biochemistry and Physiology" (C. F. Wilkinson, ed.) pp. 229–270. Plenum Press, New York

Smallman, B. N. and Fisher, R. W. (1958). *Can. J. Biochem. Physiol.* **36**, 575–586

Smissaert, H. R. (1964). *Science* **143**, 129–131

Smissaert, H. R., Abd El Hamid, F. M. and Overmeer, W. P. J. (1975). *Biochem. Pharmacol.* **24**, 1043–1047

Steele, R. W. and Maneckjee, A., (1979). *Pestic. Biochem. Physiol.* **10**, 322–332

Steele, R. W. and Smallman, B. N. (1976a). *Biochim. Biophys. Acta* **445**, 147–157

Steele, R. W. and Smallman, B. N. (1976b). *Biochim. Biophys. Acta* **445**, 131–146

Steele, R. W. and Smallman, B. N. (1976c). *Life Sci.* **19**, 1937–1942

Su, C. T. and Lien, E. J. (1980). *Res. Commun. Chem. Pathol. Pharmacol.* **29**, 403–415

Tripathi, R. K. and O'Brien, R. D. (1973). *Pestic. Biochem. Physiol.* **2**, 418–424

Tripathi, R. K. and O'Brien, R. D. (1977). *Biochim. Biophys. Acta* **480**, 382–389

Tripathi, R. K., Telford, J. N. and O'Brien, R. D. (1978). *Biochim. Biophys. Acta* **525**, 103–111

Uchida, M., Irie, Y., Kurihara, N., Fujita, T. and Nakajima, M. (1975). *Pestic. Biochem. Physiol.* **5**, 258–264

Viratelle, O. M. and Bernhard, S. A. (1980). *Biochemistry* **19**, 4999–5007

Weiden, M. H. J. (1971). *Bull. Wld. Hlth. Org.* **44**, 203–213

Wilson, I. B. (1967). In "Drugs Affecting the Peripheral Nervous System" (A. Burger, ed.) pp. 381–397. Dekker, New York

Wilson, I. B. (1971). In "Cholinergic Ligand Interactions" (D. J. Triggle, ed.) pp. 1–18. Academic Press, London and New York

Wilson, I. B., Hatch, M. A. and Ginsburg, S. (1960). *J. Biol. Chem.* **235**, 2312–2315

Worthing, C. R. (ed.) (1979). "The Pesticide Manual', 6th edn. British Crop Protection Council, London

Wustner, D. A. and Fukuto, T. R. (1974). *Pestic. Biochem. Physiol.* **4**, 365–376

4 | Insecticides Acting Elsewhere in the Nervous System

I. COMPOUNDS WHICH INTERFERE WITH AXONAL TRANSMISSION
A. Biochemical background

As a prelude to discussing the mode of action of the compounds in this section it is necessary to examine the events involved in axonal conduction; this description owes much to the accounts of Dowson (1977) and Shankland (1976). The extracellular fluid surrounding the axonal membrane contains a high concentration of sodium ions and a low concentration of potassium ions. The nerve cell, on the other hand, contains a high concentration of potassium ions and a low concentration of sodium ions; the membrane is relatively permeable to potassium ions but only slightly permeable to sodium ions. It is necessary to maintain the correct concentrations of ions against their concentration gradients and this is achieved by a sodium/potassium exchange pump driven by hydrolysis of ATP.

At rest the inside of the cell is negative with respect to the outside and the potential difference (inside with respect to outside) is usually around

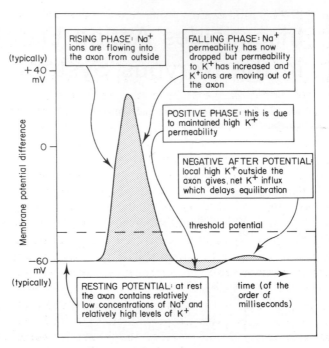

Fig. 4.1 An idealised action potential

−60 mV. When a propagated impulse passes along the axon, the action potential, i.e. the localised change in potential between the inside and the outside of the membrane, can be measured by a recording electrode. An idealised action potential is shown in Fig. 4.1. As the action potential approaches the recording electrode, the potential difference across the membrane becomes more positive (this is called depolarisation). There is a critical membrane potential difference (called the threshold potential) which is around 10 to 20 mV more positive than the resting potential, and when this threshold is reached propagation of the action potential is irreversible. The rising phase of the action potential is followed by a peak and a falling phase as the membrane becomes repolarised and the potential difference returns to the resting value (Fig. 4.1). Let us now consider the ion movements (i.e. current) associated with the action potential. These have been determined by the voltage clamp technique in which the membrane potential difference is changed from the resting value and held at some different pre-set value. More or less current has to be supplied to maintain this new voltage and the direction of the current shows whether ions are flowing into or out of the axon, indicating the ion movements which would take place at this voltage in the normal action potential.

Shankland (1976) illustrates the use of the method as follows. Suppose an axon is depolarised, resulting in an inward current. If this current is abolished when sodium ions are removed from the bathing medium, or in the presence of a chemical known to block the sodium channel, then the inward current must have been carried by sodium ions.

Experiments of this type have been interpreted in the following terms. When the threshold potential is reached, sodium channels in the membrane open, sodium ions rapidly enter the cell down their concentration gradient and thus the potential difference becomes more positive. This rise in potential serves to close the sodium channels and to increase the permeability of the axon to potassium ions, which therefore flow out of the cell down their concentration gradient. This brings about the falling phase of the action potential, in which the potential difference swings back towards the resting value. Inspection of the falling phase reveals that the membrane potential difference drops a little below the resting value. This is because the membrane retains enhanced permeability to potassium ions even when it has been repolarised to the resting potential. However, toward the end of the action potential, when net potassium efflux has slowed considerably, the high local concentration of potassium ions in the area outside the membrane results in increased net potassium influx. This leads to a slight rise in potential difference called (confusingly, but for historical reasons) the negative (or depolarising) after-potential. Eventually the local concentration of potassium ions is dissipated and the membrane potential difference returns to the resting value.

The ionic concentration gradients that usually exist across the membrane are large enough to sustain the passage of many impulses, for example, during periods of high activity. However, the exchange pump probably cannot keep up with sodium influx at such times and must continue to operate during quiet periods to restore resting conditions.

B. Pyrethroids
1. *Introduction*

The synthetic pyrethroids shown in Table 4.1 have emerged from prolonged efforts to improve the biological activity and chemical stability of the natural pyrethrins, which comprise a mixture of insecticidal esters obtained from the flowers of *Chrysanthemum cinerariaefolium*. The natural materials are expensive and unstable to light and air, and early synthetic analogues such as allethrin and bioallethrin were also expensive, rather unstable chemically and rapidly metabolised by insects. They were therefore mainly used in mixtures with a synergist (see p. 330) as household insecticides. However, recent research, both by Elliott and his colleagues and in industry, has led to the discovery of simpler compounds (e.g. cypermethrin and fenvalerate) which are more stable under field conditions. Although the synthetic pyrethroids are also relatively expensive they can often be applied at lower rates than other insecticides (Worthing, 1979) and their use in agriculture is already widespread.

Table 4.1   Pyrethroids*

Name	Structure
Allethrin	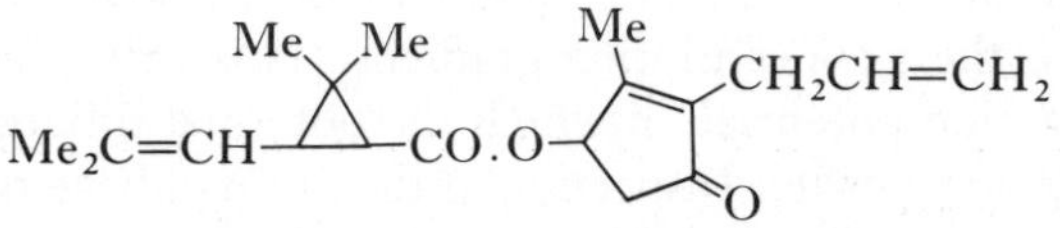
5-Benzyl-3-furylmethyl (E)-(1R)-cis-2,2-dimethyl-3-(tetrahydro-2-oxo-3-thienylidenemethyl)cyclopropanecarboxylate	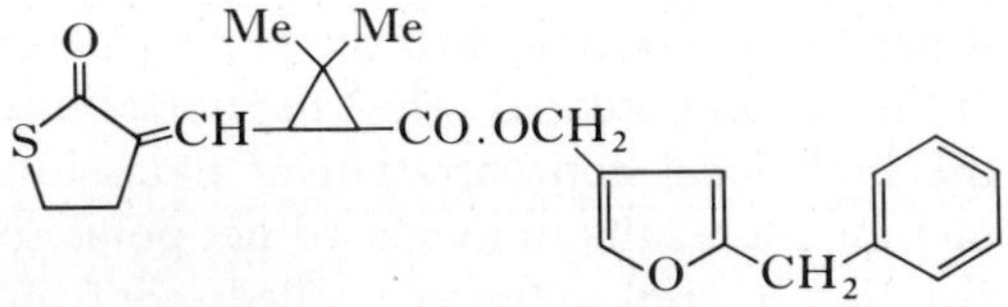
Bioallethrin	

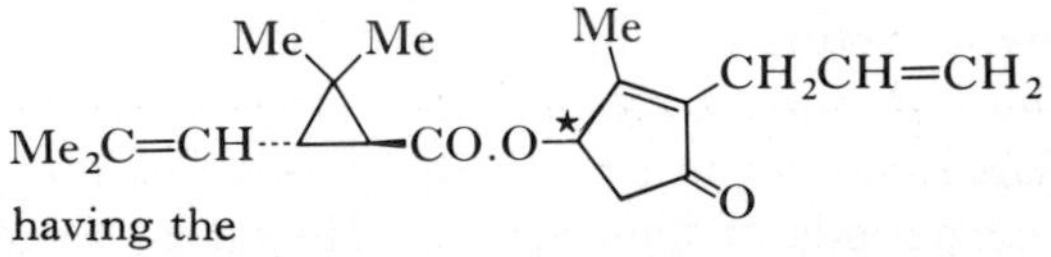

★(S)-Bioallethrin, an isomer having the (S) configuration at the carbon atom indicated, is also used commercially

Bioresmethrin	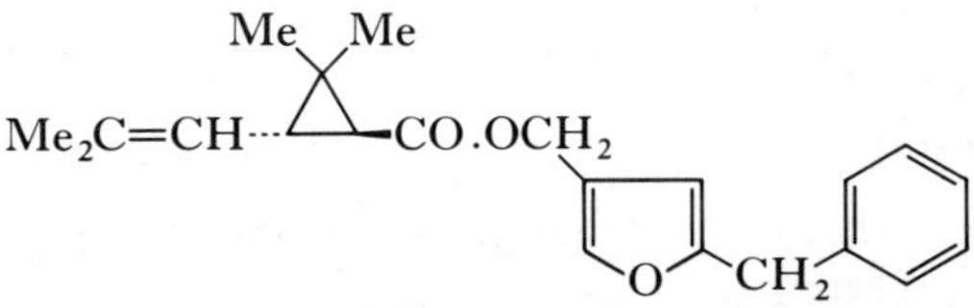

*In some cases the configuration at the double bonds and at chiral centres is known (Worthing, 1979), but the latter has only been indicated here in the cases of bioallethrin (cf. allethrin) and also bioresmethrin (cf. resmethrin)

Name	Structure
(S)-α-Cyano-3-phenoxy-benzyl (1R)-cis-(1'RS)-2,2-dimethyl-3-(1,2,2,2-tetrabromo-ethyl) cyclopropanecarboxylate	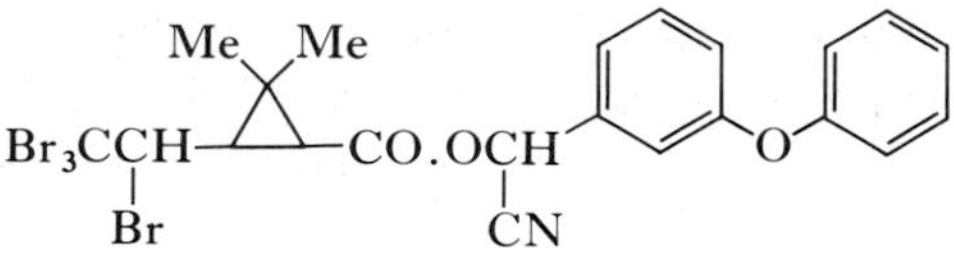
Cypermethrin	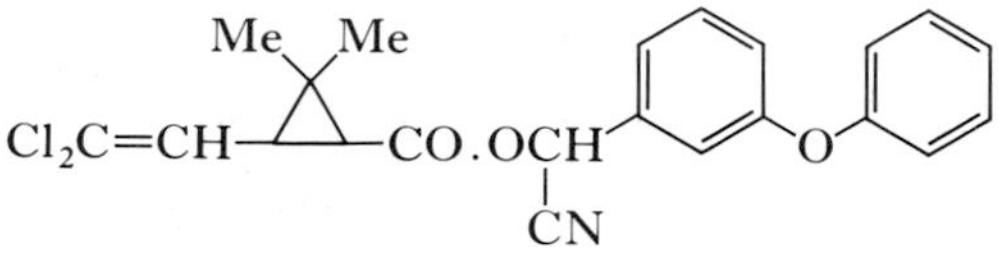
Deltamethrin	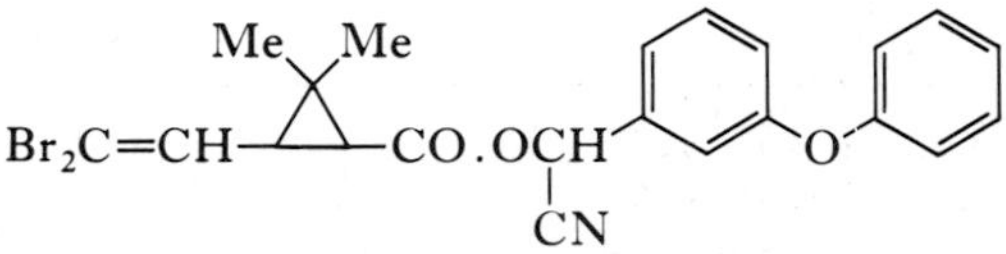
Fenvalerate	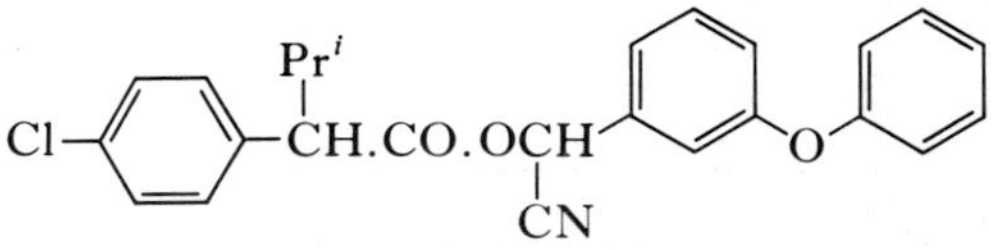
Flucythrinate	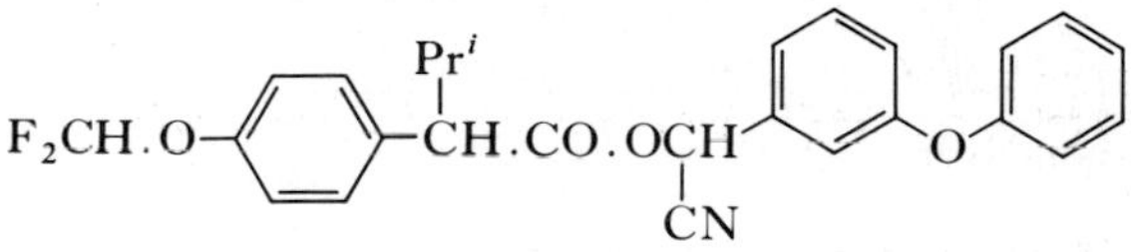
Fluvalinate	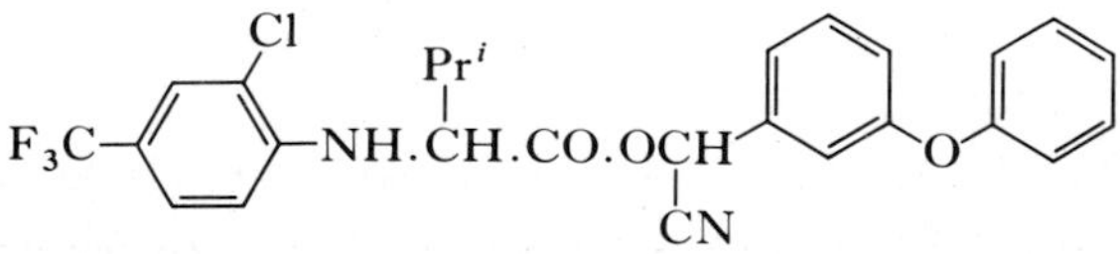
Permethrin	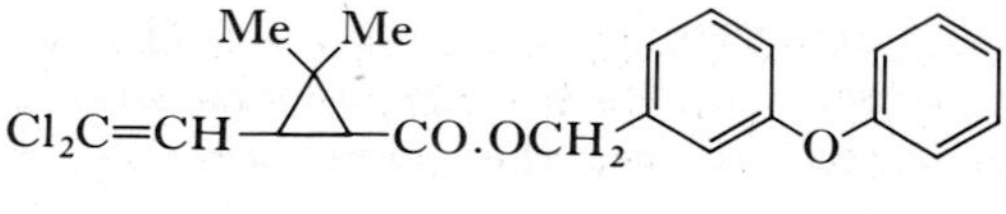

Name	Structure

Phenothrin

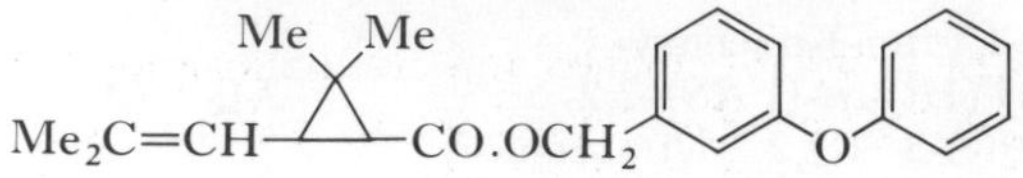

$$\text{Me}_2\text{C=CH} \overset{\text{Me}\quad\text{Me}}{\triangle} \text{CO.OCH}_2 \text{—}\bigcirc\text{—O—}\bigcirc$$

Pyrethrins or Pyrethrum

Esters from:

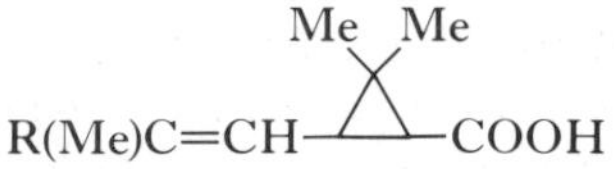

$$\text{R(Me)C=CH} \overset{\text{Me}\quad\text{Me}}{\triangle} \text{COOH}$$

and

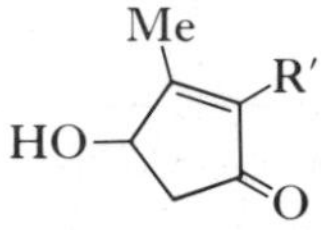

where R = Me and —COOMe
and   R' = —CH$_2$ CH=CH.CH=CH$_2$
—CH$_2$ CH=CH.Me
and
—CH$_2$ CH=CH.Et

Resmethrin

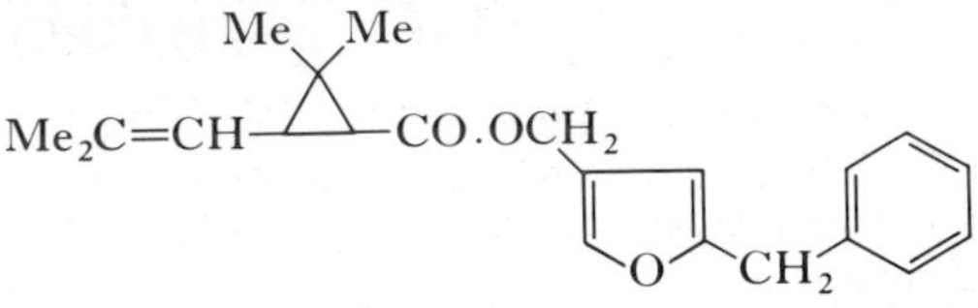

$$\text{Me}_2\text{C=CH} \overset{\text{Me}\quad\text{Me}}{\triangle} \text{CO.OCH}_2$$

Tetramethrin

$$\text{Me}_2\text{C=CH} \overset{\text{Me}\quad\text{Me}}{\triangle} \text{CO.OCH}_2\text{—N}$$

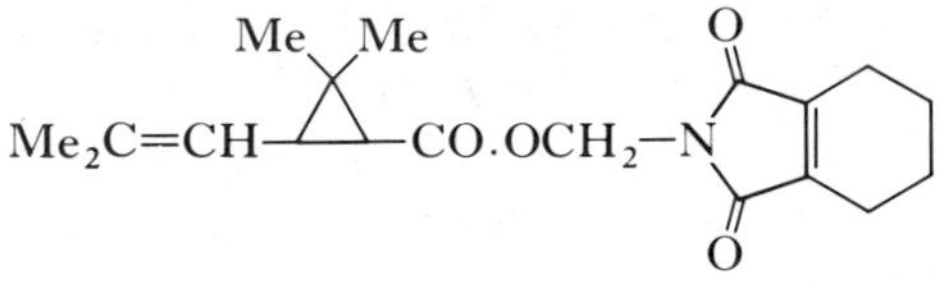

Progress in the synthesis of these materials (reviewed by Elliott, 1980) has been accompanied by a great deal of research into their mode of action (reviewed by Wouters and van den Bercken, 1978). Although the exact molecular basis for the mode of action is not yet known it is nevertheless clear that the pyrethroids interfere with the insect's normal nervous activity. Insects treated with pyrethroids show restless behaviour, become uncoordinated and are then paralysed; flying insects are, in general, rapidly knocked down.

2. *The effects of pyrethroids on nervous transmission*

Over the years Narahashi and his colleagues have made significant contributions to the elucidation of the mode of action of the pyrethroids. In an early paper (Narahashi, 1962) it was reported that application of 1 μM* allethrin to giant axons in the nerve cord of the cockroach (*Periplaneta americana*; this species is used in most insect neurophysiological studies) led to an increase in the negative after-potential and eventually blocked conduction. When the concentration was reduced to 0.3 μM the negative after-potential was again increased, the conduction block was not observed, and it was found that above a critical temperature (26°C) stimulation of the axon led to the appearance of several action potentials, rather than the one normally observed.

Repetitive discharges have subsequently been recorded from various parts of the nervous system of many insects treated with pyrethroids, e.g. from the cockroach cerci (anal sense organs) (Gammon, 1978a), motor neurones of the housefly (*Musca domestica*) (Adams and Miller, 1979), neurosecretory cells of both the stick insect (*Carausius morosus*) (Orchard and Osborne, 1979) and the blood-sucking bug (*Rhodnius prolixus*) (Orchard, 1980a), the central nervous system of the leech (*Hirudo medicinalis*) (Leake, 1977) and the peripheral nervous system of the desert locust (*Schistocerca gregaria*) (Clements and May, 1977). The nerve cord of the crayfish (Nishimura and Narahashi, 1978) and frog sensory organs (van den Bercken and Vijverberg, 1980a) behave similarly.

It should be noted, however, that the nature of the response of nervous tissue to pyrethroids depends on a number of factors. These include the properties of the neurone in question (van den Bercken *et al.*, 1979), the structure of the insecticide (e.g. Clements and May, 1977; Gammon *et al.*, 1981) and the exact experimental conditions including the temperature. Thus Narahashi and Lund (1980) reported that four different axons located in the same area of the crayfish nervous system showed four distinct responses to 0.1 μM (+)-*cis*-phenothrin, ranging from no effect to repetitive discharge.

In some test systems repetitive activity has been induced by particularly low levels of insecticide; for example Orchard and Osborne (1979) reported that permethrin at concentrations as low as 0.05 nM induced

---

*Pyrethroids, like almost all of the compounds described in this chapter, are very insoluble in water. They are usually dissolved in ethanol and then suspended in saline for testing. Concentration figures refer to the resulting suspension. It should be noted that pyrethroids could be absorbed from the bathing medium by nervous tissue, so that the amount actually present in the nerve may be higher than that indicated by the concentration figures.

repetitive activity in the neurosecretory cells of the stick insect, while Gammon *et al.* (1981) have found that pyrethroid levels of 0.3 pM led to repetitive activity in cockroach sensory nerves.

An explanation for the repetitive activity induced by pyrethroids was originally advanced by Narahashi (1962), who showed (as we have seen) that application of allethrin to the cockroach giant axon altered the evoked action potential so that the depolarising after-potential was prolonged and increased. Narahashi concluded that allethrin altered the properties of the nerve membrane in such a manner that, when this enhanced depolarisation exceeded the threshold potential, a second action potential was generated, then another and so on. More recent work by Starkus and Narahashi (1978) on the effect of allethrin on the squid giant axon supported this interpretation since it was found that repetitive discharges were evoked only when the negative after-potential exceeded the threshold membrane potential.

The altered ionic permeabilities which could underly the changes in the action potential have been investigated using the voltage clamp method and have been reviewed by Narahashi (1976a) and Wouters and van den Bercken (1978). Further examples are provided by Pelhate *et al.* (1980), Narahashi and Lund (1980) and van den Bercken and Vijverberg (1980b). From these and other studies it has been concluded that, although relatively high concentrations of pyrethroids have a number of effects on nervous tissue, their main effect, as demonstrated by experiments at low concentrations, is to prolong the inward sodium current (Narahashi and Lund, 1980).

Lund and Narahashi (1981) and Vijverberg *et al.* (1982) (using nerve preparations from squid and frog respectively) have proposed that pyrethroids bring about this effect by delaying the closing of a small percentage of the sodium channels which open on depolarisation. The majority of channels behave normally. The time course of the closing of modified channels depends on the structure of the compound under test (Vijverberg *et al.*, 1982), being particularly slow in the case of the cyano compounds deltamethrin and fenvalerate (Narahashi, 1982). The magnitude of the residual current also depends on the structure of the chemical. When the residual current is small, as again is the case with pyrethroids containing the cyano group, the threshold potential is not reached and repetitive activity is not seen. The continuous slow depolarisation leads eventually to blockage of nervous conduction (Narahashi, 1982). These findings are in agreement with those of Gammon *et al.* (1981) who found that several pyrethroids containing the cyano group did not induce repetitive activity in cockroach sensory neurones.

To summarise the events in the order in which they occur, the pyrethroids first delay the closing of sodium channels. This means that, at the end of the action potential, sodium ions still move into the axon. Thus the membrane potential difference becomes more positive, so that the negative after-potential is increased. In many cases, when the after-potential exceeds the membrane threshold another action potential is generated, and this process continues leading to the repetitive activity frequently recorded from the nervous system of treated insects. Since little information is available on the structure of the sodium channel it is not surprising that the molecular detail of pyrethroid action is, as yet, unknown. It is possible, however, that the sodium ion gating mechanism is located towards the inside of the nerve membrane (Hille, 1977).

Pyrethroid action may not, however, be limited to an effect on sodium channels. For example, Orchard and Osborne (1979) have shown that several pyrethroids increased electrical activity in isolated neurosecretory neurones of the stick insect, neurones in which the inward current is carried wholly or partly by calcium ions. In the light of this and subsequent studies (Orchard, 1980b) it can be stated that pyrethroids either interfere with the natural pathway of activation of these cells or act directly on their membrane so as to increase the spontaneous activity which is observed in the absence of insecticide.

As a second example Clements and May (1977), using locusts as a test species, found that, in addition to acting on neurones, many pyrethroids caused sustained muscle contraction without any clearly associated nervous activity; the authors suggested that the compounds could have been acting directly on the muscle itself. They also found that many pyrethroids blocked neurally evoked muscle contractions, a result which may also be consistent with a direct effect on the muscle. The mechanisms underlying these effects are not yet known.

Gammon et al. (1981) classified pyrethroids as Types I or II, depending on their effects both on sensory neurones and treated cockroaches. Type I compounds produced repetitive firing in sensory neurones in vitro whereas Type II pyrethroids, which typically contain the cyano group, did not. It was suggested that Type II compounds might have a synaptic action (see also Ford, 1979), perhaps at synapses where the neurotransmitter is glutamic acid or 4-aminobutyric acid (GABA).

3. Effect of temperature on pyrethroid action

An intriguing feature of the action of pyrethroids is that they become more toxic to insects as the temperature is lowered; this effect is reversible. It has also been found that the blockage of axonal conduction induced by allethrin becomes more pronounced as temperature is lowered (see, for example, Wang et al., 1972). Therefore Narahashi and his colleagues have suggested that nerve blockage is a critical factor in the insecticidal effect (Narahashi, 1976b).

However, Gammon (1978a) has argued that symptoms of insects treated with allethrin at low temperatures included hyperactivity, a finding which could only be explained on the nerve blocking hypothesis if inhibitory neurones were selectively inhibited. Since there appears to be no evidence for this, Gammon (1978a) has re-examined the temperature dependence of the neural effects of allethrin. He applied LD_{95} doses at 15°C and at 32°C (the LD_{95} at 15°C is approximately a tenth of that at 32°C) and, using implanted electrodes, he recorded responses from peripheral and central neurones of free-walking cockroaches. This procedure reduces problems which may arise in interpreting results obtained using physically restrained (i.e. stressed) insects, and allows symptoms observed in treated insects to be compared with responses recorded from the nervous system. Temperatures of 32°C and 15°C were chosen as they are above and below the critical temperature (≈26°C) above which treated cockroach giant axons fire repetitively in vitro. Good correlations were observed between the symptoms of treated insects and effects on the nervous system. In particular, restless behaviour, evident soon after treatment, was correlated with prolonged discharges of sensory peripheral neurones. Uncoordinated behaviour became evident later as abnormal discharges arose in motor neurones and the central nervous system. Gammon concluded that at 32°C hyperexcitation of peripheral and central neurones were primary effects but at 15°C only peripheral excitation was primary; nerve blockage was considered to be a secondary consequence of allethrin action. Thus, the negative temperature coefficient of allethrin poisoning was considered to be due to the greater sensitivity of peripheral neurones to repetitive firing at lower temperatures.

In related work, Wouters et al. (1977) found that repetitive activity in the allethrin-treated neuromuscular junction of the frog also increased as temperature was lowered. The molecular mechanisms underlying these temperature effects remain to be clarified. One possibility is that the insecticides form a complex with their receptor and that the complex is more stable at reduced temperatures, as has been suggested by Holan (1969) for DDT. The existence of a discrete receptor is indicated by the fact that the stereochemistry of a pyrethroid molecule is important in determining its neurotoxicity and insecticidal activity (see, for example, Gammon et al., 1981). A second possibility is that reduction in temperature itself delays the closing of the sodium channels so that a combination of reduced temperature and the presence of the insecticide delays closing even more, leading to an even more prolonged sodium current (van den Bercken *et al.*, 1979; Vijverberg *et al.*, 1982).

4. *Critical site of pyrethroid action*

As we have seen it is generally agreed that a major biochemical effect of pyrethroids is to interfere with sodium channels and hence disrupt axonal transmission. Nevertheless there is as yet no agreement on which axons are most sensitive in the whole insect. Some attempts to answer this question will now be described.

Gammon's experiments, referred to above, showed that, depending on temperature, both peripheral and central neurones were affected by allethrin. In another study, in which the effect of temperature was not investigated, Clements and May (1977) reported that there was a good, but not absolute, relationship between the ability of pyrethroids to cause hyperactivity in peripheral sensory axons in locusts and their ability to cause knockdown in houseflies. They therefore suggested that knockdown was caused by excessive sensory input to the central nervous system. This is consistent with the possibility that the site of action of pyrethroids may not be very deep within the insect, since quick knockdown demands ready access to the site. A correlation between increased sensory activity and knockdown by pyrethroids has also been demonstrated by Miller and Adams (1977) but the authors did not rule out effects on the central nervous system. It should be noted, however, that knockdown and toxicity in terms of kill are not themselves related (see, for example, Clements and May, 1977). In commenting on a number of investigations in this area, Burt (1980) noted that poor correlations had usually been found between the toxicity of pyrethroids to insects and their effects on an *in vitro* nerve preparation from the same insect, a discrepancy which he did not ascribe to differential penetration, distribution and elimination. He concluded that either the particular type of neurone constituting the critical site remained untested or that interactions within the nervous system make the testing of single units an irrelevant approach. Burt's own preliminary findings from intact cockroaches with implanted electrodes supported the view that the peripheral nervous system was strongly affected by pyrethroids (in this case bioresmethrin) but did not indicate whether or when the central nervous system was affected.

In summary it seems likely that pyrethroids have the capacity to affect all types of neurone but that the peripheral sensory nervous system may be more vulnerable to these contact insecticides.

5. Consequences of pyrethroid poisoning in insects

Clearly the abnormal nervous activity described above would have immediate and serious consequences for the insect. In addition, widespread release of neurohormones as a result of a direct effect on neurosecretory cells (Orchard and Osborne, 1979; Orchard, 1980a,b) or by increasing nervous activity generally would presumably result in a wide variety of secondary disruptive effects. It is well known that many insecticides cause a release of hormones in treated insects (Samaranayaka-Ramasamy, 1978; Maddrell, 1980), and this may well contribute to their toxic effect. Like other insecticides pyrethroids induce water-loss in treated insects (p. 137).

6. Pyrethroid resistance

Although pyrethroids have only recently achieved widespread use as pesticides several insect species are known to exhibit pyrethroid resistance at least in the laboratory, and several instances of resistance serve to corroborate the conclusion that the primary site of pyrethroid action is in the nervous system.

A major type of DDT resistance (*kdr* or knockdown resistance) is cause for great concern since insects with this type of resistance are cross resistant (p. 330) to pyrethroids. It has recently been shown (Miller *et al.*, 1979) that the nervous system of a knockdown-resistant strain of houseflies is less susceptible to pyrethroids than the nervous system of susceptible insects. This finding has been confirmed and extended in further experiments on houseflies (Nicholson *et al.*, 1980a; De Vries and Georghiou, 1981), the cattle tick *Boophilus microplus* (Nicholson *et al.*, 1980b) and larvae of the Egyptian cotton leafworm (*Spodoptera littoralis*) (Gammon and Holden, 1980). This suggests that DDT and pyrethroids share a common site of action and this is discussed below (p. 157).

C. DDT and its analogues
1. Introduction

DDT (a contracted form of *d*ichloro*d*iphenyl*t*richloroethane) is probably the world's best known pesticide. In spite of saving millions of lives by killing the insects responsible for the transmission of malaria and other diseases, it has been the subject of an enormous amount of adverse publicity, mainly because its persistence in the environment, so helpful to its pesticidal utility, led to the accumulation of it or its metabolites in the tissues of man and other species. For this reason and because resistance to DDT has arisen in many species, strenuous efforts have been made to find a substitute for the chemical. However, these have not been particularly success-

Table 4.2   DDT and related compounds

Name	Structure
Bromopropylate	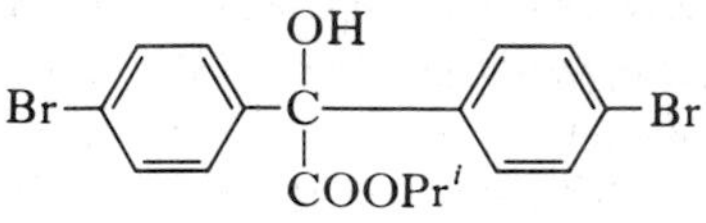 Br—⟨C₆H₄⟩—C(OH)(COOPrⁱ)—⟨C₆H₄⟩—Br
Chlorfenethol	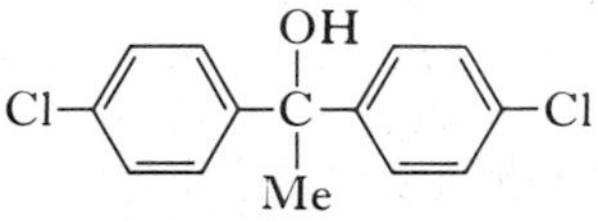 Cl—⟨C₆H₄⟩—C(OH)(Me)—⟨C₆H₄⟩—Cl
Chlorobenzilate	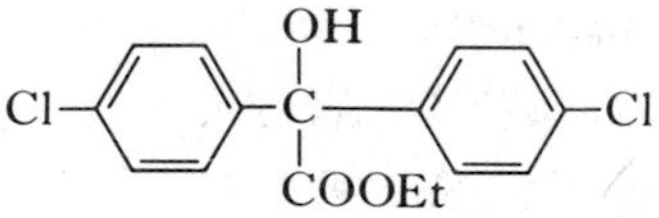 Cl—⟨C₆H₄⟩—C(OH)(COOEt)—⟨C₆H₄⟩—Cl
Chloropropylate	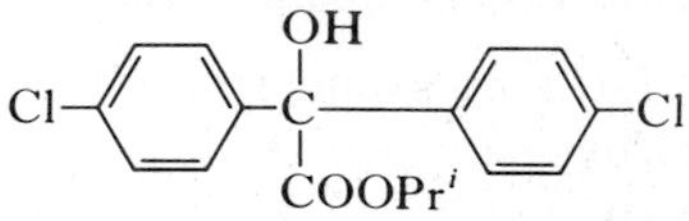 Cl—⟨C₆H₄⟩—C(OH)(COOPrⁱ)—⟨C₆H₄⟩—Cl
DDT	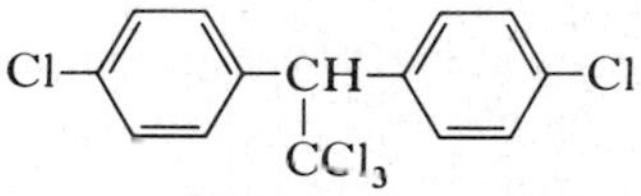 Cl—⟨C₆H₄⟩—CH(CCl₃)—⟨C₆H₄⟩—Cl
1,1-Dichloro-2,2-bis-(4-ethylphenyl) ethane	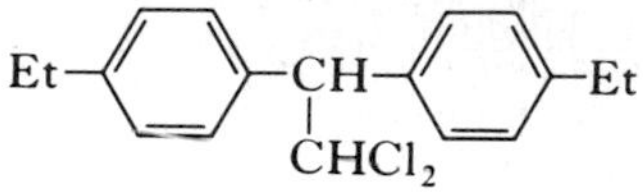 Et—⟨C₆H₄⟩—CH(CHCl₂)—⟨C₆H₄⟩—Et
Dicofol	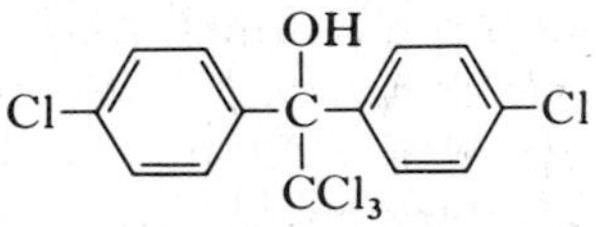 Cl—⟨C₆H₄⟩—C(OH)(CCl₃)—⟨C₆H₄⟩—Cl
Methoxychlor	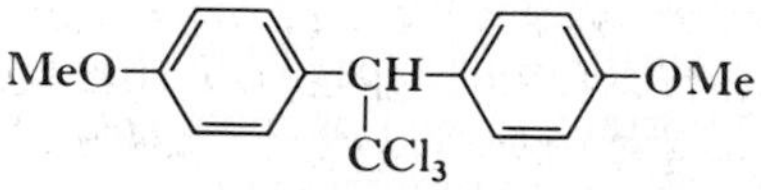 MeO—⟨C₆H₄⟩—CH(CCl₃)—⟨C₆H₄⟩—OMe

Name	Structure
2,2,2-Trichloro-1- (3,4-dichlorophenyl) ethyl acetate	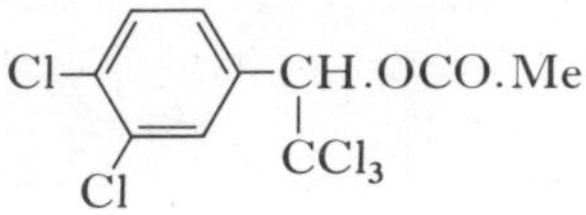 Cl—⟨benzene ring with Cl⟩—CH.OCO.Me | CCl_3

ful so that, although DDT has been largely banned for agricultural use in the developed countries, it is still used to combat insect vectors of diseases in other parts of the world.

There is continuing interest in the mode of action of DDT. One reason for this is the hope that any data obtained might help in the search for a new compound with the desirable properties of DDT.

The parent compound and its analogues and derivatives are shown in Table 4.2.

2. *Action of DDT on nervous transmission*

The effects of DDT on axonal transmission have been studied extensively over the years by Narahashi and his colleagues and have recently been reviewed (Narahashi, 1976a, 1979; Woolley, 1982). Typical findings are those of Narahashi and Yamasaki (1960) who studied the effect of DDT on the action potential of the cockroach giant axon and found that 100 μM DDT slowed the falling phase of the action potential and increased the negative after-potential. They also found that repetitive discharges could be induced in treated axons by a single stimulus and such repetitive activity has been recorded from various parts of the nervous system of treated insects, e.g. sensory neurones of the cockroach leg (Roeder and Weiant, 1946), labellar hair receptors of the housefly (Woodward and Sternburg, 1979a,b) and the blowfly (*Lucilia cuprina*) (Holan *et al.*, 1978), housefly motor neurones (Adams and Miller, 1979), neurosecretory neurones of the stick insect (Orchard and Osborne, 1979) and peripheral and central neurones of intact cockroaches (Gammon, 1978b).

The effects of DDT on the action potential have been analysed using the voltage clamp method (Narahashi, 1976a) and it has been shown that the compound prolongs the inward sodium current and also suppresses the increase in potassium permeability. A combination of these effects leads to the prolonged falling phase, increases the negative after-potential, and leads, in turn, to repetitive activity. Further studies have led to the conclusion (Narahashi and Lund, 1980) that the effect on sodium current is the more important and it is now generally agreed that the chemical delays the closing of a proportion of sodium channels. In this respect DDT resembles the pyrethroids, and Narahashi (1982) and Vijverberg *et al.* (1982) have proposed that the two classes of compound have the same type of action, but that the kinetics of closing of the insecticide-modified channels are different.

3. Effect of temperature on DDT action

DDT, like pyrethroids, is more toxic to insects as the temperature is reduced, probably because the nervous system is more susceptible to DDT at lower temperatures (Gammon, 1978b). Gammon treated free-walking cockroaches having implanted electrodes with topically applied LD_{95} doses of DDT at 16.5°C and 32°C and with an estimated LD_{95} dose at 25°C. The chemical had excitatory effects on both peripheral and central neurones. Although peripheral effects were not quantified, the effects on central neurones were measured and became more pronounced as the temperature was reduced. Possible explanations for the effect of reduced temperature on insecticide activity have already been discussed in relation to the pyrethroids (p. 149) and are equally applicable here.

4. Structure-activity relationships of DDT analogues

A hypothesis to explain DDT action has been put forward by Holan (1969), based on an earlier theory of Mullins (1956). Holan synthesised a range of diaryl halocyclopropane analogues of DDT, compared projections of steric models of these, and other analogues, with their insecticidal activity, and deduced that all the active compounds resembled 'molecular wedges'. The apex of the wedge is, in DDT, the $-CCl_3$ moiety, but the actual chemical nature of the group appears to be largely irrelevant, provided that certain dimensional requirements are met. The base of the wedge, comprising in DDT the two substituted phenyl rings, may be more variable in size than the apex. It was postulated that the base of the wedge interacts with a protein region of an axonal membrane, while the apex keeps open a gate for sodium ions, thus interfering with the ionic basis of normal axonal conduction.

Recent work has largely been interpreted in terms of this model, but with some revision. Thus, Goodford *et al.* (1976) have used statistical methods to analyse the effect of structure on insecticidal activity in a series of methoxychlor analogues. They found that bulky substituents in the *ortho* position in the aromatic rings reduced activity markedly (possibly by interfering with the fit to the receptor of the apex of the wedge) and that, for high activity, the ring substituents had to be electron-releasing* possibly to aid formation of a charge-transfer complex (p. 119) with the receptor. Studies by Metcalf and his colleagues (see, for example, Brown et al., 1981) have led to modification of Holan's model to the extent that they regard the DDT-receptor as being more flexible than that previously proposed. Brown et al. also concluded on the basis of results with chiral analogues that, perhaps not surprisingly, the conformation of the molecule was important in optimising the fit to the receptor.

The simplest representation of Holan's model is that DDT affects the sodium channel by binding to the outside of the nerve membrane. However, as we have seen (p. 149), a recent hypothetical model of the sodium channel (Hille, 1977) suggests that the gating mechanism is actually close to the inside of the membrane. Woolley (1982) has reviewed evidence to support the hypothesis that the lipophilic insecticide is able to penetrate deep into the channel and then interfere with this mechanism.

Turning now to the relationship between structure and in vitro activity, Uchida et al. (1974) demonstrated a good correlation between the ability of DDT and several analogues to produce repetitive discharges in the nerve cords of the cockroach, and their ability to induce convulsions in adult cockroaches. The ability to induce repetitive activity in cockroach nerves also correlated well with the insecticidal action of the compounds against the azuki bean weevil (*Callosobruchus chinensis*) and the housefly, suggesting a common mode of action in these species. On the other hand many DDT analogues which affect in vitro nerve preparations are not good insecticides (Narahashi, 1979).

5. *Consequences of DDT poisoning in insects*

In discussing consequences of DDT poisoning, Gammon (1978b) noted the probability that the widespread disruption of the insect nervous system could lead to changes in hormone levels. In this connection Orchard and Osborne (1979) have reported that DDT (5–50 μM) increased the electrical activity of neurosecretory cells of the stick insect. It is likely, therefore, that DDT causes a general release of hormones either by a direct action on neurosecretory cells or as a result of hyperactivity in axons which activate them. Since the acetylcholine antagonist nereistoxin (see p. 161) did not affect the neurotoxic action of DDT, Uchida et al. (1978) concluded that DDT attached non-cholinergic neurones. On the other hand Bodnaryk (1976) suggested that an elevated

---

*Although chlorine is an electronegative atom causing net electron withdrawal from an aromatic ring, it can nevertheless release electrons to specific positions in that ring.

level of the cyclic nucleotide, cyclic GMP, present in treated insects, was due to release of excess acetylcholine at cholinergic synapses. Finally, there may well be a connection between likely hormone release and the observation that the blood of DDT-treated cockroaches contained a factor which killed flies and affected the nerve cords of untreated cockroaches (Sternburg *et al.*, 1959 and references therein).

6. Comparison of DDT and pyrethroids

The above discussion shows that DDT and pyrethroids have a broadly similar biochemical mode of action and indeed, in some cases, the effects of the compounds on the sodium channel are strikingly similar (Vijverberg *et al.*, 1982). Insects with the *kdr* resistance factor, whose resistance to DDT may be due to an altered site of action, are also resistant to pyrethroids (p. 152), again consistent with the view that the compounds act in a similar way. It is therefore possible that they could bind to the same site on the nerve membrane. Although the two classes have considerable structural differences, Holan *et al.* (1978) have pointed out that some DDT isosteres and many pyrethroids contain a substituted cyclopropane ring. The hypothesis that there is a common binding site could possibly be explored by investigating whether or not the binding of radiolabelled DDT to nerve membrane could be competitively inhibited by pyrethroids and vice versa. This type of experiment has already been described for inhibitors of the Hill reaction (p. 70).

7. Effect of DDT on enzymes which utilise adenosine triphosphate

In recent years it has been found that DDT inhibits a number of enzymes which use the energy obtained by hydrolysis of ATP to mediate ion transport (see Ghiasuddin and Matsumura, 1979 for references). An example of such a system is the sodium/potassium exchange pump involved in maintaining axonal ion concentrations, but inhibition of the pump, or of mitochondrial ATPases as found by Desaiah *et al.* (1974), would lead to nerve failure through dispersion of the usual ionic gradients or lack of energy respectively and so cannot account for the repetitive activity caused by DDT. However, Ghiasuddin and Matsumura (1979) have recently reported that DDT inhibits a calcium ATPase which may be involved in regulating calcium levels at the axon surface. The enzyme, obtained from axons of the American lobster, was 50% inhibited by *c.* 1 nM DDT and inhibition was more pronounced at low temperatures. It has previously been concluded (Frankenhaeuser and Hodgkin, 1957) that reduced levels of calcium decrease the axonal membrane threshold, i.e. the axon is more prone to fire. It was, therefore, proposed that in-hibition of the ATPase could result in the repetitive activity usually seen in DDT-treated tissue (Ghiasuddin and Matsumura, 1979). Consistent with this it should be noted that, in isolated lobster nerve, the usual neuro-physiological responses to DDT are enhanced by low levels and reversed by high levels of calcium in the bathing medium (Matsumura and Narahashi, 1971).

It will be of great interest to see if these findings can be duplicated in insect systems. In particular it remains to be seen if a good correlation can be demonstrated between the ability of DDT analogues to cause hyperactivity in isolated nervous tissue and to inhibit the ATPase. In this connection Ghiasuddin and Matsumura (1979) have reported that DDE, the non-insecticidal dehydrochlorination product of DDT, caused 50% inhibition of the calcium ATPase at about 0.1 μM. In many areas of pesti-cide biochemistry such powerful inhibition of a biochemical system by a pesticide would constitute good evidence that the system represented the major site of action of the pesticide. That this degree of inhibition was caused by an inactive analogue (DDE) therefore argues against the proposal that the ATPase is the site of action of DDT.

8. *Mode of action of dicofol and its analogues*

Dicofol (Table 4.2) is a metabolite of DDT in some species and dicofol and related compounds are used mainly as acaricides rather than insecti-cides. Using DDT-resistant houseflies, Sawicki (1978) has shown that, in comparison with DDT, dicofol and its analogues induced different symp-toms of poisoning, did not exhibit the negative temperature coefficient of toxicity typical of DDT, and killed flies resistant to DDT, including the knockdown-resistant (*kdr*) strain.

Thus DDT and dicofol appear to have different actions in the housefly but this does not rule out the possibility that the mode of action of dicofol in acarines is the same as that of DDT in the fly.

Sawicki's experiments included chlorobenzilate, another acaricide related to DDT. On the basis of its structural similarity to quinuclidinyl-benzilate (QNB), a drug known to bind to muscarinic acetylcholine

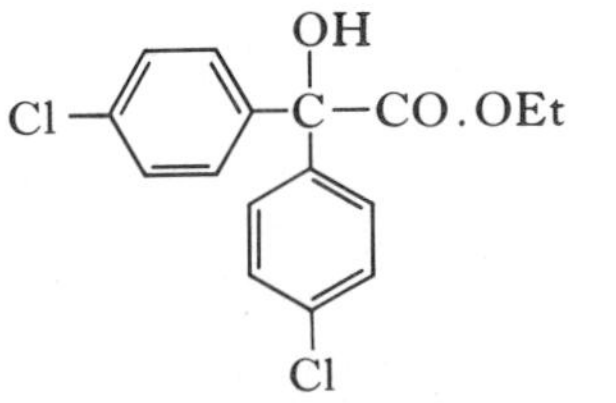

chlorobenzilate 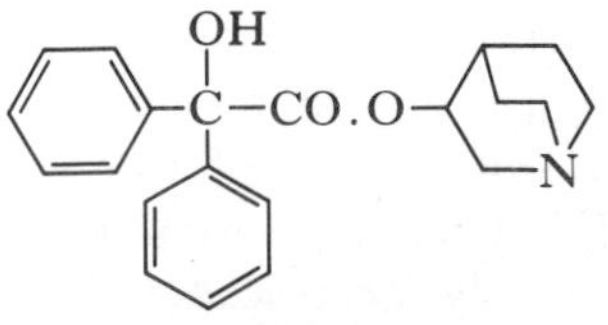   QNB receptors, Shaker and Eldefrawi (1981) tested chlorobenzilate as an inhibitor of the binding of radiolabelled QNB to a putative housefly receptor. The compound was a good inhibitor ($K_i = 14$ μM) but, as the authors acknowledge, this does not necessarily account for its acaricidal activity. Further experiments with mites will be required.

Interestingly, several compounds known to stimulate muscarinic receptors were found to be toxic to the tick *Boophilus microplus* (Bigg and Purvis, 1976). The authors suggested that other such compounds might be found which would also be acaricidal.

II. COMPOUNDS THAT BIND TO THE ACETYLCHOLINE RECEPTOR
A. Biochemical background

Arrival of an impulse at the pre-synaptic side of the synapse causes the release of a transmitter from the pre-synaptic membrane (p. 99). Where the liberated transmitter is acetylcholine, it passes across the synapse and binds to the acetylcholine receptor on the post-synaptic membrane, altering the ionic permeability of this membrane so that it becomes depolarised. By this means a new impulse is generated in the next nerve or effector cell. The molecular nature of the receptor and the mechanism by which transmitter binding induces ion flux have been the subject of intense research. Much of the work has been done in the experimentally convenient tissue from electric fish but an increasing amount of information on insect receptors is becoming available and has recently been reviewed briefly by Dudai (1979) and in detail by Sattelle (1980). In the electric fish it appears that the receptor consists of two identical linked protein complexes. Each of these has a molecular mass of *c.* 250,000 daltons and is made up of five subunits, two of which are the same (Raftery *et al.*, 1980). The present view, at least in electric fish, is that the five subunit complex both recognises the transmitter and mediates ion flux (Raftery *et al.*, 1980).

Mammalian tissue contains different types of receptor which bind acetylcholine, though they differ in their ability to respond to other molecules structurally related to the transmitter, particularly nicotine (see below) and muscarine. Sattelle (1980) has lucidly reviewed the evidence which shows that insects possess three putative receptors, having nicotinic, muscarinic and mixed (nicotinic/muscarinic) specificity, and has presented evidence that the nicotinic receptor has a physiological role in synaptic transmission.

nicotine   muscarine   nereistoxin

It should be noted in the following discussion that a compound which binds to the receptor so as to cause the normal ion flux is termed an agonist while a compound which binds but does not cause ion flux is called an antagonist. The compounds discussed are shown in Table 4.3.

Table 4.3   Insecticides which bind to the acetylcholine receptor

Name	Structure
Cartap	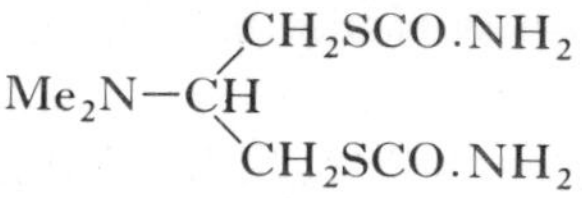 $Me_2N-CH \begin{array}{c} CH_2SCO.NH_2 \\ CH_2SCO.NH_2 \end{array}$
Nicotine	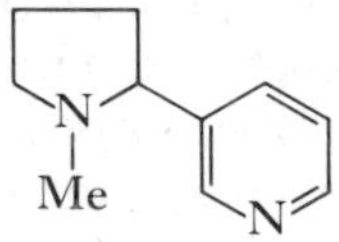
Thiocyclam	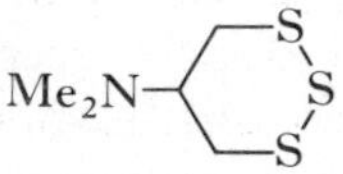 Me_2N- (ring with S, S, S)

B. Nicotine

Nicotine has been used as an insecticide for at least 200 years, but this naturally-occurring compound is both hazardous in use and lacks persistence, and has now largely been replaced by safer and more effective synthetic insecticides. Insects treated with nicotine develop tremors followed by convulsions and paralysis (O'Brien, 1967) and death usually occurs within an hour.

It has been known for a long time that nicotine stimulates nicotinic acetylcholine receptors at low concentrations and causes blockage at higher concentrations. In recent years direct binding studies have provided additional evidence on the mode of action. Thus, experiments with cell-free preparations of the 'mixed' receptor from housefly heads showed that nicotine binds reversibly to the receptor with a K_i of 3.2 μM, and that the binding of radiolabelled nicotine was inhibited by toxic nicotinoids but not by non-toxic ones (Eldefrawi et al., 1970).

Data on the nicotine receptor have also been obtained, for example by Gepner et al. (1978). These authors measured the ability of nicotine to inhibit the binding of radiolabelled α-bungarotoxin (a compound known to bind to nicotinic receptors) to tissue extracts from both fruit fly (*Drosophila melanogaster*) and cockroach nerve cord, and compared this with the ability of nicotine to suppress evoked transmission at an α-bungarotoxin-sensitive synapse in the cockroach terminal abdominal ganglion. The insecticide inhibited toxin binding by 50% at a concentration of 1 μM (*Drosophila*) and 3 μM (cockroach), and inhibited synaptic transmission in the cockroach ganglion by 50% at a concentration of 0.2 μM, results consistent with the view that the insecticide acts by binding to the acetylcholine receptor.

Electrophysiological experiments with nicotine have been reviewed by Narahashi (1976a) and include voltage clamp experiments showing that it affects post-synaptic sodium and potassium currents when applied to a frog neuromuscular junction preparation (Wang and Narahashi, 1972).

Extensive studies of the relationship between the structure and the insecticidal activity of nicotine and related compounds show that active compounds must have a certain structural similarity to acetylcholine and possess the 'highly basic nitrogen, which is protonated in the insect body and anchors the molecule to the anionic site of the receptor; a carbon atom, which arranges the pyridine ring and the nitrogen in a definite position; and the pyridine ring, which effects some unidentified influence on the nerve membrane' (Yamamoto, 1965). Beers and Reich (1970, 1971) have suggested that the nitrogen of the pyridine ring acts as a hydrogen bond acceptor at a defined distance from the charged nitrogen of the pyrrolidine ring. This distance is equivalent to that between the charged nitrogen and the hydrogen bond formed when the carbonyl oxygen of acetylcholine itself is the acceptor.

C. Cartap

Cartap was derived from a programme of chemical synthesis based on nereistoxin, a poison found in the common marine worm, *Lumbriconereis heteropoda*. It is perhaps the first example of a commercial pesticide developed from a natural product not already used as a pesticide. Early work, mainly by Sakai and his colleagues, on the insecticidal effect and mode of action of both nereistoxin and cartap has been reviewed by Eldefrawi (1976).

Direct neurophysiological measurements indicated that both nereistoxin and cartap block synaptic, but not axonal, transmission in cockroach nerves, and that the major site of nereistoxin action is the nicotinic post-synaptic membrane. Further work by Bettini *et al.* (1973) with cartap (which they considered would be metabolised to nereistoxin in their preparation) and

Callec *et al.* (1980) with nereistoxin are consistent with this interpretation; Callec *et al.*, for example, found that concentrations of nereistoxin as low as 0.05 μM blocked the receptor without depolarising the post-synaptic membrane.

Whether the toxin binds to the transmitter binding site (as discussed by Eldefrawi *et al.* (1980)) or elsewhere on the receptor complex is not certain.

Insects poisoned with nereistoxin are quickly immobilised without convulsive symptoms; the symptoms are different from those of nicotine, possibly because nicotine depolarises the receptor, which would lead to excitation, whereas nereistoxin does not induce depolarisation except at high concentrations.

Because it is an acetylcholine antagonist nereistoxin has been used in experiments designed to identify those regions of the insect nervous system with which other insecticides interfere (pp. 156, 166).

There appears to be no information on the mode of action of thiocyclam but its close structural similarity to nereistoxin suggests a similar mode of action. It is intriguing to speculate on the role of the S–S bonds in these compounds in view of the fact that, at least in the electric fish (*Torpedo californica*), an S–S group appears to be present near the acetylcholine binding site of the receptor (Damle and Karlin, 1980).

III. COMPOUNDS WHICH ARE THOUGHT TO CAUSE EXCESSIVE RELEASE OF ACETYLCHOLINE

Insects treated with cyclodiene insecticides, related compounds and *gamma*-HCH (Table 4.4) develop violent quivering of the body followed by convulsions and paralysis, symptoms consistent with an effect on the nervous system. Experiments described below have provided good evidence that the compounds cause excessive release of acetylcholine and hence lead to hyperactivity at cholinergic synapses.

A. Cyclodienes

The mode of action of dieldrin and related compounds has been studied in more detail than that of other cyclodienes. It is well established that, in some insects, aldrin can be epoxidised to dieldrin, and dieldrin can be hydrated to give the *cis* and *trans* isomers of aldrin diol. In the course of early studies (reviewed by Narahashi, 1976a) attempts were made to establish whether dieldrin or one of its metabolites constitutes the actual toxicant. Wang *et al.* (1971) found that, in dieldrin-treated cockroaches which had developed symptoms of poisoning, synapses in the metathoracic,

Table 4.4 Compounds thought to cause excessive release of acetylcholine*

Name	Structure
Aldrin	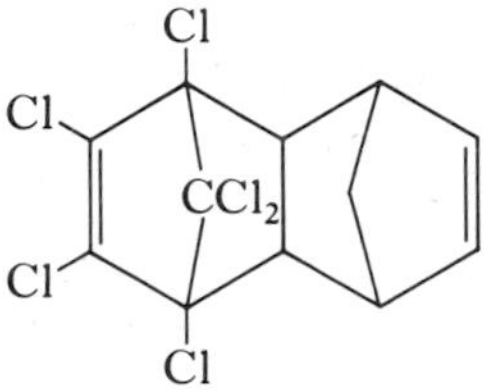
Bromocyclen	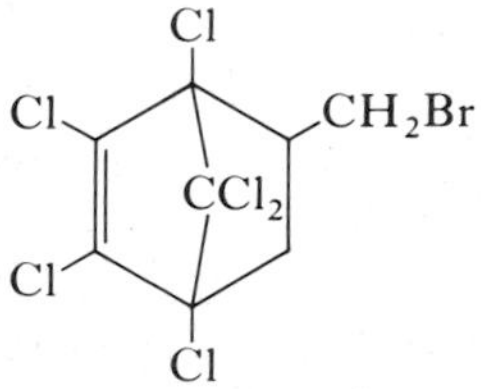
Camphechlor An isomeric mixture of chlorinated camphenes:	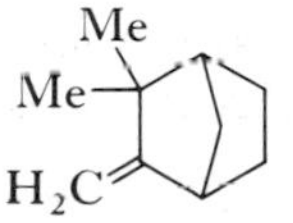
Chlordane	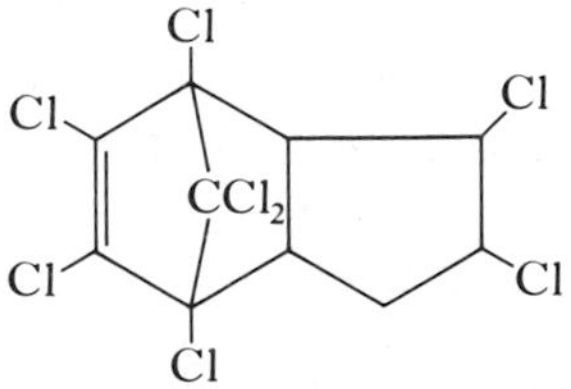
Dieldrin	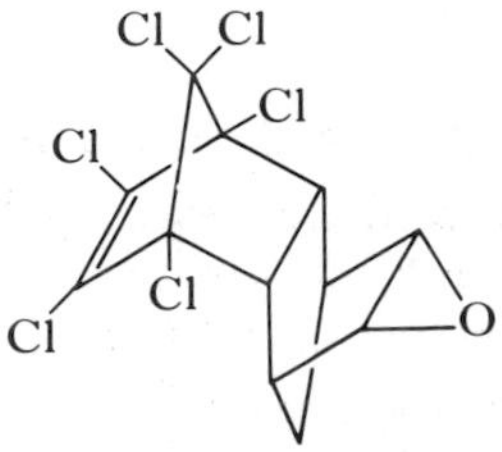

*Stereochemistry is indicated for dieldrin and endrin which are isomeric.

Name	Structure
Dienochlor	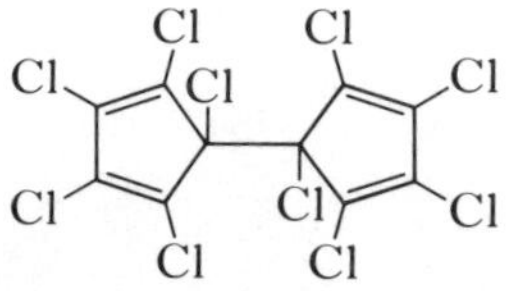
Endosulfan	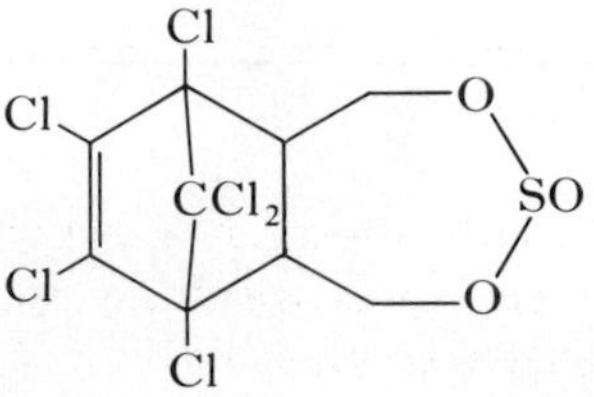
Endrin	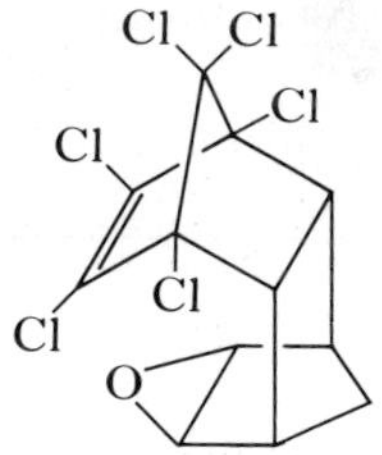
Gamma-HCH	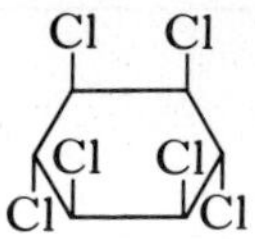
HCH	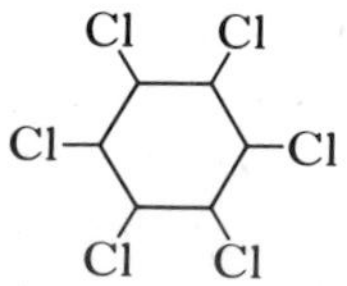

Name	Structure

Heptachlor

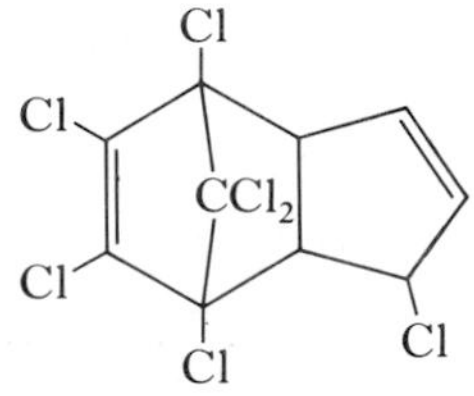

but not the sixth abdominal, ganglion were severely affected, a single pre-synaptic stimulus resulting in prolonged repetitive post-synaptic activity. Similarly when dieldrin and a variety of metabolites were applied directly to the exposed metathoracic ganglion all compounds caused this enhanced post-synaptic activity. Dieldrin was very slow to act whereas aldrin *trans*-diol exerted its effect very quickly. The diol was also faster in

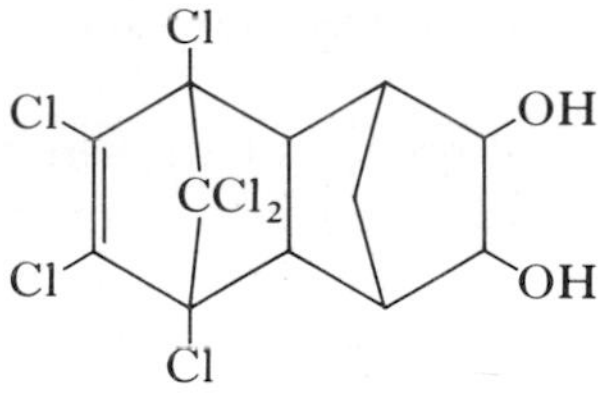

aldrin diol

producing repetitive trains of impulses from sensory neurones in the insect leg. These authors claimed that the diol was more potent than dieldrin itself although it is not clear how potency, as opposed to speed of action, had been assessed. Wang *et al.* (1971) suggested that the diol could represent one of the toxic forms of dieldrin but stressed that proof of this would require a demonstration that dieldrin poisoning did, in fact, result in the presence of the diol in the nervous system in sufficient quantity to exert a toxic action.

This and other aspects of the insecticidal action of dieldrin have been re-investigated by Shankland and his colleagues (reviewed by Shankland, 1979) with different results. Firstly, it was found that dieldrin, although slow acting, was much more toxic to the cockroach (by injection) than either aldrin *trans*- or *cis*-diol although all three compounds penetrated the nerve cord to roughly the same extent. Secondly, dieldrin was much more toxic than either diol to isolated nerve cords, affecting both the desheathed sixth abdominal ganglion and the desheathed and intact metathoracic ganglion, causing multiple post-synaptic discharges and an increase in the spontaneous activity of the preparations. Dieldrin did not affect ganglia which had previously been depleted of acetylcholine and its effects were overcome by added magnesium ions; this latter finding is relevant since transmitter release depends on calcium ions and magnesium antagonises calcium in this respect (Shankland 1979). On the basis of these results the authors concluded that dieldrin was the toxic species and that it enhanced calcium-dependent release of acetylcholine at cholinergic synapses.

The discrepant neurotoxicity results with aldrin *trans*-diol and dieldrin are, as yet, unresolved, but dieldrin appears to be the more lethal compound and the effects of the actual toxicant on acetylcholine release are generally agreed. This interpretation has been supported by Uchida *et al.* (1978) who also examined the effects of dieldrin on the cockroach nerve cord. The typical hyperactivity described above was observed and this effect was overcome by nereistoxin (p. 161) indicating that the hyperactivity was due to the presence of acetylcholine.

Although most studies have been carried out with dieldrin and the aldrin diols, it seems probable that the related chlorinated cyclic compounds have a similar mode of action (see, for example, Bloomquist and Shankland, 1983).

As in the case of DDT, several groups of workers have examined the effects of cyclodienes on enzymes using adenosine triphosphate as the energy source for ion transport. In a representative example Yamaguchi *et al.* (1979) found that heptachlorepoxide, which may be the toxic species in heptachlor action, inhibited a calcium/magnesium ATPase from the synaptic regions of rat brain, with an I_{50} around 5 μM. The authors suggested that inhibition might lead to increased calcium levels in the presynaptic region. Since rises in calcium are known to enhance transmitter release a link between inhibition of the ATPase and excessive transmitter release was suggested. However, as the authors themselves point out, much more data are required before any conclusion is possible on the relevance of these findings to the mode of action.

B. Gamma-*HCH*

The neurotoxicity of this insecticide has been examined by Uchida and his colleagues (reviewed by Uchida *et al.*, 1978) and *gamma*-HCH was found to be much more active than other HCH isomers in inducing repetitive activity in the cockroach nerve cord. Nereistoxin suppressed the symptoms and it was concluded that *gamma*-HCH behaved in the same way as dieldrin, i.e. it caused excessive acetylcholine release. These authors then went on to examine a number of *gamma*-HCH analogues and found a good correlation between their neurotoxic effect and both their ability to induce convulsions in treated cockroaches and their lethal activity. Interestingly, when neuro-excitatory activity was plotted against convulsive activity for both *gamma*-HCH and its analogues, and a series of carbamates, it was found that the points fell virtually on the same line, suggesting that convulsions are caused by high levels of acetylcholine, however these arise. In this connection *gamma*-HCH treatment was shown by Collins *et al.* (1979) to cause the presence of vacuoles in the thoracic ganglia of houseflies, an effect caused (p. 137) by the carbamate carbofuran.

In the course of further structure-activity studies of *gamma*-HCH by the Japanese group (Kiso *et al.*, 1978) the authors noted that, since cyclodienes and *gamma*-HCH appear to have the same mode of action, it would be of interest to try to identify and characterise a common pre-synaptic site to which the compounds may bind. Perhaps competition experiments of the type indicated above for DDT and the pyrethroids would also be useful in this case.

IV. CHLORDIMEFORM AND RELATED COMPOUNDS WHICH ARE THOUGHT TO ACT ON THE OCTOPAMINE RECEPTOR

Chlordimeform (Table 4.5) and its analogues have a number of behavioural effects on insects and acarines, and several proposals have been made as to the mode of action of the compounds; for example it has been suggested that they could act by uncoupling oxidative phosphorylation (p. 29) or by inhibiting monoamine oxidase, an enzyme involved in the (mammalian) metabolism of neuroactive amines (for a review see Beeman, 1982). Although such effects could play some part in the toxic action of the compounds recent findings favour the view that chlordimeform and its analogues interfere with the action of octopamine, the neurophysiological role of which has begun to emerge in recent years.

There is compelling evidence from several sources that chlordimeform is demethylated *in vivo* to a compound which then binds to the octopamine receptor. Much of this evidence has been provided by Hollingworth and his colleagues and their early work (and that of others) has been considered by Hollingworth and Murdock (1980). These workers have investigated the effects of chlordimeform and related compounds on the firefly (*Photinus pyralis*) light organ (Hollingworth and Murdock, 1980), in which light production appears to be controlled by neurones which release octopamine. Chlordimeform, applied topically to the firefly, caused the light organ to light up after a lag time of a few hours. The *N*-demethylated analogue acted more quickly and was much more potent, consistent with previous evidence that it was more active *in vivo* than chlordimeform, from which it

Table 4.5   Chlordimeform and related compounds

Name	Structure
Amitraz	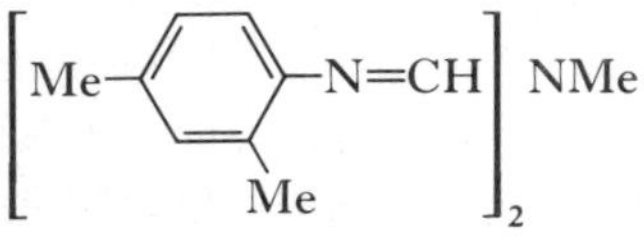 $\left[Me\text{—}\underset{Me}{\bigcirc}\text{—}N{=}CH \right]_2 NMe$
Chlordimeform	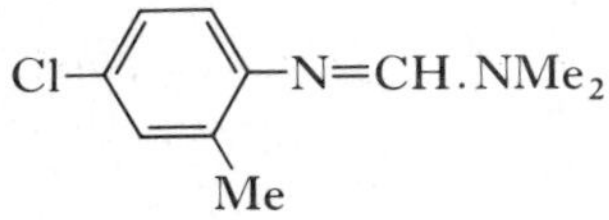 $Cl\text{—}\underset{Me}{\bigcirc}\text{—}N{=}CH.NMe_2$
Chloromethiuron	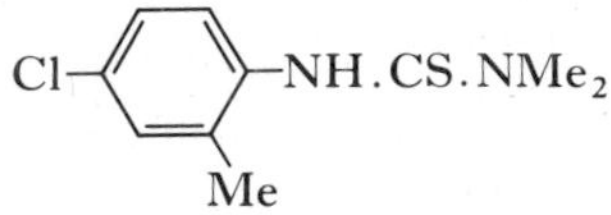 $Cl\text{—}\underset{Me}{\bigcirc}\text{—}NH.CS.NMe_2$
Clenpyrin	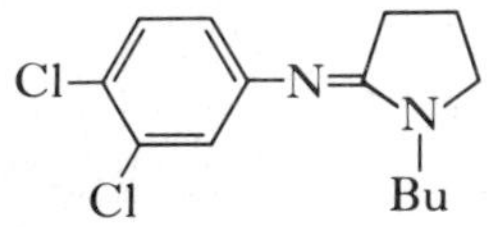 $Cl\text{—}\underset{Cl}{\bigcirc}\text{—}N{=}\underset{Bu}{\bigcirc}$
Cymiazole	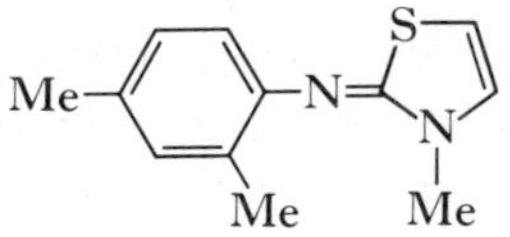 $Me\text{—}\underset{Me}{\bigcirc}\text{—}N{=}\underset{Me}{\overset{S}{\bigcirc}}$

may be produced by demethylation (Atkinson and Knowles, 1974). The action was considered to be at the post-synaptic octopamine receptor (rather than, for example, on octopamine release), because the compounds were active under conditions in which the nerve terminals to the light organ had degenerated, i.e. there was no pre-synaptic activity. Drugs which antagonise octopamine in other systems reduced glowing. These results led to the hypothesis that chlordimeform is converted *in vivo* to the *N*-demethylated derivative which then acts as an octopamine agonist (Hollingworth and Murdock, 1980).

Experiments with isolated firefly tails (Murdock and Hollingworth, 1980) support this view. Chlordimeform itself caused no light emission, while *N*-demethychlordimeform was very active, being at least as potent as octopamine.

Cl—⟨benzene ring⟩—N=CH.NHMe with Me substituent   HO—⟨benzene ring⟩—CH.CH$_2$NH$_2$ with OH

N-demethylchlordimeform   octopamine

In further experiments both *N*-demethylchlordimeform and octopamine stimulated the enzyme adenylate cyclase which is probably involved in mediating the octopamine response; octopamine antagonists inhibited this effect on the cyclase. Chlordimeform itself did not stimulate the cyclase, in agreement with previous work in which it failed to stimulate a cyclase present in the cockroach nervous system (Matsumura and Beeman, 1976). A more recent study on the effects of chlordimeform and *N*-demethylchlordimeform on adenylate cyclase from the firefly gave substantially the same results (Nathanson and Hunnicutt, 1981).

The compounds also interfere with other systems in which octopamine is either known or thought to be involved. Firstly, Evans and Gee (1980) found that the chemicals duplicated the effects of octopamine in a nerve-muscle preparation from the locust (*Schistocerca americana gregaria*). The demethylated compound was more active by an order of magnitude, the effects of the chemicals were more prolonged than those of octopamine and octopamine antagonists counteracted the effects of the compounds.

Secondly, Singh *et al.* (1981) found that both compounds stimulated release of hyperlipemic hormone (a hormone regulating lipid levels) from the isolated corpus cardiacum of the locust (*Locusta migratoria*). The normal response probably involves octopamine. In this study chlordimeform and its demethylated derivative had the same order of activity.

Turning now to other compounds in Table 4.5, the triazapentadiene, amitraz, has been found to stimulate the motor output from insect ganglia (Lund *et al.*, 1979) and to give rise to the simple formamidine, ArN=CH.NHMe(Ar=2,4-dimethylphenyl) *in vivo* (Schuntner and Thompson, 1978). There is evidence that this compound causes transient increased nervous activity in a nerve preparation from the tick, *Boophilus microplus*, and it was suggested that the increased activity could be linked with the tendency of ticks to detach from their hosts after treatment (Binnington and Rice, 1977). It would be of interest to know if octopamine is involved in this particular nerve preparation.

The acaricidal thiourea, chloromethiuron, has recently been investigated by Schuntner and Thompson (1979) who concluded that chloromethiuron has the same mode of action as chlordimeform and its analogues.

Clenpyrin and cymiazole are acaricides structurally related to chlordimeform. General biochemical effects of clenpyrin have been reported (Andrews and Stendel, 1975) but the primary site of action does not seem to have been identified. Since the compounds are structurally related to chlordimeform it seems possible that they could be octopamine agonists.

V. MISCELLANEOUS COMPOUNDS
A. Trifenmorph

Although the mode of action of the molluscicide trifenmorph (Table 4.6) is not established the compound is included here because it is neurotoxic and, although this is probably not the cause of its molluscicidal activity, it represents a clue for further work. In a series of papers Brezden and Gardner (1980 and references therein) have examined the effects of trifenmorph and a number of its analogues on the aquatic snail (*Lymnaea stagnalis*) and found that, although the exposed central nervous system was sensitive to compounds applied directly (see also Plummer and Banna, 1979), brain activity was normal in poisoned snails. Nevertheless only those analogues which were neurotoxic were molluscicidal and further experiments showed that both neurotoxicity and (with one exception) lethal potency were correlated with the hydrolysis rate of the materials. In this connection Boyce and Milborrow (1965) had previously concluded that the molluscicidal activity of trifenmorph analogues was associated with nucleophilic attack at the aliphatic carbon atom, and Brezden and Gardner (1980) took a similar view, namely that neurotoxicity and lethal action could both be the result of a nucleophilic displacement reaction but that the reaction occurred at different biological receptors.

However, on chemical grounds, hydrolysis of such compounds would be expected to proceed via the carbonium (positive carbon) ion (Ph_3C^+); it is therefore possible that the activities described could be associated with the ease of production of this ion (possibly by protonation of the nitrogen atom by a hydrogen donor on the putative receptor, followed by departure of the leaving group) rather than the nucleophilic displacement referred to above.

It is interesting to speculate here on whether there is any connection between the mode of action of trifenmorph and that of the trisubstituted tin compounds, particularly fentin acetate (p. 40). Both triphenyl tin compounds and trifenmorph could give rise to the Ph_3X^+ cation ($X = Sn$ and C respectively) and both types of compound have been considered to affect chloride ion movement across membranes. Interestingly, trisubstituted tin compounds, including fentin acetate, are active against snails (Smith *et al.*, 1979).

Table 4.6 Miscellaneous compounds

Name	Structure
Trifenmorph	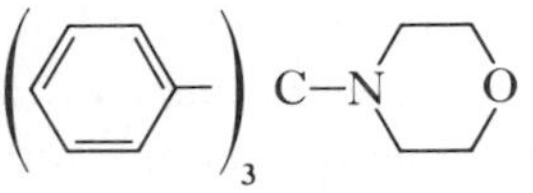
Ryanodine	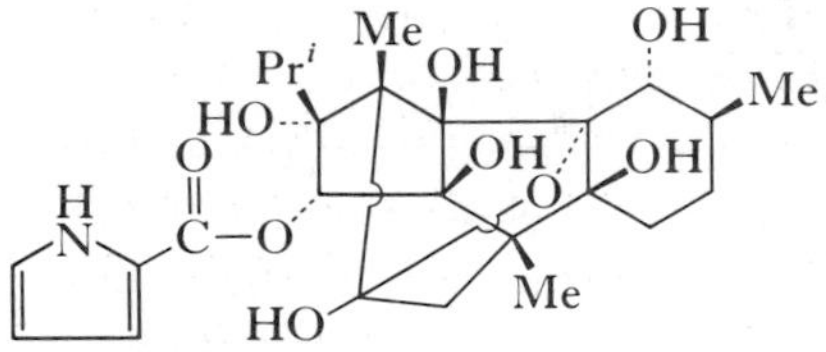
Avermectin B_{1a}	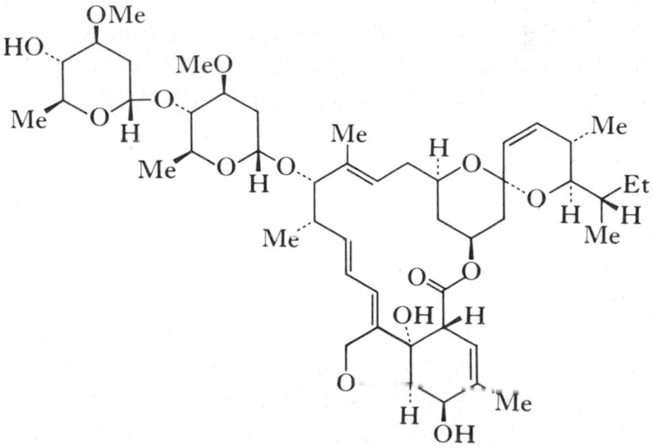

As an alternative to the nervous system Brezden and Gardner have proposed the snail musculature as a possible site of action of trifenmorph. Support for this suggestion is found in a report by Banna and Plummer (1978) that the compound slows the molluscan heart. However, there is as yet no indication in molecular terms of how these effects may occur.

B. Ryanodine

Ryanodine (Table 4.6) is a stomach-acting insecticide isolated from the wood of the shrub *Ryania speciosa*. Early work on its mode of action (summarised by O'Brien, 1967) led to the conclusion that it acted on insect muscle. More recent work has shown that the compound affects calcium transport in a number of systems and has provided good evidence that, at low concentrations, ryanodine causes muscle contraction by interfering with calcium movement across the specialised membranes of muscle cells (see Pezzementi and Schmidt, 1981).

C. The avermectins

The avermectins, isolated from the soil organism *Streptomyces avermitilis*, are a very interesting class of compounds having high activity as acaricides, insecticides and particularly anthelmintics (Putter *et al.*, 1981). One of the most active members of the family is avermectin B_{1a} (Table 4.6) and evidence is accumulating that this compound interferes in some way with the proper function of synapses at which 4-aminobutyric acid (GABA) is the transmitter; GABA is an inhibitory transmitter (i.e. it leads to reduced excitability in the post-synaptic cell). Following a neurophysiological study of the effects of avermectin B_{1a} on the parasitic nematode *Ascaris suum*, Kass *et al.* (1980) concluded that the compound affected inhibitory synapses, possibly by acting as a GABA agonist or by stimulating GABA release. Further work suggests that the latter may be more likely (Pong *et al.*, 1980). This effect on the nervous system is consistent with the paralysing effect of the compound on nematodes. A great deal of work is being done in this area and further clarification of the mode of action of these compounds can be expected.

VI. PHEROMONES
A. Introduction

In the last two decades, there has been spectacular progress in our understanding of the way in which naturally-occurring chemicals control numerous facets of the behaviour of insects (Silverstein, 1981; Baker and Evans, 1980). The most intensively researched area is concerned with pheromones, i.e. substances released externally by insects and which elicit immediate behavioural responses from members of the same species. These chemical messengers do not exert toxic effects within a species, and as such cannot be classed as pesticides. However, it has been demonstrated that interference with natural communication mechanisms can provide effective population control of pest species.

Amongst the many types of pheromones, sex attractants have received the greatest attention. These are chemicals released by one sex of a species (usually the female) to attract the opposite sex of the same species for the purpose of mating. Other types of pheromones are involved in aggregation and colonisation of a host (population attractants), egg-laying, and foraging and maintenance of social order in colonies of insects such as bees and ants. As an example, a population of the western pine beetle *Dendroctonus brevicomis* is attracted to the pine by a combination of three components, one each from the female, the male and the tree (Silverstein, 1981).

B. Biochemistry

Insects usually detect pheromones by sweeping the air with their antennae which possess thousands of olfactory hairs. These hairs contain sensory nerve cells which have receptor sites for the chemicals. The presence of a sufficient number of molecules at their receptors leads to depolarisation (p. 142) of the membrane of the sensory cell and this generates a nerve impulse which is transmitted to the central nervous system; the greater the number of molecules arriving at the receptor, the greater is the frequency of impulses transmitted. It has been shown that a sensory cell will respond to one molecule of pheromone and it has been estimated that the presence of some two hundred molecules is sufficient to produce a behavioural response.

The effects of chemicals on the sensory receptors of the antennae can be studied in the laboratory by the technique of electroantennography, in which the impulses arising in activated cells are amplified, recorded and measured. Recordings from single receptors have revealed different olfactory cell types by observation of relative amplitudes and frequency of pheromone-induced nerve impulses (Kaissling, 1979; Priesner, 1979). Electroantennographic bioassay has also been used to analyse the complexity of multicomponent signals. For example, in the case of the silkworm moth, *Bombyx mori*, a single component pheromone had been proposed. Electroantennography clearly showed the existence of a second type of olfactory cell, and subsequently a second compound was shown to be an inhibitory component acting at the second site (Kaissling *et al.*, 1978).

In recent years, it has become apparent that pheromone secretions are most commonly multicomponent mixtures of organic compounds, each component of which may mediate a different facet of behaviour. Characteristically, sex attractants are volatile organic compounds, emitted in minute quantities, which exert powerful effects over relatively long distances. For example, females of *Heliothis virescens* produce a multifunctional seven-component mixture of homologous long-chain aldehydes, alcohols and acetates, although not all components are attractants (Klun *et al.*, 1980).

It is known in specific cases that certain components of a mixture are responsible for long-range attraction, whereas other components take over at shorter range (Roelofs and Cardé, 1977). It is important to note that the relative ratio of components is critical for full behavioural response. Synthetic lures therefore need to have exactly the right composition. The isomeric and geometric purity of alkene pheromones is also critical, as is the enantiomeric purity of chiral compounds.

C. Applications of pheromones in pest control

The attractant responses of moths and beetles presently represent the areas of greatest commercial interest (Silverstein, 1981) and four main approaches have been investigated.

Firstly, pheromones have been used for monitoring pest population levels, a technique which relies upon incorporation of a pheromone system into an immobilising trap. Trap catch data are then used to forecast pest populations, thus enabling optimisation of spraying regimes. The exact chemical composition of pheromones is not critical in this case because the method is based upon a calibrated response. The second approach involves the luring and trapping of pest insects (usually the male), so that populations are reduced to levels which do not constitute a problem. Virtually all the males have to be caught to prevent significant mating. An additional problem is that mated females could enter the treated area from outside and then lay their eggs. Furthermore, high standards of purity and accuracy of composition of the lure are generally required to ensure high catches.

The third area of pheromone use is in mating disruption. Here, a relatively very high level of sex attractant pheromone is introduced into the atmosphere in order to effect complete disruption of mating. As with mass trapping, the ingress of mated females and the requirement for high standards of pheromone purity can cause problems. In general, the specificity of pheromone responses limits effectiveness to one pest species and this is a severe disadvantage in the protection of crops which support a complex of pests, but has the advantage that beneficial species are not directly affected. Finally pheromones may be used in combination with conventional insecticides in integrated control programmes, but there has been relatively little progress reported from a commercial standpoint. However, the attraction of particular pest species to insecticide-treated traps could provide an ecologically acceptable pest control method in many instances.

The advantages and disadvantages of the methods described above, together with some case histories, are described by Silverstein (1981) and Roelofs and Cardé (1977).

REFERENCES

Adams, M. E. and Miller, T. A. (1979). *Pestic. Biochem. Physiol.* **11**, 218–231
Andrews, P. and Stendel, W. (1975). *Pestic. Sci.* **6**, 129–143
Atkinson, P. W. and Knowles, C. O. (1974). *Pestic. Biochem. Physiol.* **4**, 417–424
Baker, R. and Evans, D. A. (1980). *Chemistry in Britain* **16**, 412–415
Banna, H. B. and Plummer, J. M. (1978). *Comp. Biochem. Physiol* **61 C**, 33–36
Beeman, R. W. (1982). *Ann. Rev. Entomol.* **27**, 253–281
Beers, W. H. and Reich, E. (1970). *Nature (Lond.)* **228**, 917–922
Beers, W. H. and Reich, E. (1971). *Nature (Lond.)* **232**, 422–423
Bettini, S., D'Ajello, V. and Maroli, M. (1973). *Pestic. Biochem. Physiol.* **3**, 199–205
Bigg, D. C. H. and Purvis, S. R. (1976). *Nature (Lond.)* **262**. 220–222
Binnington, K. C. and Rice, M. J. (1977). *J. Aust. Entomol. Soc.* **16**, 80
Bloomquist, J. R. and Shankland, D. L. (1983). *Pestic. Biochem. Physiol.* **19**, 235–242
Bodnaryk, R. P. (1976). *Can. J. Biochem.* **54**, 957–962
Boyce, C.B.C. and Milborrow, B. V. (1965). *Nature (Lond.)* **208**, 537–539
Brezden, B. L. and Gardner, D. R. (1980). *Pestic. Biochem. Physiol.* **13**, 189–197
Brown, D. D., Metcalf, R. L., Sternburg, J. G. and Coats, J. R. (1981). *Pestic. Biochem. Physiol.* **15**, 43–57
Burt, P. E. (1980). In "Insect Neurobiology and Pesticide Action", pp. 407–414. Society of Chemical Industry, London
Callec, J. J., Sattelle, D. B., Hue, B. and Pelhate, M. (1980). In "Insect Neurobiology and Pesticide Action", pp. 93–100. Society of Chemical Industry, London
Clements, A. N. and May, T. E. (1977). *Pestic. Sci.* **8**, 661–680
Collins, C., Kennedy, J. M. and Miller, T. (1979). *Pestic. Biochem. Physiol.* **11**, 135–158
Damle, V. N. and Karlin, A. (1980). *Biochemistry* **19**, 3924–3932
Desaiah, D. Cutkomp, L. K. and Koch, R. B. (1974). *Pestic. Biochem. Physiol.* **4**, 232–238
De Vries, D. H. and Georghiou, G. P. (1981). *Pestic. Biochem. Physiol.* **15**, 242–252
Dowson, R. J. (1977). *Pestic. Sci.* **8**, 651–660
Dudai, Y. (1979). *Trends Biochem. Sci.* **4**, 40–44
Eldefrawi, A. T. (1976). In "Insecticide Biochemistry and Physiology" (C. F. Wilkinson, ed.) pp. 297–326. Plenum Press, New York
Eldefrawi, M. E., Eldefrawi, A. T. and O'Brien, R. D. (1970). *J. Agric. Food Chem.* **18**, 1113–1116
Eldefrawi, A. T., Bakry, N. M., Eldefrawi, M. E., Tsai, M.-C. and Albequerque, E. X. (1980). *Molec. Pharmacol.* **17**, 172–179
Elliott, M. (1980). *Pestic. Sci.* **11**, 119–128
Evans, P. D. and Gee, J. D. (1980). *Nature (Lond.)* **287**, 60–62
Ford, M. G. (1979). *Pestic. Sci.* **10**, 39–49
Frankenhaeuser, B. and Hodgkin, A. L. (1957). *J. Physiol.* **137**, 218–244
Gammon, D. W. (1978a). *Pestic. Sci.* **9**, 79–91
Gammon, D. W. (1978b). *Pestic. Sci.* **9**, 95–104
Gammon, D. W. and Holden, J. S. (1980). In "Insecticide Neurobiology and Pesticide Action", pp. 481–488. Society of Chemical Industry, London
Gammon, D. W., Brown, M. A. and Casida, J. E. (1981). *Pestic. Biochem. Physiol.* **15**, 181–191.
Gepner, J. I., Hall, L. M. and Sattelle, D. M. (1978). *Nature (Lond.)* **276**, 188–190

Ghiasuddin, S. M. and Matsumura, F. (1979). *Pestic. Biochem. Physiol.* **10**, 151–161

Goodford, P. J., Hudson, A. T., Sheppey, G. C., Wootton, R., Black, M. H., Sutherland, G. J. and Wickham, J. C. (1976). *J. Med. Chem.* **19**, 1239–1247

Hille, B. (1977). *BioSystems* **8**, 195–199

Holan, G. (1969). *Nature (Lond.)* **221**, 1025–1029

Holan, G., O'Keefe, D. F., Virgona, C. and Walser, R. (1978). *Nature (Lond.)* **272**, 734–736

Hollingworth, R. M. and Murdock, L. L. (1980). *Science* **208**, 74–76

Kaissling, K. E. (1979). In "Chemical Ecology: Odour Communication in Animals" (F. J. Ritter, ed.) pp. 43–56. Elsevier/North Holland, Amsterdam

Kaissling, K. E., Kasang, G., Bestmann, H. J., Stransky, W. and Vostrowsky, O. (1978). *Naturwissenschaften* **65**, 382–384

Kass, I. S., Wang, C. C., Walrond, J. P. and Stretton A. O. W. (1980). *Proc. Natl. Acad. Sci. U.S.A.* **77**, 6211–6215

Kiso, M., Fujita, T., Kurihara, N., Uchida, M., Tanaka, K. and Nakajima, M. (1978). *Pestic. Biochem. Physiol.* **8**, 33–43

Klun, J. A., Bierl-Leonhardt, B. A., Plimmer, J. R., Sparks, A. N., Primiani, M., Chapman, O. L., Lepone, G. and Lee, G. H. (1980). *J. Chem. Ecol.* **6**, 177–183

Leake, L. D. (1977). *Pestic. Sci.* **8**, 713–721

Lund, A. E. and Narahashi, T. (1981). *J. Pharmacol. Exp. Ther.* **219**, 464–473

Lund, A. E., Hollingworth, R. M., Murdock, L. L. (1979). In "Advances in Pesticide Science" (H. Geissbühler, ed.) Pt 3, pp. 465–469. Pergamon Press, Oxford

Maddrell, S. H. P. (1980). In "Insect Neurobiology and Pesticide Action", pp. 329–334. Society of Chemical Industry, London

Matsumura, F. and Beeman, R. W. (1976). *Environ. Health Perspect.* **14**, 71–82

Matsumura, F. and Narahashi, T. (1971). *Biochem. Pharmacol.* **20**, 825–837

Miller, T. A. and Adams, M. E. (1977). In "Synthetic Pyrethroids" (M. Elliott, ed.) pp. 98–115. American Chemical Society, Washington, D.C.

Miller, T. A., Kennedy, J. M. and Collins, C. (1979). *Pestic. Biochem. Physiol.* **12**, 224–230

Mullins, L. J. (1956). In "Molecular Structure and Functional Activity of Nerve Cells" (R. G. Grenell and L. J. Mullins, eds) pp. 123–166. American Institute of Biological Sciences, Washington, D.C.

Murdock, L. L. and Hollingworth, R. M. (1980). In "Insect Neurobiology and Pesticide Action", pp. 415–422. Society of Chemical Industry, London

Narahashi, T. (1962). *J. Cell. Comp. Physiol.* **59**, 61–65

Narahashi, T. (1976a). In "Insecticide Biochemistry and Physiology" (C. F. Wilkinson, ed.) pp. 327–352. Plenum Press, New York

Narahashi, T. (1976b). *Pestic. Sci.* **7**, 267–272

Narahashi, T. (1979). In "Neurotoxicology of Insecticides and Pheromones" (T. Narahashi, ed.) pp. 211–243. Plenum Press, New York

Naharashi, T. (1982). In "Neuropharmacology of Insects: Ciba Foundation Symposium 88" pp. 291–306. Pitman, London

Narahashi, T. and Lund, A. E. (1980). In "Insecticide Neurobiology and Pesticide Action", pp. 497–505. Society of Chemical Industry, London

Narahashi, T. and Yamasaki, T. (1960). *J. Physiol.* **152**, 122–140

Nathanson, J. A. and Hunnicutt, E. J. (1981). *Molec. Pharmacol.* **20**, 68–75

Nicholson, R. A., Hart, R. J. and Osborne, M. P. (1980a). In "Insect Neurobiology and Pesticide Action", pp. 465–471. Society of Chemical Industry, London.

Nicholson, R. A., Chalmers, A. E., Hart, R. J. and Wilson, R. G. (1980b). In "Insect Neurobiology and Pesticide Action", pp. 289–295. Society of Chemical Industry, London

Nishimura, K. and Narahashi, T. (1978). *Pestic. Biochem. Physiol.* 8, 53–64

O'Brien, R. D. (1967). "Insecticides: Action and Metabolism". Academic Press, London and New York

Orchard, I. (1980a). *Pestic. Biochem. Physiol.* 13, 220–226

Orchard, I. (1980b). In "Insect Neurobiology and Pesticide Action", pp. 321–328. Society of Chemical Industry, London

Orchard, I. and Osborne, M. P. (1979). *Pestic. Biochem. Physiol.* 10, 197–202

Pelhate, M., Hue, B. and Sattelle, D. B. (1980). In "Insect Neurobiology and Pesticide Action", pp. 65–71. Society of Chemical Industry, London

Pezzementi, L. and Schmidt, J. (1981). *J. Biol. Chem.* 256, 12651–12654

Plummer, J. M. and Banna, H. B. (1979). *Comp. Biochem. Physiol.* 62c, 9–18

Pong, S.-S., Wang, C. C. and Fritz, L. C. (1980). *J. Neurochem.* 34, 351–358

Priesner, E. (1979). In "Chemical Ecology: Odour Communication in Animals" (F. J. Ritter, ed.) pp. 57–71. Elsevier/North Holland, Amsterdam

Putter, I., MacConnell, J. G., Preiser, F. A., Haidri, A. A., Ristich, S. S. and Dybas, R. A. (1981). *Experientia* 37, 963–964

Raftery, M. A., Hunkapiller, M. W., Strader, C. D. and Hood, L. E. (1980). *Science* 208, 1454–1457

Roeder, K. D. and Weiant, E. A. (1946). *Science* 103, 304–306

Roelofs, W. L. and Cardé, R. T. (1977). *Ann. Rev. Entomol.* 21, 377–405

Samaranayaka-Ramasamy, M. (1978). In "Pesticide and Venom Neurotoxicity" (D. L. Shankland, R. M. Hollingworth and T. Smyth, Jr., eds) pp. 83–93. Plenum Press, New York

Sattelle, D. B. (1980). In "Advances in Insect Physiology" (M. J. Berridge, J. E. Treherne and V. B. Wigglesworth, eds) Vol. 15, pp. 215–315. Academic Press, London and New York

Sawicki, R. M. (1978). *Nature (Lond.)* 275, 443–444

Schuntner, C.A. and Thompson, P. G. (1978). *Aust. J. Biol. Sci.* 31, 141–148

Schuntner, C. A. and Thompson, P. G. (1979). *Pestic. Sci.* 10, 519–526

Shaker, N. and Eldefrawi, A. (1981). *Pestic. Biochem. Physiol.* 15, 14–20

Shankland, D. L. (1976). In "Insecticide Biochemistry and Physiology" (C. F. Wilkinson, ed.) pp. 229–270. Plenum Press, New York

Shankland, D. L. (1979). In "Neurotoxicology of Insecticides and Pheromones" (T. Narahashi, ed.) pp. 139–153. Plenum Press, New York

Silverstein, R. M. (1981). *Science* 213, 1326–1332

Singh, G. P. J., Orchard, I. and Loughton, B. G. (1981). *Pestic. Biochem. Physiol.* 16, 249–255

Smith, P. J., Crowe, A. J., Das, V. G. K. and Duncan, J. (1979). *Pestic. Sci.* 10, 419–422

Starkus, J. G. and Narahashi, T. (1978). *Pestic. Biochem. Physiol.* 9, 225–230

Sternburg, J., Chang, S. C. and Kearns, C. W. J. (1959). *J. Econ. Entomol.* 52, 1070–1076

Uchida, M., Naka, H., Irie, Y., Fujita, T. and Nakajima, M. (1974). *Pestic. Biochem. Physiol.* 4, 451–455

Uchida, M., Fujita, T., Kurihara, N. and Nakajima, M. (1978). In "Pesticide and Venom Neurotoxicity" (D. L. Shankland, R. M. Hollingworth and T. Smyth, Jr., eds) pp. 133–151. Plenum Press, New York

van den Bercken, J. and Vijverberg, H. P. M. (1980a). In "Insect Neurobiology and Pesticide Action", pp. 391–397. Society of Chemical Industry, London

van den Bercken, J. and Vijverberg, H. P. M. (1980b). In "Insect Neurobiology and Pesticide Action", pp. 79–85. Society of Chemical Industry, London

van den Bercken, J., Kroese, A. B. A. and Akkermans, L. M. A. (1979). In "Neurotoxicology of Insecticides and Pheromones" (T. Narahashi, ed.) pp. 183–210. Plenum Press, New York

Vijverberg, H. P. M., van der Zalm, J. M. and van den Bercken, J. (1982). *Nature* (Lond.) **295**, 601–603

Wang, C. M. and Narahashi, T. (1972). *J. Pharmacol. Exp. Ther.* **182**, 427–441

Wang, C. M., Narahashi, T. and Yamada, M. (1971). *Pestic. Biochem. Physiol.* **1**, 84–91.

Wang, C. M., Narahashi, T. and Scuka, M. (1972). *J. Pharmacol. Exp. Ther.* **182**, 442–453

Woodward, W. E. and Sternburg, J. G. (1979a). *Pestic. Biochem. Physiol.* **10**, 287–298

Woodward, W. E. and Sternburg, J. G. (1979b). *Pestic. Biochem. Physiol.* **10**, 299–305

Woolley, D. E. (1982). In "Mechanisms of Actions of Neurotoxic Substances" (K. N. Prasad and A. Vernadakis, eds) pp. 95–141. Raven Press, New York

Worthing, C. R. (ed.) (1979). "The Pesticide Manual", 6th edn. British Crop Protection Council, London

Wouters, W. and van den Bercken, J. (1978). *Gen. Pharmacol.* **9**, 387–398

Wouters, W., van den Bercken, J. and van Ginneken, A. (1977). *Eur. J. Pharmacol.* **43**, 163–171

Yamaguchi, I., Matsumura, F. and Kadous, A. A. (1979). *Pestic. Biochem. Physiol.* **11**, 285–293

Yamamoto, I. (1965). *Adv. Pest Control Res.* **VI**, 231–260

5 | Compounds Interfering with Cell Growth, Division and Development

I. COMPOUNDS INTERFERING WITH PLANT HORMONE ACTION
A. Introduction

The pattern of cell growth and division within a plant is under the influence of five well established hormones or classes of hormone: auxin, cytokinins, gibberellins, ethylene and abscisic acid. The last two are relative late-comers to the scene and it is possible that other hormonally active compounds may yet be discovered.

A hormone is a compound which is produced at one site, and which, at low concentrations, produces physiological effects at another. In the case of the plant hormones, it is the relative levels of all the hormones which dictate the behaviour of a given cell though in certain instances fluctuations in the levels of a single hormone trigger particular processes. Another way of expressing this is to say that a particular hormone is, at a given time, limiting for a certain process. It is important to realise that even if the application of a hormone causes a positive response it does not necessarily follow that it is the sole mediator of the process under all conditions, but

179

only that it can be made the controlling factor in the particular circumstances of that test.

This traditional view of hormone action has been challenged recently by Trewavas (1982) who suggests instead that cells may come under the influence of hormones as a result of changes in the levels of sensitive sites within the cells. According to either view one would expect the plant to be damaged by an insufficient supply of a hormone important for a particular process, or the application of extremely large doses of hormonally active compounds. As we shall see, influences of this type account for the activity of the great majority of herbicides which interfere with hormone action.

Herbicides interfere with the action of at least three of the five classes of hormone. We shall consider the classes in turn, indicating something of the biochemical and physiological action of the hormone, followed by an account of the compounds which interfere with this action. General reviews of the field of plant hormones have been written (see, for example, Letham *et al.*, 1978).

B. Auxins
1. *The natural hormone*

The term 'auxins' applies to compounds which are able to stimulate the growth of excised coleoptile (p. 186) tissue. When used in the singular and without further qualification 'auxin' usually refers to the naturally-occurring compound indol-3-yl-acetic acid (IAA; indoleacetic acid) which

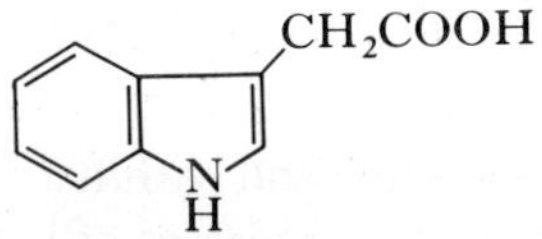

$$CH_2COOH$$

IAA

probably occurs universally in plant tissue together with closely related molecules and derivatives which may be implicated in auxin action (Bearder, 1980).

Auxin is involved in many aspects of the development and function of plant cells and tissues. These include phototropism, apical dominance, and cell growth and differentiation; auxins can also, for instance, promote lateral root formation in cuttings and delay senescence (Jacobs, 1979).

The biochemical mode of action is not known in detail, though a huge literature attests to the many effects which might be anticipated following treatment of plant tissue with a hormone. In recent years a good deal of attention has been focused on auxin-binding components of plant tissues. This work is based on the assumption that there must be a site (or sites) to which auxin has to bind in order to trigger the correct cell responses. Both soluble and membranous auxin-binding components have been identified and there may well be binding sites with different properties (and perhaps different functions) located on the various membranes of the cell (for reviews see Dodds and Hall, 1980; Rubery, 1981; Venis, 1978, 1979).

The secondary question of what changes are brought about when auxin binds to a site has not been answered. It is recognised that the growth stimulation shown by excised stem sections falls into two phases. First, there is a rapid response (within minutes), simulated by low pH and perhaps due to auxin stimulating the pumping of protons into the cell wall and loosening it. The second phase of the response occurs 35–45 min after treatment. Inhibitor work shows it to be dependent on macromolecule synthesis (see Venis, 1978 for original references), and ribonucleic acid (RNA) synthesis in particular is implicated (Jacobsen, 1977). The various theories on the mode of action of auxin are discussed by Rubery (1981).

It is worth noting that indoleacetic acid administered to tissues can be destroyed by an oxidase whereas this does not usually occur with synthetic chemicals. Therefore, in many studies on auxin binding and action, auxin mimics are used on the assumption that the modes of action are identical; as we shall see this is very likely to be the case.

2. Synthetic auxins mimicking indoleacetic acid

These will be considered in broad chemical classes. A wide-ranging account of the synthetic auxin herbicides is given by Garraway and Wain (1976). Some general properties of synthetic auxins will be illustrated by reference to the phenoxyalkanoic acids, their relatives, and related compounds. For convenience we have collated compounds of quite diverse structure under this one heading (Table 5.1). The compounds are used principally as herbicides, though auxins can be used to modify fruit abscission (Addicott, 1976) and this is the major use of ethyl 5-chloro-1H-3-indazolyl-acetate (etychlozate; Kamuro, 1981). In addition 4-(indol-3-yl) butyric acid is used to promote rooting of cuttings and this is also one of several growth-regulating applications of 1-naphthylacetic acid (NAA). It is assumed that all these uses reflect the properties of the compounds as auxin mimics, so that the biochemical mode of action is the same in each case. It should be mentioned, however, that auxin herbicides do stimulate ethylene production by treated tissues (Abeles, 1968; Morgan, 1976), and, while this is not involved in the herbicidal action (Abeles, 1968), it can account for some of the sub-toxic effects (Morgan, 1976) and it is stated to be involved in the fruit-thinning action of etychlozate (Kamuro, 1981).

Table 5.1 Synthetic auxins mimicking indoleacetic acid

Name	Structure
A. Phenoxyalkanoic acids, related compounds and their precursors	
Chlorfenac	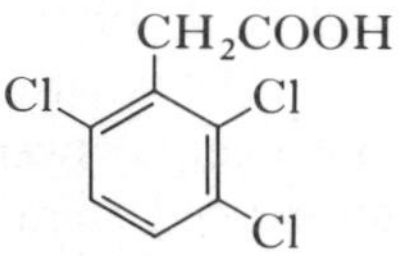CH₂COOH Cl Cl Cl
2,4-D	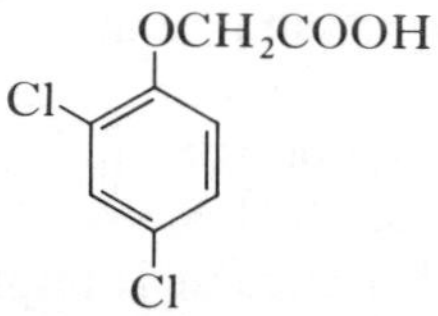OCH₂COOH Cl Cl
2,4-DB	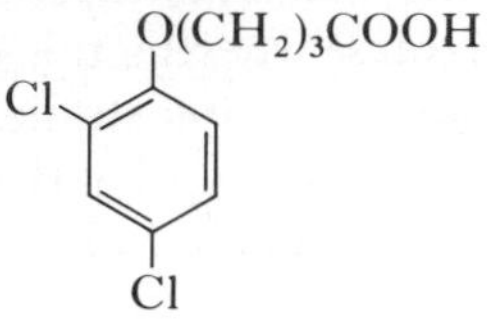O(CH₂)₃COOH Cl Cl
2,4-DES-sodium (disul-sodium)	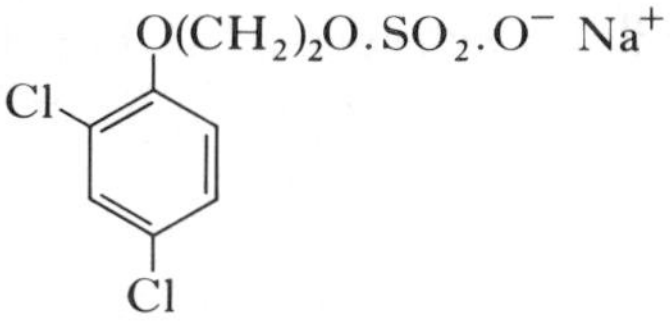O(CH₂)₂O.SO₂.O⁻ Na⁺ Cl Cl
Dichlorprop	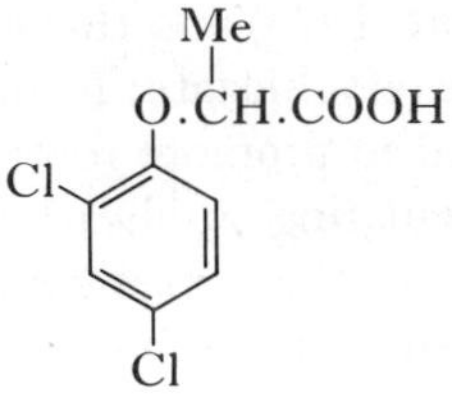Me O.CH.COOH Cl Cl
Ethyl 5-chloro-1*H*-3-indazolyl-acetate (Etychlozate)	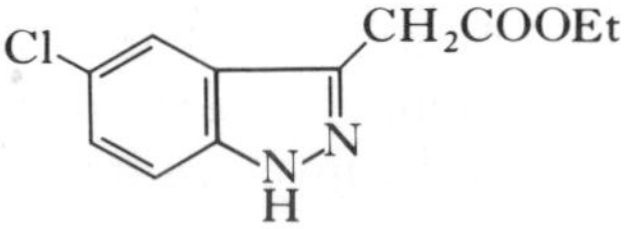Cl CH₂COOEt N N H

Name	Structure
Fenoprop	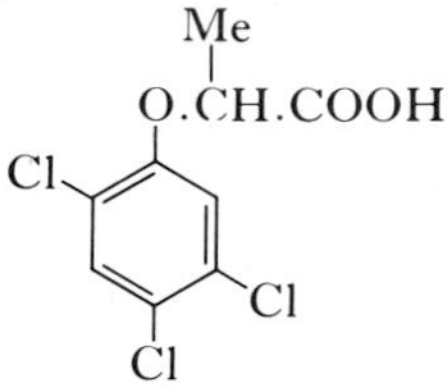 Me O.CH.COOH Cl Cl Cl
4-(Indol-3-yl)butyric acid	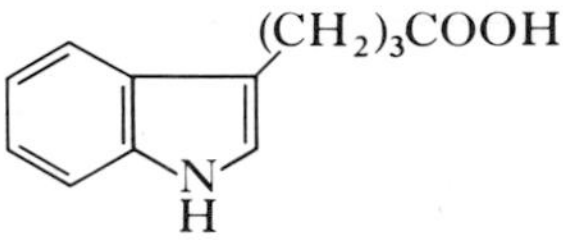 (CH2)3COOH N H
MCPA	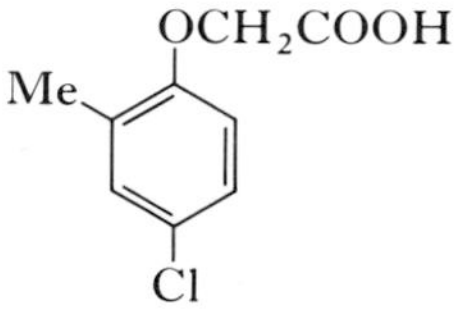 OCH2COOH Me Cl
MCPB	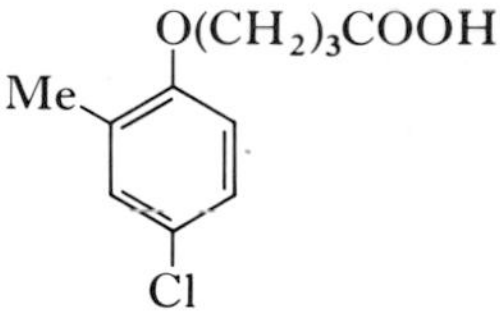 O(CH2)3COOH Me Cl
Mecoprop	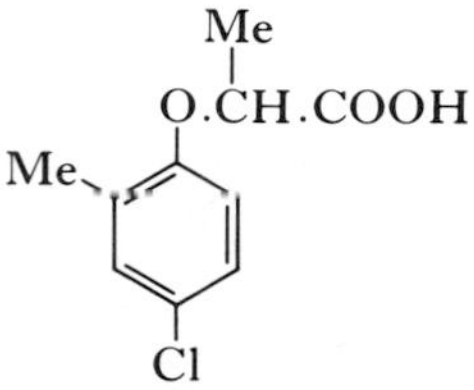 Me O.CH.COOH Me Cl
1-Naphthylacetamide	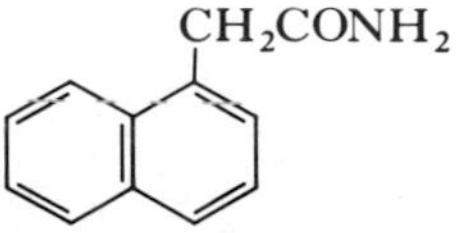 CH2CONH2

Name	Structure
1-Naphthylacetic acid	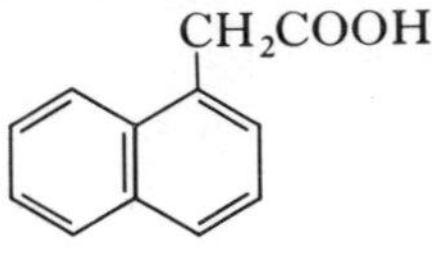 CH_2COOH
2-Naphthyloxyacetic acid	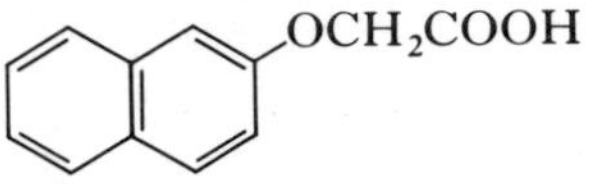 OCH_2COOH
2,4,5-T	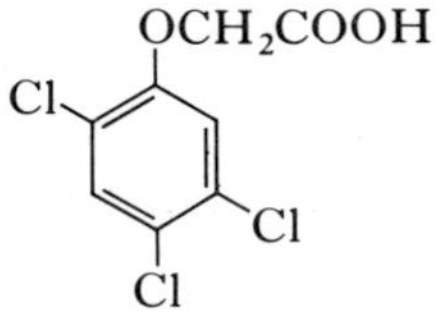 OCH_2COOH Cl, Cl, Cl
Triclopyr	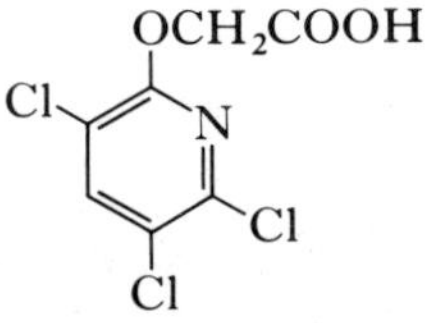 OCH_2COOH Cl, N, Cl, Cl
B. *Benzoic acids and related compounds* Chloramben	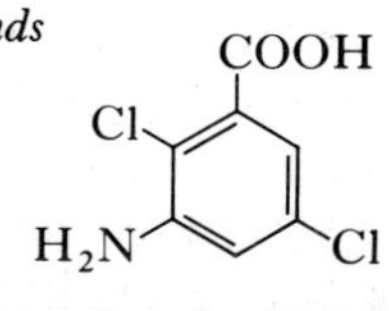 COOH Cl, H_2N, Cl
Dicamba	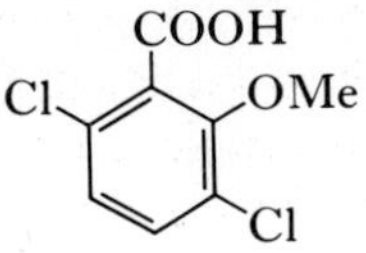 COOH Cl, OMe, Cl
3,6-Dichloropicolinic acid	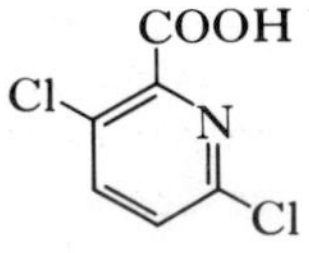 COOH Cl, N, Cl

Name	Structure
Picloram	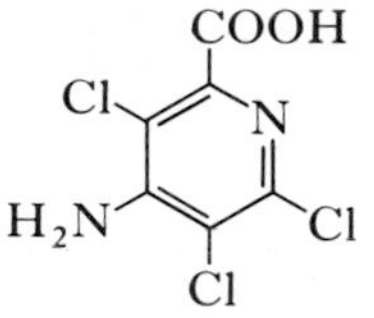
2,3,6-TBA	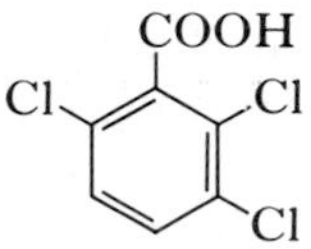
C. Benazolin	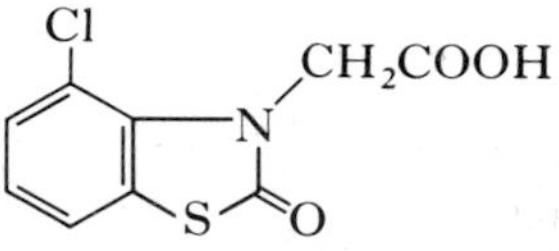

(a) Phenoxyalkanoic acids, related compounds and their precursors

(i) Introduction. It is no exaggeration to say that the discovery of 2,4-D and MCPA (Table 5.1) was revolutionary for agriculture. These hormone herbicides, together with related chemicals, were discovered independently in Britain and the U.S.A. during the Second World War and, since they were the first non-toxic organic herbicides that were effective at low levels, they completely altered agricultural techniques of weed control (Kirby, 1980). The compounds have since had an excellent record of safety to animals and man. It was probably the success of these synthetic structures that stimulated industry to invest in the research which subsequently led to the discovery of the wide variety of herbicides now available. General reviews of the phenoxyacetic acids have been compiled by Pillmoor and Gaunt (1981) and Ashton and Crafts (1981).

We can regard 2,4-D and its relatives as being persistent auxins that induce disorderly growth by their presence in excess of the normal amounts. This is supported by the fact that they produce morphological effects on plants that indicate exaggeration of normal auxin action and they also show auxin activity in a variety of standard tests on excised tissues.

Before considering their action as auxins, it should be noted that some of the compounds listed in part *A* of Table 5.1 have to be converted to an auxin before they can act. 2,4-DES-sodium and 1-naphthylacetamide both lack a carboxyl group, which appears to be a requirement for high activity as an auxin (see later). 2,4-DES-sodium is not phytotoxic itself but in moist soil is converted to the alcohol which is in turn oxidised to 2,4-D (Worthing, 1979). It seems likely that 1-naphthylacetamide is active after hydrolysis to NAA. 2,4-D itself can be applied as an ester (Worthing, 1979), and most plants are able to hydrolyse such esters to the acid (Loos, 1975). The action of 4-(indol-3-yl)butyric acid as an agent for promoting rooting of cuttings can be accounted for by its conversion to IAA (Fawcett et al., 1958) by the process of β-oxidation. This is explained below in relation to the selective activation of 2,4-DB and MCPB which are converted by the same process to 2,4-D and MCPA.

(ii) *Auxin activity measured by conventional assays.* A variety of tests, most of which are described in the handbook of Mitchell and Livingston (1968), have been used to measure the effects of plant growth regulators in general. As an example, auxin activity may be estimated by the split pea stem test, in which the curvature of slit halves of etiolated pea stems immersed in a solution of the auxin is measured under defined conditions and after a set time.

Hanson and Trewavas (1982) have argued that the fact that excised tissues grow by cell enlargement in response to auxin does not necessarily mean that auxin controls growth rate *in vivo*. Rather, they suggest that in intact tissues auxin is an essential, but non-limiting factor, and they consider that hormone-induced growth in excised tissues largely represents an accelerated recovery from a metabolic injury caused by excision. Nevertheless the assays have proved to be a satisfactory way of quantifying *in vitro* the activity of the synthetic auxins that concern us here, and it is on such data that the models of an auxin-binding site (see below) are based.

However, if different tests are used, different orders of activity emerge. In the oat coleoptile curvature test, a small agar block containing auxin is placed asymmetrically on a decapitated coleoptile. (The coleoptile is the sheath that protects the young emerging shoot, and it grows essentially by cell enlargement at the top.) The unilateral application of auxin causes the cells below the block to grow more so that the coleoptile bends, and the extent of bending indicates the activity of the auxin. Although this test is sensitive to natural auxins, it is not particularly responsive to synthetic compounds; for example, 2,4-D is one thousand times less active than indoleacetic acid, whereas in the related oat coleoptile straight growth test, in which the elongation of the coleoptile section immersed in a solution of the auxin is measured, indoleacetic acid and 2,4-D show comparable activity (Bentley, 1950).

The differing results from the various tests for auxin activity make a meaningful quantitative comparison of the activity of the various auxins rather difficult. However, it is fair to conclude that the synthetic auxins under consideration in this section have an activity which is at least comparable to that of indoleacetic acid itself. Correlation between the structure and the activity of auxins will be deferred until we have dealt with other structures having auxin activity.

The absolute amount of 2,4-D required to produce a marked effect on plant growth is small. A concentration of 0.4 to 4 μM caused 20 to 55% stimulation of wheat coleoptile segment growth when compared with an untreated control (Wain and Fawcett, 1969). It is perhaps more relevant to consider the amount of 2,4-D that it is necessary to apply to an intact plant to achieve a modification of growth. Cotton is particularly sensitive, and 40 μg of 2,4-D sprayed carefully, to avoid run-off, onto 2-month-old cotton plants will cause marked formative effects on leaves (McIlrath and Ergle, 1953).

(*iii*) *Morphological effects on plants.* Pillmoor and Gaunt (1981) have catalogued the symptoms observed in plants following treatment with phenoxyacetic herbicides. The symptoms include leaf chlorosis, altered stomatal function, stem tissue proliferation, root initiation in stem tissue, disintegration of root tissues and abnormal apical growth. In trying to define the major cause of death, they supported the long-held view that the compounds create a general hormonal imbalance that leads to abnormal cell behaviour.

The morphological effects caused by auxins are turned to advantage in the cases of 4-(indol-3-yl)butyric acid and 1-naphthylacetic acid which are used to stimulate the rooting of cuttings and also, in the case of the latter compound, to control flowering and fruiting (Worthing, 1979).

(*iv*) *Selectivity of action.* 2,4-D and MCPA are primarily used to control broad-leaved weeds (dicotyledons) in cereal crops (monocotyledons). Possible explanations for this important selective action have been reviewed in detail by Pillmoor and Gaunt (1981) but they could only conclude that it was doubtful whether any single aspect of herbicide behaviour would explain selectivity. Loos (1975) considered that the prime cause could be differences in plant structure and rates of herbicide translocation in monocotyledonous and dicotyledonous plants. The arrangement of the vascular tissue in bundles surrounded by protective tissue in monocotyledons seems to prevent the destruction of the phloem by the disorganised growth caused by the herbicide. The fact that there is no auxin-sensitive layer of cells capable of cell division in the vascular bundles of monocotyledons may also be an important factor in conferring selectivity (Hanson and Slife, 1969). As far as translocation is concerned, movement of foliar-applied auxins from the site of application is less in monocotyledons than in susceptible dicotyledons (see Loos, 1975). In addition there are differences in metabolism between monocotyledons and dicotyledons which could also contribute to selectivity (Garraway and Wain, 1976; Loos, 1975). There is no evidence for differences in the target auxin binding sites (see below) but this remains a possibility.

2,4-DB and MCPB (Table 5.1) are higher homologues of 2,4-D and MCPA, and they are more selective in their action, so that they can, for instance, be used to control weeds in legumes; the parent compounds cannot be used since they damage the crop (Wain and Smith, 1976). The history of the discovery of the herbicidal effectiveness of 2,4-DB and MCPB is interesting since it suggests a rational approach to herbicide design.

Synerholm and Zimmerman (1947) observed that, in a series of 2,4-D homologues having an increasing number of methylene groups in the side chain, only those compounds with an even number of carbon atoms in the whole of the chain showed activity on tomatoes. They suggested that this was because the plants degraded the side chain by β-oxidation (i.e. by removing two carbon atoms at a time), which would mean that compounds containing an even number of carbon atoms would generate 2,4-D, while those with an odd number would yield 2,4-dichlorophenol, which is not an auxin. This sequence of events was directly demonstrated by Wain's group, who showed that flax seedlings only produced phenol from acids of the type $C_6H_5O(CH_2)_nCOOH$ when n was even (i.e. when the number of carbon atoms in the whole of the side chain was odd). Experiments with the analogous 2,4,5-trichlorophenoxy series showed a similar pattern of activity with wheat, but not with pea and tomato. This evidence that some plants can carry out the β-oxidation while others cannot suggested to Wain that a logical approach to selective weed control could be based upon the plant's own enzyme make-up (Wain and Smith, 1976).

Recent work on the metabolism of phenoxyalkanoic herbicides has been reviewed by Loos (1975) and Pillmoor and Gaunt (1981), and it supports the β-oxidation hypothesis. However, Linscott and Hagin (1970) showed that the legume alfalfa will actually convert 2,4-D to 2,4-DB, and 2,4-DB itself to higher homologues. They suggest that the tolerance of rapidly growing alfalfa to 2,4-DB may be due to the predominance of these synthetic reactions over those responsible for the β-oxidation that produces 2,4-D.

(b) Benzoic acid and related compounds

Herbicides related to benzoic acids (Table 5.1) act in the same way as the phenoxyalkanoic acids, i.e. they are persistent auxins. Their properties have been reviewed in detail by Frear (1975) and Ashton and Crafts (1981).

The activity of 2,3,6-TBA in the oat coleoptile straight growth test was demonstrated by Bentley (1950) who showed that a 4 μM concentration caused 78% of the stimulation produced by indoleacetic acid at the same concentration. The auxin-like nature of 2,3,6-TBA was emphasised by Zimmerman and Hitchcock (1951), who showed that it caused cell elongation and tissue proliferation, induced adventitious roots, modified the arrangement of leaves and other organs and caused fruit to develop in the absence of fertilisation.

The three auxins indicated were shown by Keitt and Baker (1966) to cause, at a concentration of 1 μM, the following percentage elongation of young oat stems compared with a control incubated in the basal medium.

indoleacetic acid	369
chloramben	182
dicamba	295

Although picloram (Table 5.1) is structurally distinct from the other auxins, it has been shown by Kefford and Caso (1966) and by Eisinger and Morré (1971) to possess properties typical of an auxin. Thus, it stimulates cell elongation in a variety of plant stem tissues at a concentration of 1 μM and causes other typical auxin symptoms.

Since picloram is highly mobile in plants, yet does not undergo extensive alteration, it was used by Davis et al. (1972) to determine the actual concentration of herbicide inside the plant that is required to produce a lethal effect. They found that a concentration of 10 μM picloram in cotton leaf buds prevented the normal development of half the leaves, though the buds continued to accumulate more herbicide than this.

(c) Benazolin

Benazolin (Table 5.1) caused marked stimulation of growth of wheat coleoptile segments at 40 μM, and its herbicidal action is probably related to this auxin effect since related structures lacking such activity were not herbicidal (Brookes and Leafe, 1963).

(d) Relation between the structure and activity of auxins

Many compounds have been made and their auxin activity determined in bioassay systems of the sort outlined above. The results provided a basis for a succession of attempts to define a model of the auxin receptor which could accommodate the active auxins and account for the inactivity of others. For the agrochemical industry, of course, such models provide a basis for the synthesis of new, potentially useful chemicals.

For some time the most satisfactory suggestion was that of Porter and Thimann, who held that auxin activity depended on the presence of a fractional positive charge situated 0.55 nm away from the negative charge of a carboxyl group, as shown below for IAA, 2,4-D, 2,3,6-TBA and picloram (Porter and Thimann, 1965; Thimann, 1969).

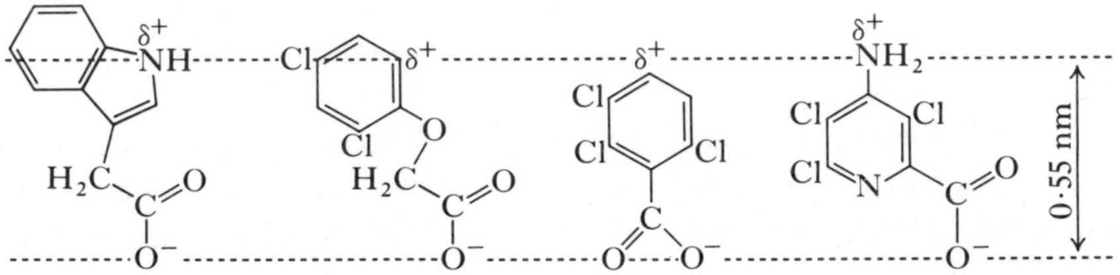

In a theoretical chemical treatment Block and Clements (1975) found no relation between the electron density at the indole nitrogen atom and auxin activity for a series of indole derivatives. A similar approach by Farrimond *et al.* (1978) appeared to support a modified form of the Porter and Thimann theory, although this suggestion was abandoned when it did not withstand the test of a more detailed analysis (see Farrimond *et al.*, 1981).

Approaches other than the charge-separation theory have been followed over the years. Following Porter and Thimann's theory Wain and Fawcett (1969) proposed that a three-point contact between auxin and receptor was required for activity. This accounted for the lack of activity of the 'wrong' isomer of 2,4-dichlorophenoxypropionic acid:

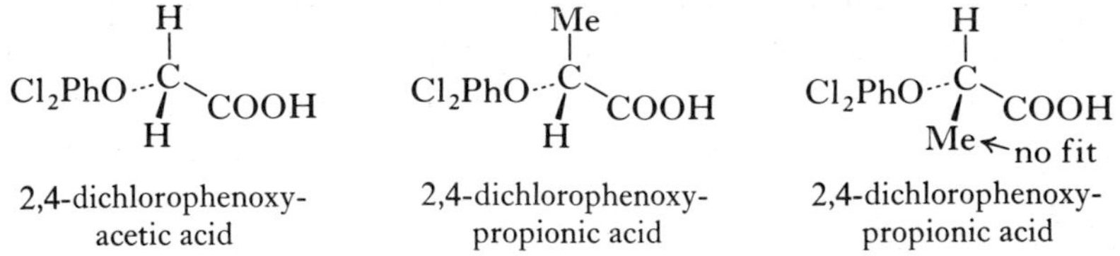

This idea moved towards the view of a chiral three-dimensional biological receptor, and is intrinsically more realistic than a simple two-dimensional charge-separation requirement. It allowed for the differences in potencies between isomers, a point which was also the major feature of work by Lehmann (1978). Using a mathematical analysis of data on the relative potencies of isomers, he also favoured a three-point attachment theory.

A quite distinct suggestion was made by Kaethner (1977) who proposed that, to be active, an auxin had to approach the receptor in a 'recognition' configuration and then go through a simultaneous change in shape with the receptor to adopt a 'modulation' conformation. Kaethner's theory also accounted for the action of anti-auxins, compounds which prevent auxin from exerting its effect in a manner that can be overcome by adding

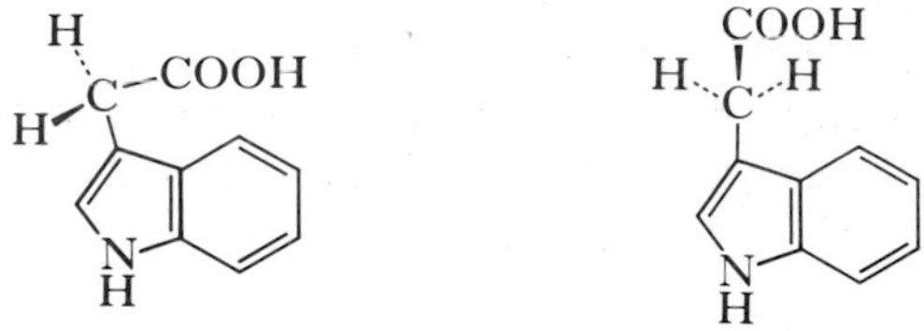

recognition
conformation

modulation
conformation

further auxin. They can be explained on this theory as compounds which can adopt the recognition conformation and therefore bind, but which cannot move with the receptor to give the modulation conformation. However, as Kaethner (1977) himself points out, there are some compounds with auxin activity which are not well accommodated by the theory. In addition Farrimond *et al.* (1978) calculated that in the case of IAA, only about 0.5% of the molecules would be in the recognition conformation.

Rakhaminova *et al.* (1978) studied the structures of active and inactive auxins reported in the literature, and formulated a model of a binding site in which the aromatic part of the auxin bound to a flat platform of restricted dimensions, with the carboxyl group lying out of the plane of the ring. They suggest that a conformational change is induced in the receptor when an auxin becomes bound.

Katekar (1979) proposed a model which is conventional in that it attempts to define the site by reference to all the compounds known to act there. Katekar considers that the delocalised electrons in the aromatic portion of auxin molecules interact with a relatively positive area of the auxin binding site. He also suggests that chlorine substituents on the aromatic nucleus might mimic to some extent the binding capabilities of the electrons of aromatic rings and thus interact directly with the site rather than through modification of the properties of the aromatic nucleus.

The site was depicted thus:

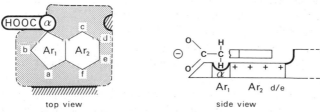

top view   side view

Reproduced from Katekar (1979)

A good auxin must possess the carboxyl group but not have groups which would hinder binding in the α, c–d or a–f regions. With these criteria satisfied, the auxin activity would be determined by the extent to which Ar_1 and Ar_2 were covered. Note that the auxin is considered to bind with the side-chain extended away from the ring, a conformation favoured by Farrimond *et al.* (1978). A recent study of the activities of a series of chlorinated IAA derivatives as stimulators of pea stem section elongation gave results consistent with the model (Katekar and Geissler, 1982).

3. *Flurecols and naptalam—compounds that may act by interfering with auxin transport*

The flurecols (Table 5.2) are also known as morphactins (*morph*ogenetically *acti*ve substances) and can be used as herbicides or growth regulators. Naptalam can be used as a pre-emergence herbicide and commercial formulations are also available for peach thinning (W.S.A Handbook, 1979).

The flurecols and naptalam abolish the normal curvature of roots towards the ground and of shoots toward the light. The flurecols are active

Table 5.2   Compounds which interfere with indoleacetic acid transport

Name	Structure
1. *Flurecols* (*morphactins*) Chlorflurecol-methyl	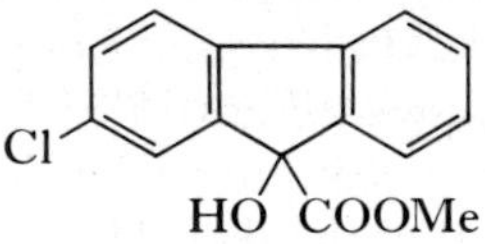
Flurecol-butyl	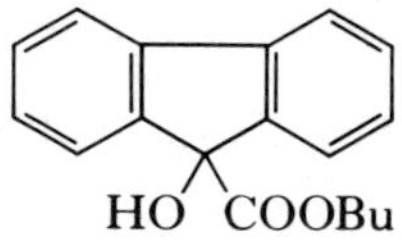
2. *Naptalam*	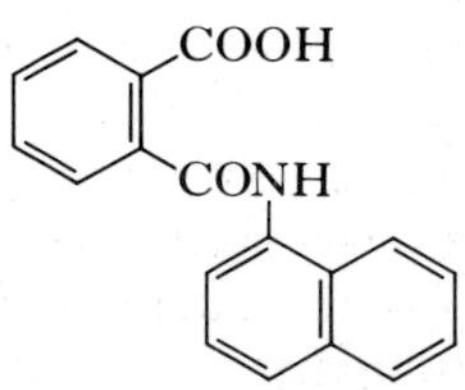

at 60 μM or less (Schneider, 1970; Ziegler, 1970) and naptalam is effective at 17 μM (Jones *et al.*, 1954). As the bending of plant organs in response to gravity or light requires correct movement of auxin within the tissue, the symptoms are consistent with a disruption of auxin transport.

Intact plants transport auxin downwards (transport is said to be basipetal) and this is thought to be caused by the preferential efflux of auxin from the basal ends of the cells, perhaps due to differences in cellular auxin permeability and/or the distribution of carrier systems for the indoleacetic acid anion (Goldsmith, 1977). Inhibitors of auxin transport such as naptalam and morphactins bring about their effect by preventing cellular auxin efflux so that auxin accumulates within the cell (see Goldsmith, 1977, for original references). As this efflux may be carrier-mediated (Jacobs and Hertel, 1978), it is possible that, for example, naptalam could interfere with the carrier's correct function (Sussman and Goldsmith, 1981).

Such a carrier would presumably be located in the outer cell membrane (plasmalemma). Lembi *et al.* (1971) showed that, after separation of a total cell membrane fraction on a sucrose density gradient, naptalam binding coincided with a staining reaction thought to be characteristic of the plasmalemma. The view that naptalam binding sites are probably located here has been supported by more recent membrane fractionation and binding studies (Normand *et al.*, 1975; Ray, 1977).

Naptalam binding to cellular membrane preparations has been reported to be competitively inhibited by naptalam derivatives (Thomson and Leopold, 1974). Indoleacetic acid, NAA or 2,4-D did not compete in this way (Thomson, 1972) though it has since been shown that, when the naptalam receptor is rendered soluble by detergent treatment, naptalam binding does become susceptible to competition by IAA though not benzoic acids (Sussman and Gardner, 1980).

Other work has demonstrated the existence of a class of auxin-binding membrane fragments to which IAA binding is actually increased by naptalam and another auxin-transport inhibitor, 2,3,5-triiodobenzoic acid (Jacobs and Hertel, 1978). It has been suggested that both this site and the one mentioned in the previous paragraph may represent particular conformations of an IAA carrier (see Rubery, 1981).

Whilst the emphasis above has been on naptalam, it has been shown that the morphactins and naptalam competed for a common binding site in homogenates from maize coleoptiles (Thomson and Leopold, 1974).

A recent attempt was made to define common structural characteristics amongst auxin transport inhibitors (Katekar and Geissler, 1977, 1980). These chemicals also inhibit root geotropic responses and the compounds appeared to possess the structural features previously postulated to be necessary for this effect (Katekar, 1976). These features were (*i*) a carboxyl group (or a masked one), attached to (*ii*) an aromatic ring, connected at the *ortho-* position to (*iii*) a second aromatic ring. Further, (*iv*) the two rings could be separated by a conjugated or planar system of atoms and (*v*) high activity required that the distance between the centres of the two extreme aromatic rings was at least 0.73 nm. It was thought that substituents on the non-carboxylated ring could lead to increased activity if they assisted in the achievement of the required molecular size; otherwise they had little effect. These ideas represent a working hypothesis which is still being refined and more recent results do not indicate a requirement for coplanarity of the rings (Katekar and Geissler, 1981). Katekar and Geissler (1980), proposed the term 'phytotropin' for compounds which interfere with the tropic responses of plants and broadly satisfy these structural criteria. There may perhaps be a natural ligand whose effects are being mimicked by these chemicals (see Katekar and Geissler, 1980).

The model accommodates naptalam (Table 5.2) and various phytotropins have since been shown to prevent binding of naptalam to membrane preparations from maize coleoptiles (Katekar *et al.*, 1981). On the other hand the model will not accommodate the morphactins nor, in fact, 2,3,5-triiodobenzoic acid. Katekar and Geissler (1977) speculate that morphactins might be active after conversion to the corresponding 2-phenylbenzoic acids. However, Katekar and Geissler (1980) point out that the morphactins do induce some morphological changes which are not caused by the other compounds on which the model is based. They conclude that there must be a relationship between the action of the morphactins and the other compounds, but the nature of this relationship is not yet clear.

In summary, therefore, structure/activity studies have led to a description of a 'phytotropin'-binding site although another binding site may exist for other compounds interfering with auxin transport. Occupation of a site may lead to inhibition of auxin efflux from the basal ends of cells and thereby to an inhibition of basipetal auxin movement.

C. Gibberellins
1. *The natural hormones*

The gibberellins were the second group of hormones known to Western scientists, who became aware of them from 1950, although they were discovered in Japan in 1926. The chemistry and metabolism of gibberellins have been reviewed by Hedden *et al.* (1978) who provide an entry to the earlier literature. A comprehensive review of the entire field is provided by Graebe and Ropers (1978).

The morphological effects of gibberellins on plant growth are varied, and differ from organ to organ and plant to plant. A dramatic response to gibberellin is the conversion of dwarf plants to tall ones as demonstrated originally with beans. Gibberellins can affect the dormancy of buds and seeds where they can substitute for various cold or light treatments normally required to induce sprouting or germination. Also they can induce fruit development in the absence of fertilisation. Gibberellins do not affect root growth, though root tissues do synthesise gibberellins (Torrey, 1976).

At the time of the review by Bearder (1980) there were 57 gibberellins known, all closely related to gibberellic acid (GA_3), but thought to be active in their own right. Gibberellic acid itself has found practical application as a growth regulator.

GA_3

It is not known how gibberellins work at the molecular level and much less progress has been made towards identifying their possible receptors than in the case of auxins (Kende and Gardner, 1976). In biochemical terms perhaps the best characterised sequence of events caused by gibberellins occurs during barley seed germination. According to the traditional view, gibberellins are synthesised in the germinating embryo and diffuse to the aleurone layer (a layer of cells towards the outer part of the seed) where they induce the production of hydrolytic enzymes. Trewavas (1982), however, considers that, at a particular stage of germination, the aleurone cells become sensitive to gibberellin that was present all along. The enzymes, particularly α-amylase, break down the starch stored in the endosperm to give sugar which is then available to the developing embryo. The increase in α-amylase activity is due to new protein synthesis (Filner and Varner, 1967) and it is therefore possible that the gibberellins may act at the gene level; however, this is not proven (Amrhein, 1979).

2. Compounds that may interfere with gibberellin synthesis

The compounds in this section (Table 5.3) are growth retardants – i.e. they reduce plant growth without permanently stunting or malforming the plants. Their action has been reviewed recently by Dicks (1980).

In the discussion which follows, reference will be made to the outline of gibberellin biosynthesis shown in Fig. 5.1. Occasional reference will be made to the superseded growth retardant Amo 1618, though the properties of this compound will not be comprehensively reviewed.

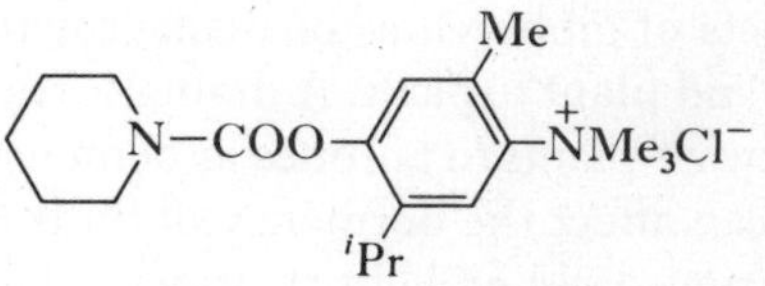

Amo 1618

(a) Chlormequat chloride

This compound (Table 5.3), discovered by Tolbert (1960), produces compact plants with shortened internodes and leaf stems and is used to prevent cereals from lodging (i.e. falling over). It may also make plants more resistant to attack by insects and fungi (Worthing, 1979).

Table 5.3　Plant growth regulators that may interfere with gibberellin synthesis

Name	Structure
Ancymidol	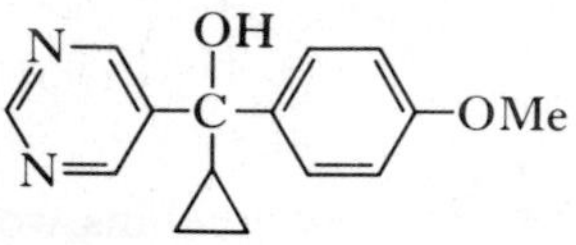
Chlormequat chloride	$\overset{+}{ClCH_2CH_2NMe_3}Cl^-$
Chlorphonium chloride	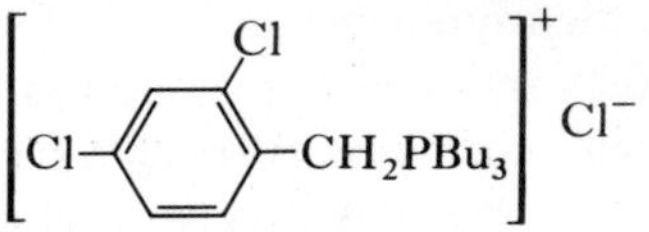
Daminozide	$\underset{CH_2COOH}{CH_2CONH.NMe_2}$
Mepiquat chloride	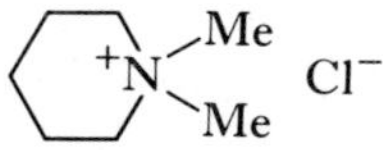
Piproctanyl bromide	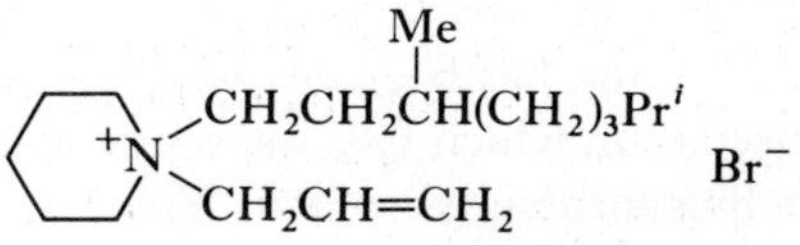

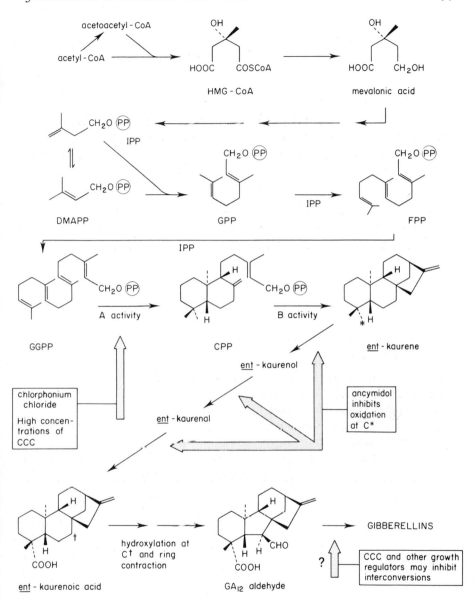

Fig. 5.1 Outline of gibberellin biosynthesis. HMG-CoA, 3-hydroxy-3-methyl-glutaryl CoA; IPP, isopentenyl pyrophosphate; DMAPP, dimethylallyl pyrophosphate; GPP, geranyl pyrophosphate; FPP, farnesyl pyrophosphate; GGPP, geranylgeranyl pyrophosphate; CPP, copalyl pyrophosphate; GA, gibberellic acid. In the structures ⓅⓅ indicates pyrophosphate. A activity and B activity are explained in the text

The relationship between the action of chlormequat chloride (CCC) and that of gibberellin was reviewed by Lang (1970). Gibberellin competitively counteracted the effect of CCC on higher plants, and CCC inhibited production of gibberellin by the fungus *Gibberella* (the gibberellins were first discovered as fungal metabolites). There was a close parallel between the ability of a series of CCC analogues to inhibit gibberellin production by the fungus and to inhibit growth in higher plants, and CCC lowered the production of gibberellin-like substances in seeds.

However, work on partially purified enzyme preparations *in vitro* has shown that, whereas inhibition of kaurene synthesis represents a likely mode of action for some growth retardants, this is not so for CCC. The step between geranylgeranyl pyrophosphate and *ent*-kaurene (Fig. 5.1) is catalysed by two enzymes operating together and known (together as) 'kaurene synthetase'. The individual enzyme activities are known as A (geranylgeranyl pyrophosphate → copalyl pyrophosphate) and B (copalyl pyrophosphate → *ent*-kaurene) activities (Fig. 5.1 and Duncan and West, 1981). Frost and West (1977) studied a preparation from immature wild cucumber (*Marah macrocarpus*) endosperm but found a negligible effect of CCC on both A and B activities at concentrations up to 500 μM. Some inhibition did take place, however, at millimolar concentrations (West, 1973). Further, kaurene synthetase from a *Gibberella* strain in which synthesis of gibberellin was 93% inhibited by 10 μM CCC was only significantly inhibited by concentrations of the growth regulant of 1 mM or above (West and Fall, 1972). Additional evidence conflicting with the view that CCC inhibits gibberellin biosynthesis was obtained by Reid and Crozier (1972), who showed that the application of CCC to intact higher plants could actually increase the amount of gibberellin in the plant, and that application of the compound at high enough levels to cause growth inhibition did not necessarily reduce the gibberellin level. This has also been found in other systems (see Dicks, 1980).

In an attempt to rationalise these confusing reports on this and other growth regulators, Dicks (1980) highlights the suggestion that CCC may, perhaps in common with other growth regulators, influence gibberellin interconversions. A similar suggestion was made to account for part of the action of ancymidol by Coolbaugh and Hamilton (1976; see below). This would lead to increased levels of some gibberellins and decreased concentrations of others and Dicks (1980) cites evidence that this does in fact occur. Inhibitory effects of CCC on kaurene formation at higher concentrations are not ruled out. The physiological consequences of effects on gibberellin interconversions would depend on the exact changes which took place, and the precise role in plant development of the gibberellins concerned. As Dicks points out, this hypothesis is open to test.

It has also been reported that CCC can inhibit sterol synthesis in tobacco seedlings (Douglas and Paleg, 1974). We shall be considering sterol synthesis later in relation to the mode of action of various fungicides but it is sufficient to point out at the moment that the pathways of sterol and gibberellin synthesis are common in part, diverging at farnesyl pyrophosphate, (Fig. 5.1). Douglas and Paleg (1974) found a correlation between growth retardation by Amo 1618, chlorphonium chloride and CCC, and inhibition of incorporation of radioactivity from (DL-2-^{14}C)-mevalonic acid into sterol (particularly desmethylsterol) fractions, though concentrations of CCC greater than 1 mM were required for 50% inhibition. In the case of Amo 1618 and CCC growth retardation could be overcome not only by gibberellic acid, but also by added sterols. However, the pattern of build-up of intermediates following inhibition varied between compounds, and the follow-up work has been done only with Amo 1618 (Douglas and Paleg, 1978). It is, therefore, too early to assess the relationship of these observations to the work on inhibition of gibberellin synthesis.

To summarise briefly it seems unlikely that inhibition of kaurene synthesis is the only effect of CCC since inhibition *in vitro* requires relatively high concentrations of compound. Perturbations in the later stages of the gibberellin pathway may be involved.

(b) Chlorphonium chloride

The evidence that chlorphonium chloride (Table 5.3) works by inhibiting gibberellin synthesis was reviewed by Lang (1970) and MacMillan (1971). West and Fall (1972) showed that kaurene synthetase from *Gibberella* was half inhibited by 0.1–1 μM compound, whereas the figure for the *Marah macrocarpus* preparation was 0.5–5 μM. It was the A activity (Fig. 5.1) that was sensitive (Frost and West, 1977).

Other effects have been demonstrated, some at somewhat higher concentrations. One such effect is on sterol synthesis, as mentioned in discussion of CCC (Douglas and Paleg, 1974). A second is a complete inhibition of wound healing in potato tuber tissue by 60 μM chlorphonium chloride (which could not be overcome by gibberellin); this could have been caused by inhibition of protein synthesis (Borchert *et al.*, 1974). Another reported effect is uncoupling of photophosphorylation at 2–5 μM in a reconstituted spinach chloroplast system (Lendzian *et al.*, 1978). Although this effect was observed at low concentrations, the fact that chlorphonium chloride is used as a growth retardant suggests that it exerts its principal effect *in vivo* through inhibition of gibberellin synthesis.

(c) Ancymidol

Ancymidol (Table 5.3) was introduced in 1971 for reducing internode elongation in commercial greenhouse plants (Worthing, 1979). It caused dwarfing in dark-grown beans and this could be overcome by gibberellin (Shive and Sisler, 1976). In addition ancymidol inhibited gibberellin-stimulated growth of lettuce hypocotyls (Leopold, 1971) and *Avena* stem segments (Montague, 1975), and the effect on lettuce could also be reversed by gibberellic acid. Coolbaugh and Hamilton (1976) confirmed that this was also true for pea shoots and went on to show that ancymidol at 1 μM completely blocked the conversion of *ent*-kaurene to *ent*-kaurenol from both pea shoots and *Marah oreganus* (wild cucumber) endosperm. In *M. oreganus* endosperm extracts formation of *ent*-kaurene was also inhibited as the concentrations approached millimolar values.

It was subsequently shown that *ent*-kaurene, *ent*-kaurenol and *ent*-kaurenal oxidations (Fig. 5.1) were equally sensitive to ancymidol (I_{50} values *c.* 0.01 μM) and that the growth regulator was able to form a complex with the cytochrome P-450 (p. 331) which is part of the oxidase system. Ancymidol was not a general mixed function oxidase inhibitor however (Coolbaugh *et al.*, 1978).

A structure-activity study of compounds related to ancymidol and tri-arimol (a structurally-related compound used as a fungicide and probably acting by inhibiting sterol synthesis – p. 256) revealed that ability to inhibit the elongation of pea internodes was correlated with the ability to inhibit kaurene oxidation in *Marah macrocarpus* extracts (Coolbaugh *et al.*, 1982). This confirms once more that the growth retardant activity of ancymidol is very probably due to its effects on gibberellin biosynthesis.

An action between *ent*-kaurene and *ent*-kaurenoic acid is not entirely in accord with the observations that ancymidol can inhibit effects brought about by exogenous gibberellic acid. Coolbaugh and Hamilton (1976) suggest that gibberellin interconversions may be required for some effects and ancymidol might interfere with these.

Shive and Sisler (1976) investigated the possibility that ancymidol could act through inhibition of sterol formation; since growth retardation took place in tissues in which there was no net sterol synthesis they considered inhibition of gibberellin synthesis to be the most likely primary effect.

(d) Daminozide

Daminozide (Table 5.3) is used commercially to control vegetative growth of fruit trees, and to modify the shape and height of ornamental plants (Worthing, 1979).

It is included here because it affects plant growth in much the same way as chlormequat chloride. For instance, growth-inhibitory effects caused by daminozide can be overcome by gibberellic acid (Menhenett, 1979). However, it did not inhibit gibberellin production by the fungus *Gibberella* (Lang, 1970), nor did it have any inhibitory effect on kaurene synthetase isolated from *Gibberella* (West and Fall, 1972) or *Marah macrocarpus* (Frost and West, 1977).

(e) *Piproctanyl bromide and mepiquat chloride*

Although there do not seem to have been any biochemical studies on the effects of piproctanyl bromide (Table 5.3), some of its effects as a growth retardant can be overcome by gibberellins (Knypl, 1977; Menhenett, 1978). Mepiquat chloride is structurally related to piproctanyl bromide and could exert its growth retardant effect in the same way. Menhenett makes the pertinent observation that the extent to which the effects of a number of growth retardants can be overcome by a gibberellin depends on which gibberellin is used. Although this may not seem surprising it does mean that the failure of a particular gibberellin to overcome growth retardation cannot prove that the growth retardant does not work via inhibition of gibberellin biosynthesis.

D. Ethylene: the natural hormone and synthetic precursors

Ethylene ($H_2C=CH_2$) is recognised as a plant hormone involved in a wide range of aspects of plant growth, development and senescence. Like all the hormones its action is tied up with the levels of the other hormones, but ethylene can influence seed germination, seedling, root and leaf growth, and various stress situations; it can also be the limiting factor for ripening, aging and senescence (reviewed by Lieberman, 1979).

The effect of ethylene is reasonably specific since closely related chemicals such as propylene are much less effective (Pratt and Goeschl, 1969). It is envisaged that the first stage in ethylene action is binding to a specific receptor (Lieberman, 1979).

Ethylene is, of course, a gas, and so it is applied for agricultural use as a precursor, two of which are known:

$$\underset{\text{ethephon}}{ClCH_2CH_2\overset{\displaystyle O}{\overset{\displaystyle \|}{P}}(OH)_2} \qquad \underset{\text{etacelasil}}{ClCH_2CH_2Si(OCH_2CH_2OMe)_3}$$

Ethephon is used, for instance, to stimulate pre-harvest and end-of-season ripening of fruit, to shorten stems of forced daffodils and to stimulate branching and basal bud initiation for geraniums and roses (Worthing, 1979). Etacelasil is an abscission agent for olives (Worthing, 1979).

The rate of release of ethylene from ethephon was measured by Warner and Leopold (1969), who found that a 200 μM solution containing pea seedling stems released approximately one third of the available ethylene in 24 hours. It seems that the reaction, which is accelerated by high pH, is not catalysed by enzymes inside the plant, since Dennis *et al.* (1970) showed that autoclaved or boiled homogenates of green tomato fruits released ethylene from ethephon at the same rate as untreated homogenate. However, the plant may contain constituents that stimulate the reaction non-enzymatically, since the homogenates released ethylene at a greater rate than a buffer solution of similar pH.

Etacelasil is a member of a class of silicon compounds long known to be capable of breakdown to give ethylene (Sommer *et al.*, 1948).

E. Other hormones

Cell division is stimulated by the cytokinin class of hormone whilst, on the other hand, abscisic acid is a hormone with an inhibitory type of action so that, for instance, it promotes dormancy in buds and seeds and abscission in cotton. There is no evidence that any pesticide has its principal effect through interference with the action of these hormones and we will therefore not consider them further.

II. COMPOUNDS INTERFERING WITH CELL DIVISION
A. Biochemical background
1. *General description of cell division*

Development in a higher organism consists of a continuous process of cell division and differentiation by which new cells able to carry out specialised functions are produced. Before a cell can divide, it must copy its genetic information, which is maintained in the form of deoxyribonucleic acid (DNA) in the chromosomes of the cell. The information in DNA is encoded in a linear sequence of two purine (adenine and guanine) and two pyrimidine (cytidine and thymidine) nucleotide bases arranged in two strands that form a double helix. In this helix the strands interact by 'base-pairing', that is they form hydrogen bonds between the strands so that adenine always pairs with thymidine, and cytidine always with guanine. When a copy of the nuclear chromosomal DNA is to be made, the strands separate, and a new complementary strand is made for each, according to the base pairing specificity.

When replication is complete (and by this time the cell will have increased in mass by extra synthesis of all the cellular components) the chromosomes, now in pairs, line up in a plane across the centre of the cell ('metaphase'). They then separate to opposite poles of the cell along a mitotic spindle consisting of microtubules which are assembled from sub-units in the cytoplasm. Cell division is completed by the synthesis of wall and membrane material dividing the two daughter cells.

The diagram given in Fig. 5.2 is intended to aid interpretation of the following accounts of the disruptive actions of pesticides.

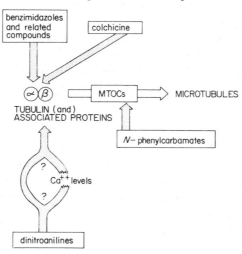

Fig. 5.2 Diagrammatic summary of the influence of pesticides on microtubule formation. Note that, although colchicine does act on tubulin, it is not thought to bind to the same site as any pesticide. MTOCs, microtubule organising centres

2. Role of microtubules in cell division

The subject of microtubules has been reviewed by Gunning and Hardham (1979, 1982) and Sabnis and Hart (1982), particularly with reference to plants, and books devoted to the subject have been published (Borgers and de Brabander, 1975; Dustin, 1978; Roberts and Hyams, 1980). The following account is based on these publications.

Microtubules are proteinaceous organelles, present in nearly all the cells of higher organisms, and made of subunits, termed tubulin, assembled into elongated tubular structures. The lengths of the tubules are variable, and changes are effected by assembly or dis-assembly of the subunits. Each tubulin subunit is itself a dimer of α and β monomers which have a high affinity for one another. Various other proteins co-purify with tubulin and are probably essential for correct structure and function. These may contribute to variations between microtubules from different sources or with different cellular roles. Microtubules form the mitotic spindle as mentioned above, and the effect of chemicals on their function in chromosome separation has received much attention. However, microtubules also have other roles, particularly in controlling deposition of wall material during cell division and in directing secondary thickening of plant cells. They also comprise the major structural component of cilia and flagella of motile organisms.

The formation of microtubules from tubulin within both animal and plant cells appears to initiate from 'nucleating regions', or microtubule organising centres (MTOCs). The appearance of these MTOCs varies greatly, but their correct function is required for proper assembly of microtubules and some compounds in this section may impair this function.

B. Fungicidal benzimidazole derivatives and related compounds
1. Biological activity

Benomyl and carbendazim are the best known of the fungicidal benzimidazole derivatives and their relatives, which are shown in Table 5.4. Some compounds of similar structure are used as anthelmintics and others have anti-tumour properties (see Davidse and Flach, 1977). Certain compounds have more than one use; thiabendazole for instance, is currently used both as a fungicide and as an anthelmintic (Worthing, 1979).

2. Chemical breakdown of benomyl and the thiophanates

Benomyl is easily converted in aqueous solution to the fungitoxic carbendazim (Clemons and Sisler, 1969), which is found in plants treated with benomyl (Sims et al., 1969), and which is used as a fungicide in its own right (Hampel and Löcher, 1973). Thiophanate-methyl in solution undergoes degradative cyclisation to carbendazim, while thiophanate gives the corresponding ethyl ester (Selling et al., 1970; Vonk and Kaars Sijpesteijn, 1971). Thiophanate-methyl cyclisation is also known to occur in vivo (Buchenaur et al., 1973; Noguchi et al., 1971).

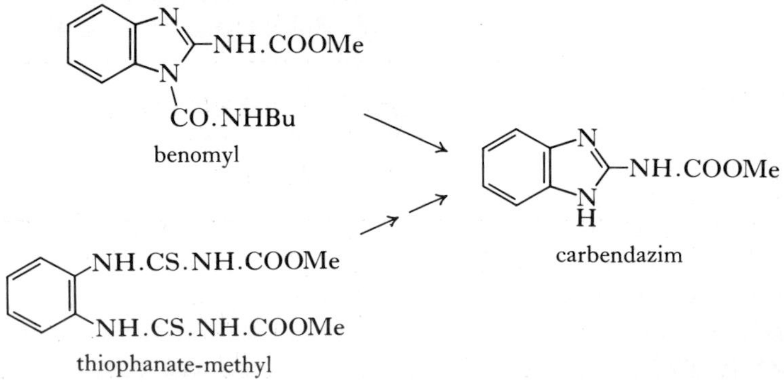

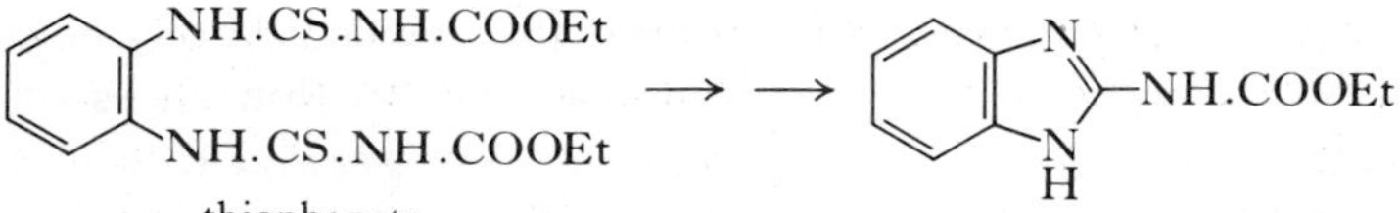

NH.CS.NH.COOEt

NH.CS.NH.COOEt

thiophanate

$\longrightarrow \longrightarrow$

—NH.COOEt

Table 5.4   Benzimidazole derivatives and related fungicides which inhibit microtubule subunit polymerisation

Name	Structure
Benomyl	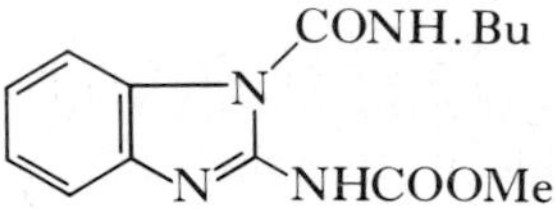 CONH.Bu N N⁀NHCOOMe
Carbendazim	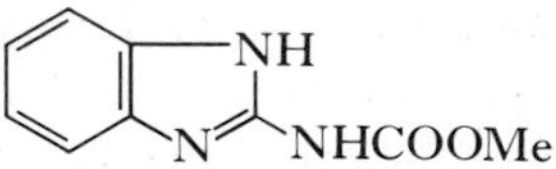 —NH N⁀NHCOOMe
Fuberidazole	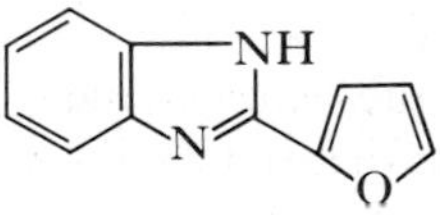 —NH N⁀(furan) O
Thiabendazole	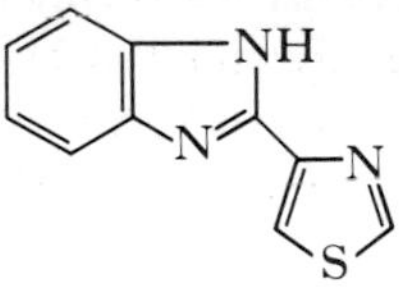 —NH N⁀(thiazole) S
Thiophanate	NH.CS.NH.COOEt NH.CS.NH.COOEt
Thiophanate-methyl	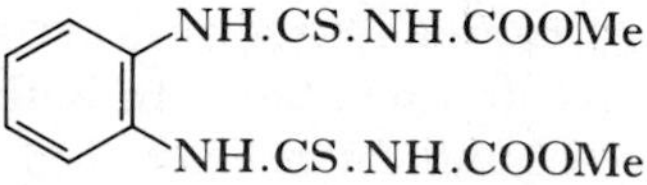 NH.CS.NH.COOMe NH.CS.NH.COOMe

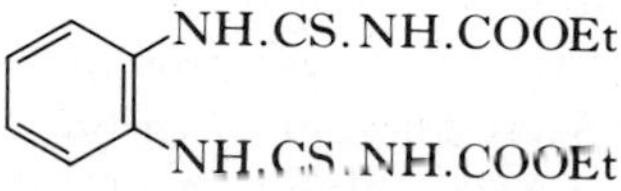

Recently Vonk *et al.* (1977) found that the conversion of thiophanate-methyl to carbendazim was greatly accelerated by thin slices of potato tuber, homogenates of apple or potato, and sap pressed from cucumber seedlings. They obtained evidence for the suggestion that the conversion was mediated by unstable quinones generated from diphenol substrates by polyphenol oxidase.

Since there is a correlation between the formation of the benzimidazole carbamate esters from thiophanate fungicides and their fungitoxicity, it seems clear that the thiophanates themselves are not toxic to fungi, but act by forming the esters (Vonk and Kaars Sijpesteijn, 1971). Similarly it is generally accepted that benomyl is only active after conversion to carbendazim (Davidse, 1973).

Benomyl and the thiophanates are protective and systemic fungicides, and there is evidence that their effectiveness depends on the retention, in or on the roots, of benomyl itself. In the case of the thiophanates, the retention of a conversion product intermediate between the thiophanate and carbendazim (or its ethyl analogue) is postulated. The precursors may act as reservoirs of the active ingredients which are gradually released to the remainder of the plant (Fuchs *et al.*, 1972). Klopping (see Edgington, 1972), has shown that 50% of the residue remaining on weathered leaf surfaces three weeks after an application of benomyl was not carbendazim, but benomyl itself, so that a similar mechanism of release might also operate on the foliage. Certainly sunlight does cause photochemical conversion of thiophanate-methyl to carbendazim (as well as degrading the ethyl analogue) on leaf surfaces and this probably contributes to the protective activity of the compound (Buchenauer *et al.*, 1973).

Eckert and Rahm (1979) have conducted a structure/activity study of carbendazim analogues. Substantial or complete loss of fungicidal activity occurred with almost any structural change in the molecule.

3. *Biochemical mode of action*

The most important early evidence relating to the mode of action of compounds of this class was provided by Clemons and Sisler (1971) who found that a 5 μM solution of carbendazim applied to *Neurospora crassa* conidia caused 85% inhibition of DNA synthesis (as indicated by the incorporation of radioactivity from labelled uridine into DNA) in 8 hours. With *Ustilago maydis* spores a 42 μM solution caused 50% inhibition of DNA synthesis after one hour. In both cases RNA and protein synthesis were not affected until later. From these observations it could be concluded that carbendazim exerted its fungitoxic effect through interference with

DNA synthesis or some closely related aspect of cell or nuclear division. Since this time, elegant and unequivocal studies principally by Davidse and collaborators have indicated that carbendazim kills fungi because it prevents assembly of tubulin into microtubules (Davidse, 1973, 1975, 1979; Davidse and Flach, 1977, 1978).

Early progress came with the recognition that carbendazim did not prevent DNA synthesis during the first cell division cycle of *Saccharomyces cerevisiae* cultures whose cell divisions had been synchronised; however, completion of cell division was inhibited (Hammerschlag and Sisler, 1973). Similarly, in asynchronous cultures of *Aspergillus nidulans* DNA synthesis became progressively inhibited after exposure to carbendazim, and complete inhibition was not achieved until the DNA content per nucleus had nearly doubled (Davidse, 1973). This result is also best explained by mitotic failure, since the cells did not become sensitive until the onset of cell division, before which the doubling of the DNA content is complete. Both sets of authors likened the effects to those of colchicine, which inhibits higher plant cell division by binding to microtubular protein and thereby preventing assembly of the subunits into spindle fibres.

Experiments were then designed to investigate whether this was the basis of the fungicidal action of carbendazim (Davidse, 1975). Cell-free extracts of 3 strains of *Aspergillus nidulans* were made; one strain was wild-type, a second was hypersensitive to carbendazim, and a third was resistant to it. Each extract was incubated with ^{14}C-carbendazim and, after 2 hours, subjected to gel filtration to determine the amount of radioactivity bound to macromolecules; this was in the order hyper-sensitive strain > wild-type > resistant strain. The binding activity was of the same order of magnitude as that obtained with cell-free extracts prepared from mycelium that had been incubated with carbendazim *in vivo*. Further characterisation with a different chromatographic separation technique indicated that the carbendazim was associated with a protein of molecular mass of approximately 100,000 daltons, having properties similar to microtubular protein from other sources. Protein with the same chromatographic properties from the resistant strain bound no carbendazim (Davidse, 1975).

Further characterisation of the protein has allowed the conclusion that it has properties unique to tubulin (Davidse and Flach, 1977). Carbendazim bound rapidly to the protein and the binding was competitively inhibited by a structurally related compound and colchicine. In this study Davidse and Flach (1977) confirmed that the resistant strain was resistant because its tubulin had a lower affinity for carbendazim; the dissociation constants were, for the hypersensitive strain 0.6 μM, for the wild-type 2 μM, and for the resistant strain 27 μM. Previous work (Davidse, 1976) had ruled out uptake and metabolism differences as causes of the resistance.

The work has been taken a stage further. A *Penicillium* strain which is hypersensitive to carbendazim, is also *less* sensitive to thiabendazole (negatively-correlated cross resistance). Davidse and Flach (1978) have shown that its tubulin has a higher affinity for carbendazim and a lower affinity for thiabendazole, and that both compounds bind to the same site. This must be due to change(s) in the part of the binding site which interacts with the side-chain by which the compounds differ (Table 5.4). The change(s) cannot affect the normal polymerisation of the subunits to form microtubules in the absence of inhibitor.

Sheir-Neiss *et al.* (1978) carried out a genetic analysis of 28 benomyl-resistant mutants, and found that 18 of these had alterations in the genes coding for the β-subunit of tubulin, whereas none had altered α-tubulin. This strongly suggests that the binding site for carbendazim is on the β-tubulin subunit.

Parallel studies have been carried out with compounds which have a similar structure to the fungicides considered in this section, but different uses. The conclusions serve to reinforce the views summarised above (see references quoted by Davidse (1979), Davidse and Flach (1977) and Laclette *et al.*, 1980).

The conclusion is inescapable that the compounds bind to tubulin and thereby interfere with microtubular subunit assembly and this undoubtedly represents the primary mode of action of carbendazim and its relatives. However, Kumari *et al.* (1977), report that in germinating conidia of *Fusarium oxysporum*, benomyl at 3 and 10 μM decreased DNA synthesis before the conidia entered mitosis. Thus they argue that there may also be a direct effect on DNA synthesis to be distinguished from the secondary reduction in DNA synthesis which follows interference with the mitotic spindle. A further additional effect may be operating in the case of benomyl, which can inhibit fungal growth at lower concentrations than does carbendazim, probably because the non-specifically fungitoxic compound butyl isocyanate is also formed when benomyl breaks down to give carbendazim (Hammerschlag and Sisler, 1973).

Since the compounds discussed in this section are systemic fungicides, we need to explain why they are not toxic to plants, whose cells divide in a similar way and which are affected by colchicine and other cell division inhibitors to be considered below. The explanation probably lies in the fact that little benzimidazole fungicide gains access to the dividing plant cell because the compounds are mainly translocated apoplastically (i.e. outside the protoplast, the living part of the cell contained by plasma-membrane) (Ben-Aziz and Ahronson, 1974 and references therein; Solel *et al.*, 1973). Plant cells can exhibit mitotic abnormalities following exposure to benomyl or carbendazim (Richmond and Phillips, 1975).

C. Herbicidal carbamates
1. *Introduction*

The carbamates listed in Table 5.5A have the general formula:

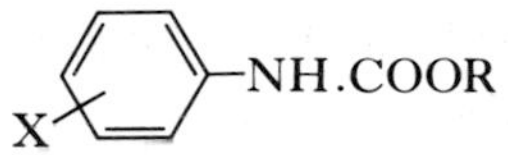

$$X\text{—}\boxed{}\text{—NH.COOR}$$

i.e. they are esters of N-phenyl carbamic acids. (Note that carbamate insecticides (Chapter 3) are esters of N-methyl or N,N-dimethyl carbamic acid.) Asulam (Table 5.5B) contains a substituted benzenesulphonyl rather than a phenyl group.

Table 5.5   Herbicides interfering with cell division

Name	Structure
A. N-Phenylcarbamates Barban	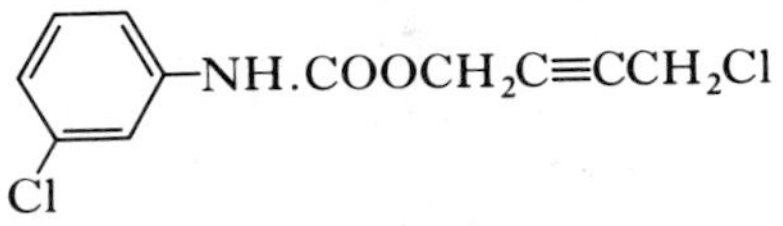 —NH.COOCH$_2$C≡CCH$_2$Cl Cl
Carbetamide	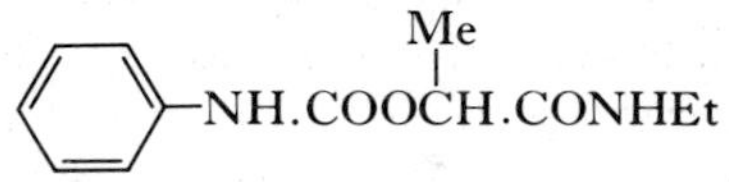 Me —NH.COOCH.CONHEt
Chlorbufam	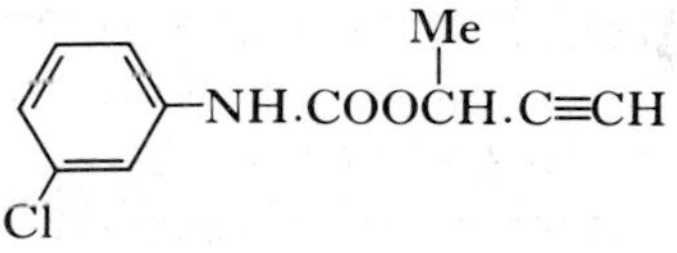 Me —NH.COOCH.C≡CH Cl
Chlorpropham	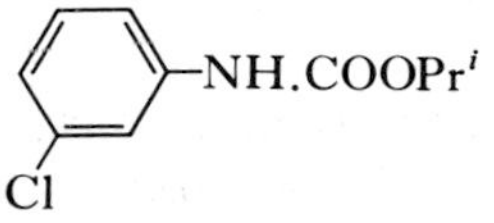 —NH.COOPr^i Cl
Propham	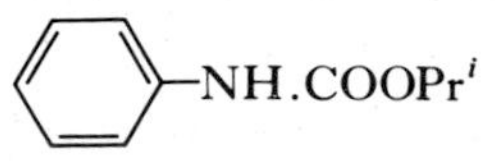 —NH.COOPr^i
B. Other carbamates Asulam	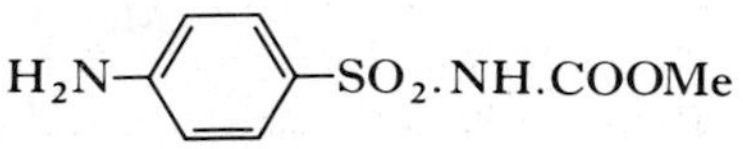 H$_2$N—⟨ ⟩—SO$_2$.NH.COOMe

Name	Structure
C. Dinitroanilines Benfluralin	F_3C—(ring with NO_2 ortho, NO_2 ortho)—N(Et)(Bu)
Butralin	^tBu—(ring with NO_2, NO_2)—NH.CH(Me)(Et)
Dinitramine	F_3C—(ring with H_2N, NO_2, NO_2)—NEt_2
Ethalfluralin	F_3C—(ring with NO_2, NO_2)—N(Et)(CH_2—C(Me)=CH_2)
Fluchloralin	F_3C—(ring with NO_2, NO_2)—N(Pr)(CH_2CH_2Cl)
Isopropalin	^iPr—(ring with NO_2, NO_2)—NPr_2
Nitralin	Me—S(=O)(=O)—(ring with NO_2, NO_2)—NPr_2

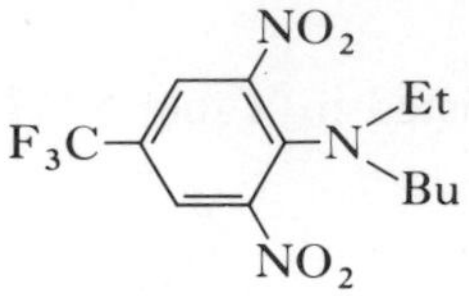

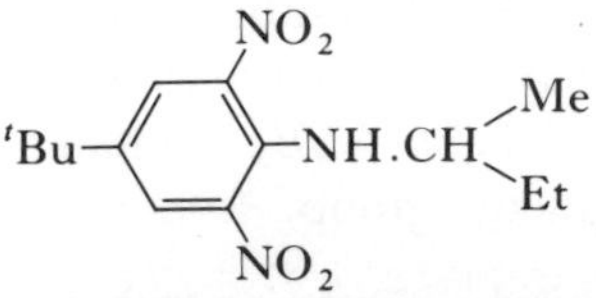

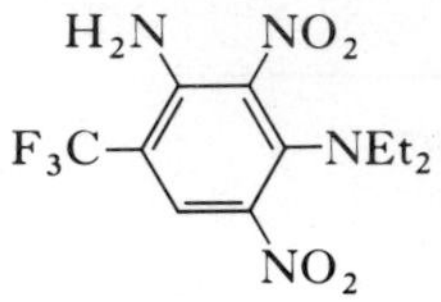

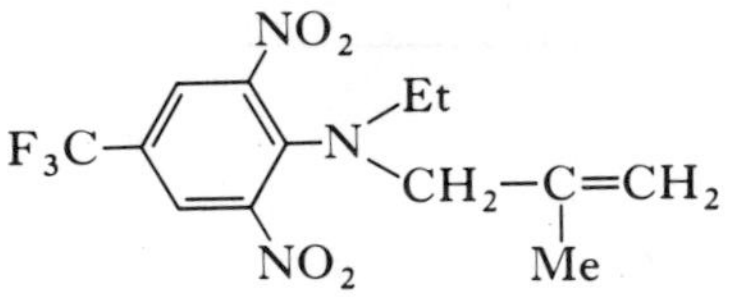

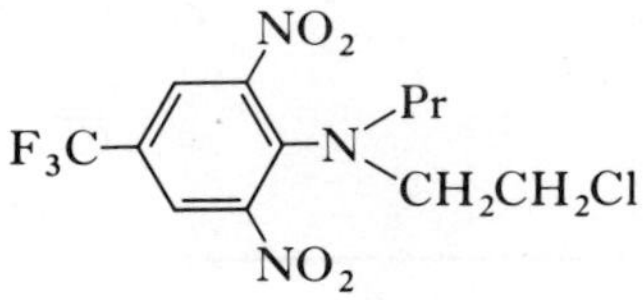

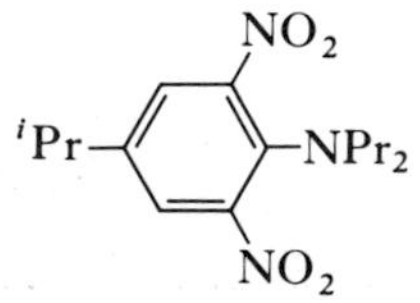

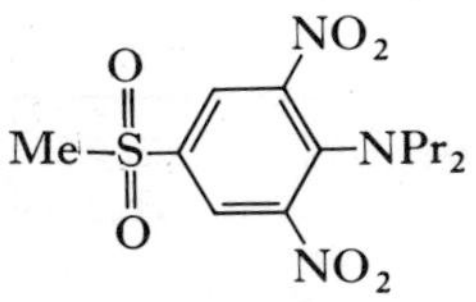

Name	Structure

Oryzalin

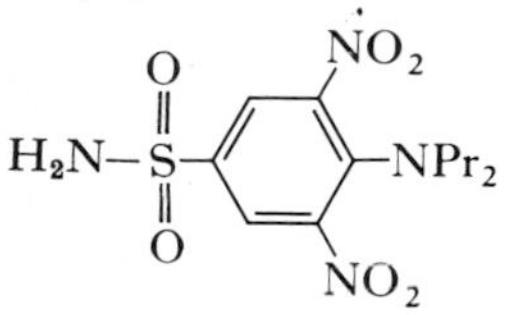

$$H_2N-\overset{\overset{O}{\|}}{\underset{\underset{O}{\|}}{S}}-\text{benzene ring}\begin{cases}NO_2\\-NPr_2\\NO_2\end{cases}$$

Pendimethalin

Me—(ring with NO_2, —NH.CHEt$_2$, Me, NO_2)

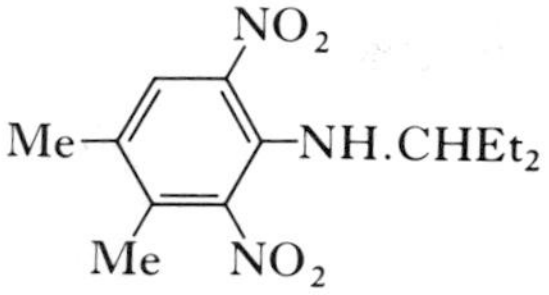

Profluralin

F_3C—(ring with NO_2, N<CH$_2$—△ / Pr, NO_2)

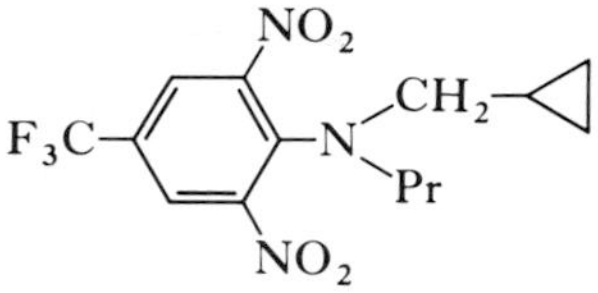

Trifluralin

F_3C—(ring with NO_2, —NPr$_2$, NO_2)

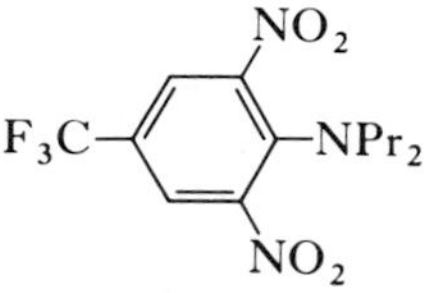

D. Phosphoric amides
Amiprophos-methyl

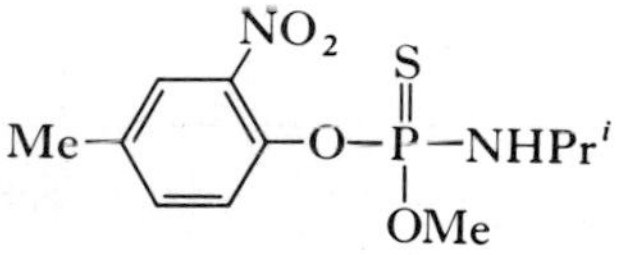

Me—(ring with NO_2)—$O-\overset{\overset{S}{\|}}{\underset{\underset{OMe}{|}}{P}}-NHPr^i$

Butamifos

(ring with NO_2, Me)—$O-\overset{\overset{S}{\|}}{\underset{\underset{OEt}{|}}{P}}-NH.CH\begin{cases}Me\\Et\end{cases}$

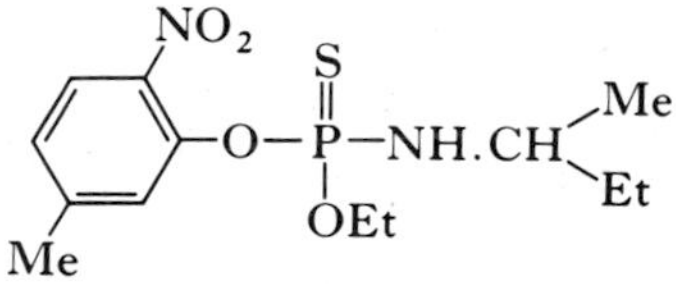

E. Sulphonylureas
Chlorsulfuron

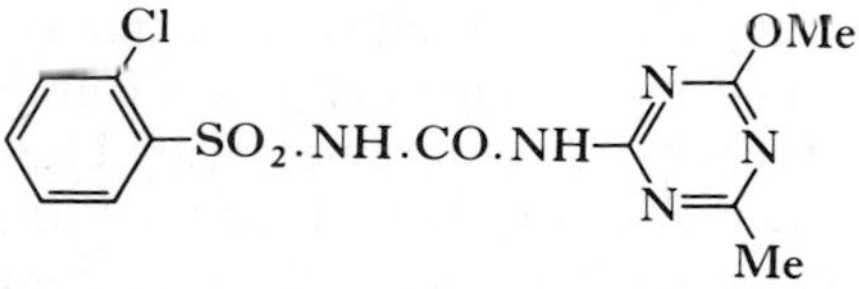

(ring with Cl)—$SO_2.NH.CO.NH$—(triazine ring with OMe, Me)

Name	Structure
Sulfometuron-methyl (proposed name)	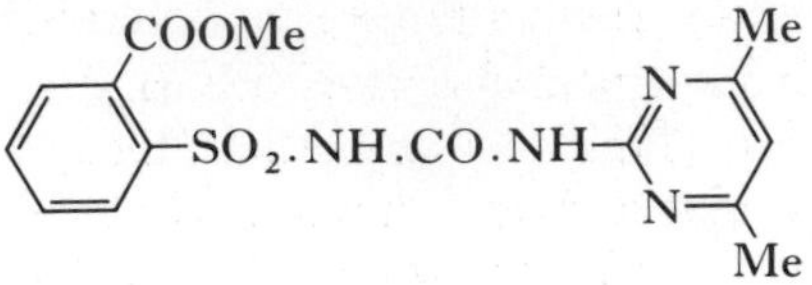
F. Miscellaneous compounds Propyzamide	
Chlorthal-dimethyl	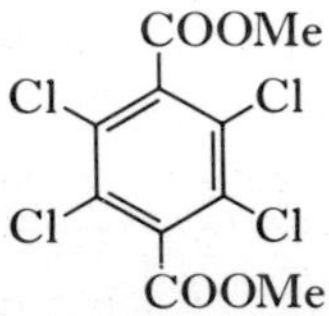
Maleic hydrazide	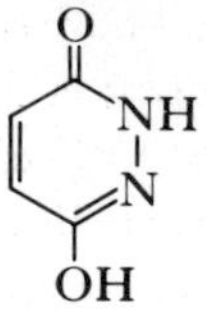

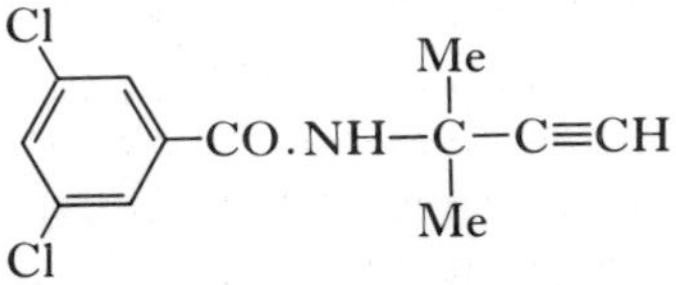

The discovery of the herbicidal effect of carbamates, particularly pro-pham, was announced by Templeman and Sexton (1945), who also pointed out that they affected cell division. This was confirmed by later workers and the compounds are referred to as 'mitotic poisons'. They are normally used as herbicides, not growth regulators, though the regulating properties of chlorpropham are noted in the WSA Handbook (1979).

2. *Effects on cell division*

The inhibiting effect of N-phenylcarbamates on mitosis, originally noted by Templeman and Sexton (1945), was confirmed by other workers, some of whose observations were reviewed by Canvin and Friesen (1959). The latter authors themselves showed the total inhibition of cell division in the roots of barley seeds germinated in an approximately 6 µM solution of propham. Root and cell structure appeared normal, but no dividing cells

·could be seen. At higher concentrations propham produced cells with an abnormally large number of chromosomes, and also multinucleate cells. The results suggested a failure in the division mechanism.

The morphological effects of barban on wild oats have been described by Dubrovin (1959) and by Abel (1962). The shoot apex of wild oats becomes swollen and development is arrested, with death usually occurring some five weeks after treatment. A cytological examination under the light microscope showed that treated apices contained swollen cells with numerous groups of chromosomes, suggesting that the herbicide was interfering with cell division (Dubrovin, 1959).

A clue to the mechanism of the inhibitory effect of N-phenylcarbamates on cell division was suggested by Hepler and Jackson (1969). As a result of a careful electron microscopic study on dividing endosperm cells of the African blood lily *Haemanthus katherinae* they observed that a 55 μM solution of propham caused disorientation of microtubules in 0.5 to 2 hours. They suggested that propham might interfere with what are now called the MTOCs.

Propham has been shown to prevent assembly of microtubules *in vivo* in the unicellular organisms *Ochromonas* (Brown and Bouck, 1974) and *Chlamydomonas* (Flavin and Slaughter, 1974); in both cases the effects were distinguishable from those of colchicine. Also, Coss and Pickett-Heaps (1974) studied the effects of propham on dividing cells of the green alga *Oedogonium cardiacum* and supported the suggestion that propham acts on the MTOCs but did not rule out effects on the polymerisation of tubulin as well. Brower and Hepler (1976) observed with the electron microscope the effect of propham on microtubules involved in secondary wall deposition in xylem of higher plants, and concluded that the effect was on the MTOCs. Other examples of effects on algae are cited by Gunning and Hardham (1982).

This view that the N-phenylcarbamates, typified by propham, do not bind to tubulin could be examined directly. *In vitro* ^{14}C-labelled chlor-propham did not bind to pig brain tubulin, nor did 1.2 mM propham affect the polymerisation of the tubulin (Bartels and Hilton, 1973). In confirmation, Coss *et al.* (1975) found that while ^{14}C-colchicine bound to chick brain microtubular protein, ^{14}C-propham did not. Further 100 μM–1 mM propham did not prevent tubulin polymerisation *in vitro*.

This corroborative evidence, plus the early observations, lends weight to the view that interference with the MTOCs, leading to disorganisation of microtubules, is the most likely primary effect of the N-phenyl carbamates. However, other possibilities have been raised. Suggestions that these herbicides may delay the transcription of new messenger ribonucleic acid (see p. 235) (Mann *et al.*, 1967; Keitt, 1967) do not seem to have been sustained. Rusness and Still (1974) speculated that inhibition of amino acid and fatty acid activation could be involved in the herbicidal action but there is no direct evidence to establish these processes as contenders for the primary target.

3. *Asulam*

Although asulam (Table 5.5*B*) is structurally distinct from the *N*-phenyl *N*-phenyl carbamates, it has been shown that 2.4 μM asulam caused onion root-tip cells to become arrested at metaphase, and the effects were likened to those caused by colchicine and propham (Sterrett and Fretz, 1975).

Asulam bears a structural resemblance to 4-aminobenzoic acid, a precursor of the cofactor tetrahydrofolic acid which is involved in one-carbon transfer reactions, particularly the methylations involved in the synthesis of nucleic acid bases. It has recently been shown that micromolar concentrations of asulam could interfere with the production of folic acid cofactors (Killmer *et al.*, 1980; Veerasekaran *et al.*, 1981), that 4-aminobenzoic acid or folic acid could reduce or in some cases overcome asulam-induced stunting of roots of various plant species (Stephen *et al.*, 1980) and that, in asulam-treated peas there was a build-up of precursors of folate-utilising reactions in purine biosynthesis (Kidd *et al.*, 1982). Certainly an effective folic acid antagonist would be expected to inhibit cell division through its effect on nucleic acid synthesis, but there does not seem to be any reason why cells deprived of tetrahydrofolic acid should be arrested at metaphase (see above).

D. Herbicidal dinitroanilines
1. *Introduction*

The first dinitroaniline herbicide, trifluralin, was introduced in 1960 (Worthing, 1979). Structures are given in Table 5.5*C*. As a class, the dinitroanilines are more toxic to monocotyledonous than dicotyledonous plants (Swanson, 1972) and they are applied, pre-emergence, for the control of annual grasses.

The most typical symptom of plants treated with herbicides of this class is the inhibition of lateral root formation and swelling of the root tips. Cells in the area of disruption are multinucleate, indicating an effect on cell division. In the stunted parts of the plant above ground, swelling and brittleness of the stem or hypocotyl is also seen (Parka and Soper, 1977; Ashton and Crafts, 1981).

Effects on cell division are most likely to be the primary changes brought about by the dinitroanilines though other possibilities have been considered. Results obtained before 1977 have been succinctly reviewed by Parka and Soper (1977).

2. Effects on cell division

Swanson (1972) reported that about 1 nM trifluralin caused mitotic aberrations in hair cells of the houseplant *Tradescantia* and in leaf cells of *Vicia faba* and concluded that a major effect of the dinitroaniline herbicides was disruption of nuclear and cell division.

Numerous observations that dinitroaniline treatment results in multinucleate cells and accumulation of other cells at metaphase are catalogued by Parka and Soper (1977). Disruptive effects on microtubules were seen. Trifluralin, for instance, was reported to bring about this effect in various tissues (Bartels and Hilton, 1973; Hess and Bayer, 1974; Jackson and Stetler, 1973; Lignowski and Scott, 1971, 1972) and this was followed by studies on isolated tubulin.

3. What causes microtubule disruption?

Bartels and Hilton (1973) found that dinitroanilines (trifluralin, oryzalin and pronamide) brought about loss of microtubules from root cells of wheat and maize, but did not prevent *in vitro* polymerisation of isolated pig brain tubulin. Since this work the question has been re-examined with the conclusion that trifluralin could prevent polymerisation of purified pig brain tubulin (Robinson and Herzog, 1977). Hess and Bayer (1977) used microtubular protein purified from the flagella of the alga *Chlamydomonas* to show that low concentrations (3 μM) of trifluralin could indeed bind to the microtubular protein isolated from the flagella but, as controls, not to membrane or matrix proteins. As a further contrast, however, Upadhyaya and Noodén (1980) could detect no binding of ^{14}C-oryzalin to any protein with a molecular weight similar to tubulin in corn root extracts.

Together these experiments are clearly inconclusive. The discrepancies may perhaps be explained by differences between the microtubular proteins from different sources, together with the difficulties associated with making reproducible solutions of these rather insoluble compounds (Hess, 1979; Strachan and Hess, 1982). Progress in methods for purifying plant tubulin offers hope for some clarification (Mizuno et al., 1981; Rikin et al., 1982).

Hess (1979) went on to investigate whether microtubule assembly was affected. This was done using a technique to deflagellate *Chlamydomonas*, followed by observations of the effect of trifluralin on the kinetics and extent of flagellar regeneration. Hess first established that trifluralin was an active cell division inhibitor in the alga (90–100% inhibition at 1 μM and above). The swelling of the cells seen in treated plants was also observed. The onset of flagellar regeneration was completely prevented by 5 μM compound and significantly inhibited by lower concentrations. When added to a system which was actively regenerating, trifluralin stopped the process and the flagella began to shorten slowly (see also Quader and Filner, 1980). The inhibiting influence could be removed by washing so that regeneration resumed. In order to demonstrate that this effect was relevant to the mechanism of herbicidal activity Hess (1979) showed that three close, but herbicidally inactive, analogues of trifluralin did not affect flagellar regeneration at 10 μM.

Much of the work described above would support a direct effect of dinitroaniline herbicides on plant cell tubulin, preventing its assembly into microtubules and also perhaps promoting its disassembly. However, an alternative suggestion is that the compounds interfere with the ability of the mitochondria to play their part in maintaining cell calcium ions at a relatively high level in the mitochondria and a relatively low level in the cytoplasm (Hertel et al., 1980; Quader and Filner, 1980). This would lead to higher levels of calcium ions in the cytoplasm, and this is known to cause microtubule depolymerisation (see Hertel et al., 1980). The idea is supported by effects of the herbicide on the ability of plant mitochondria to take up calcium ions in vitro; this was half-inhibited by 0.6 μM trifluralin (Hertel et al., 1980). It is postulated (Hertel et al., 1980, 1981) that stimulation of calcium ion release may also be involved.

To conclude, it is very likely that the dinitroanilines are herbicidal because they disrupt microtubules and therefore cell division, but the molecular mechanism of action has not yet been established with certainty.

4. Other effects of dinitroanilines

It has been shown by Moreland's group and some earlier workers (Moreland, et al., 1972a,b: Moreland and Huber, 1979 and references therein) that dinitroanilines interfere with photosynthesis and respiration, both in excised tissues and in separated mitochondria and chloroplasts (see also Parka and Soper, 1977). Some effects are seen at low micromolar concentrations, but in most studies levels of the order of 100 μM have been employed and so the relevance to the primary effect is in doubt. It is difficult also to discern much of a pattern in the effects observed, which include (sometimes for the same and sometimes for different compounds) inhibition of photosynthetic electron transport and phosphorylation, and inhibition and uncoupling of oxidative phosphorylation. Clearly the effects on chloroplasts and mitochondria could be brought about by interactions with their membranes. This topic has been reviewed by Parka and Soper

(1977) and is also considered in the papers of Robinson *et al.* (1977) and Moreland and Huber (1979).

It has been suggested that various members of the dinitroanilines exert their herbicidal action by interfering with the cell membrane. Travis and Woods (1977) found that dinitramine interfered with three plasma membrane associated activities: K^+-ATPase, glucan synthetase, and permeation of ^{86}Rb ions. Due to its low water solubility, dinitramine was sprayed from ethanolic solution on to the soybean roots used for study. These solutions were usually approximately 600 μM, but the concentration reaching the plasmalemma is unknown. The most sensitive function was the K^+-ATPase which was maximally inhibited (60%) when the strength of the solution sprayed was 150 μM. The authors suggest the possibility of a primary effect on the cell membrane, but, as they point out, the idea needs further investigation. Unfortunately the authors do not attempt to reconcile their ideas with the evidence implicating inhibition of microtubule polymerisation as the mode of action. This was attempted by Upadhyaya (1978) but much of the work which had to be accommodated at the time has been superseded by the work on microtubules already described.

E. Phosphoric amide herbicides

Amiprophos-methyl is a phosphoric amide herbicide (Kiermayer and Fedtke, 1977) of closely analogous structure to butamifos, a pre-emergence herbicide used against annual weeds, particularly grasses (Worthing, 1979). Structures of the compounds are given in Table 5.5D.

Sumida and Ueda (1974) studied the effects of butamifos on *Chlorella* in a broad, biochemical screening operation designed to focus attention on the primary effect of a new herbicide. They concluded that the observed effects could be accommodated by a primary effect on cell division and went on to examine this microscopically in onion root-tips (Sumida and Ueda, 1976). They found that, after treatment with about 30 μM butamifos for 3.5 hours, arrested mitoses were seen. Further, after centrifugation, the chromosomes of cells arrested at metaphase became displaced to one side of the cell, whereas those of control metaphases did not. Since arrested metaphases induced by colchicine had also become displaced, the authors concluded that disruption of the mitotic spindle was the primary effect. This work has recently been reviewed (Sumida and Yoshida, 1982).

Kiermayer and Fedtke (1977) conducted a microscopic examination of the effects of amiprophos-methyl on the alga *Micrasterias denticulata* and stated that it is a very active anti-microtubule agent, as judged by the effects caused by 35 nM – 35 μM compound.

Collis and Weeks (1978) have gone further in their studies of the effects of amiprophos-methyl on flagellar regeneration of *Chlamydomonas*. These experiments are similar in conception to those of Hess (1979) on dinitro-aniline herbicides, described earlier. They found that 3–10 μM compound prevented the induction of tubulin synthesis accompanying flagellar regeneration. They were able to study this by the technique of labelling newly-synthesised proteins by supplying radioactive sulphate, and, after extraction, separating them by polyacrylamide gel electrophoresis. Auto-radiographs of the gels were made, and the intensity of the bands at the known running positions of the tubulin subunits indicated the extent of synthesis during the experiment. The inhibitory effects of amiprophos-methyl were specific for tubulin, and reversible. Further investigations suggested that the messenger ribonucleic acid molecule (p. 235) coding for tubulin was not being produced.

As in the case of the dinitroaniline herbicides mentioned above, evidence has been obtained that amiprophos-methyl and butamifos may act through disrupting control of cellular calcium ion concentrations. Low concentrations (2.5 μM) of amiprophos-methyl prevented flagellar regeneration in *Chlamydomonas* and this might perhaps be due to effects on calcium ion levels as explained above for the dinitroanilines (Quader and Filner, 1980). In this connection, the ability of plant mitochondria to accumulate calcium ions was completely prevented by 100 μM amiprophos-methyl; at least part of this effect was due to enhanced calcium ion efflux (Hertel *et al.*, 1980).

F. Sulphonylurea herbicides

Chlorsulfuron (Table 5.5*E*) is a new herbicide capable of controlling weeds in cereals at the very low rates of 10–40 g/ha (Palm *et al.*, 1980). Ray (1982) studied the mode of action and concluded that cell division was much the most sensitive process, being reduced by 87% in root-tips of *Vicia faba* treated with 2.8 μM chlorsulfuron. The cells did not accumulate at any particular stage of cell division, which suggests that the compounds act on a process essential for cell division, rather than on the division process itself (Ray, 1982). Not surprisingly the related compound sulfo-meturon-methyl (proposed name; see Table 5.5*E*, for structure) is reported to work in the same way (Dodel and Carraro, 1981).

In a survey of the effects of chlorsulfuron on various biochemical processes of soybean leaf cells, Hatzios and Howe (1982) found that incorporation of radiolabelled acetate into lipids was the most sensitive of those examined, being inhibited by 38% at 0.1 μM after 2 hours. However, the response to increased dose was very shallow, so that 100 μM compound was needed to cause 90% inhibition in the same time; this depression of lipid synthesis cannot therefore challenge that on cell division as a candidate for the primary effect. The molecular basis of the effect on cell division is not known.

G. Miscellaneous compounds
1. *Propyzamide*

Propyzamide (Table 5.5*F*) is a selective herbicide which can be used in lucerne and other small-seeded legumes, and which also has pre-emergence applications (Worthing, 1979). Carlson *et al.* (1975) studied its effects on growth, ^{86}Rb uptake (as an indicator of ion uptake in general), mitosis and cell wall synthesis. Mitosis was very sensitive, so that after exposure to about 2 μM propyzamide for 30 min, there were no normal mitoses taking place in oat root-tips. The chromosomes failed to line up in the equatorial plane for division (arrested metaphase). Bartels and Hilton (1973) indicated that propyzamide disrupted microtubules in wheat and maize roots in a manner that was indistinguishable from the effects caused by the dinitroanilines trifluralin and oryzalin. Propyzamide probably, therefore, interferes with cell division.

2. *Chlorthal-dimethyl*

This herbicide (Table 5.5*F*), which acts by killing germinating seeds, was investigated by Chang and Smith (1972), who took electron micrographs of foxtail millet seeds germinated in chlorthal-dimethyl. The most striking effect was the inhibition of cell, but not nuclear division, leading to the formation of multinucleate cells, but other organelles were also damaged.

Robinson and Herzog (1977) report that chlorthal-dimethyl at about 10 mM interfered with pig brain tubulin assembly and encouraged microtubule subunit dissociation, but this concentration caused such severe damage in the alga *Oocystis solitaria* that effects on microtubules *in vivo* were not evaluated.

3. *Maleic hydrazide*

The agricultural applications of maleic hydrazide (Table 5.5*F*) whose growth-regulating activity was discovered in 1947, were described by Ashton and Crafts (1973). They include use as a herbicide and as an inhibitor of suckering in tobacco. Interactions of maleic hydrazide with biochemical systems were reviewed by Webb (1966).

McLeish (1952) showed that mitosis was permanently suppressed 2.5 days after bean roots had been exposed to a 100 μM solution of maleic hydrazide for 20 hours and it is generally accepted that the effectiveness of the compound is due to its ability to inhibit cell division (Bonaly, 1971; Hoffman and Parups, 1964; Noodén, 1972). Cell enlargement, on the other hand, appears to be unaffected (Noodén, 1969).

The mode of action is not established with molecular precision, but there is evidence to suggest that the compound interferes with DNA and/or RNA synthesis though perhaps in different ways. For instance, Noodén (1972) found that maleic hydrazide inhibited incorporation of ^{3}H-thymidine into DNA and incorporation of ^{3}H-uridine into RNA in corn roots though the former effect occurred more rapidly. On the other hand, Coupland and Peel (1971) supplied radioactive maleic hydrazide to intact willow plants and found that the radioactivity accumulated in the root-tips and that some was incorporated into RNA, but not DNA. This supported the idea that the similarity in structure between maleic hydrazide and the pyrimidine bases of DNA and RNA permitted the herbicide to exert its effect through incorporation into nucleic acids.

Approaching the problem from a different direction Cradwick (1975) has established the crystal structure of maleic hydrazide. On the basis of his results he suggests that maleic hydrazide could perhaps be incorporated into a nucleotide through either the substituent –OH or a ring N and that in the former case this nucleotide could base pair with uracil or thymine, whereas in the latter case base pairs could be formed by hydrogen bonding with adenine.

Successful attempts to overcome inhibitory effects of maleic hydrazide by the supply of excess of the natural bases were made. References to such work are given by Noodén (1969), though he could obtain no such evidence in experiments with corn or pea roots. Subsequently Mathur and Yadav (1975) have indicated that thymine could overcome inhibitory effects of maleic hydrazide on the growth of the pondweed *Spirodela polyrrhiza*.

H. The 'aromatic hydrocarbon' and dicarboximide fungicides

The compounds of this group are given in Table 5.6 and they have been left until last since the evidence for their site of action is indirect. The term 'aromatic hydrocarbon group' was proposed for compounds of the type given in part *A* of the table (Georgopoulos and Zaracovitis, 1967), though it should be noted that all but one of those currently used as pesticides carry multiple substituents. These and the compounds in part *B* of the table have been grouped together on the basis of their ability to induce mutations in fungal colonies (Georgopoulos *et al.*, 1979). Earlier work is reviewed in this paper. The test system used to establish activity is as follows.

Table 5.6   The 'aromatic hydrocarbon' and dicarboximide groups of fungicides affecting cell division

Name	Structure
A. *'Aromatic hydrocarbons'* Biphenyl	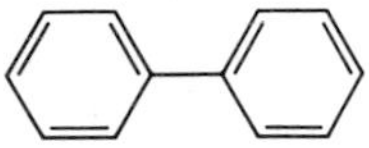
Chloroneb	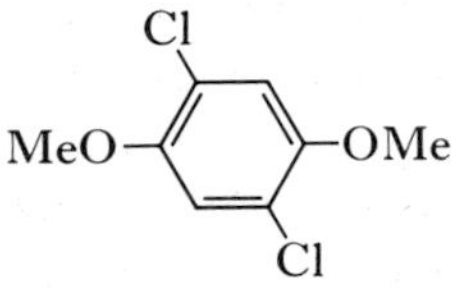
Dicloran	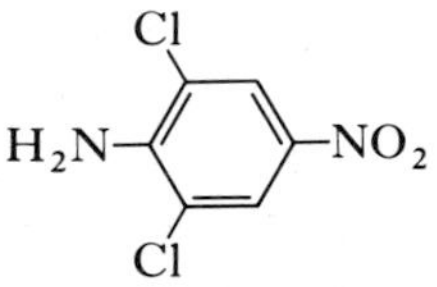
Hexachlorobenzene	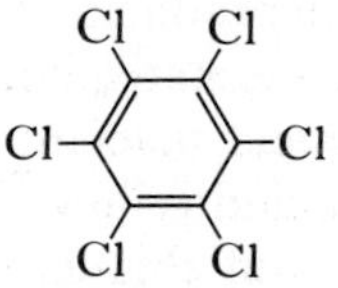
Quintozene	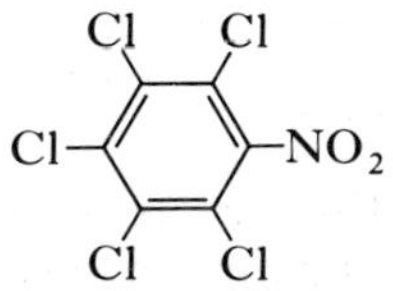
Tecnazene	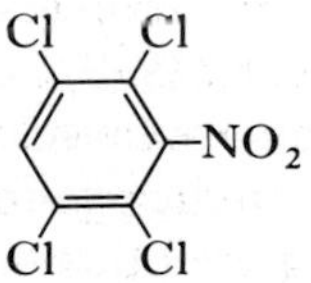

Name	Structure

B. Dicarboximides
Iprodione

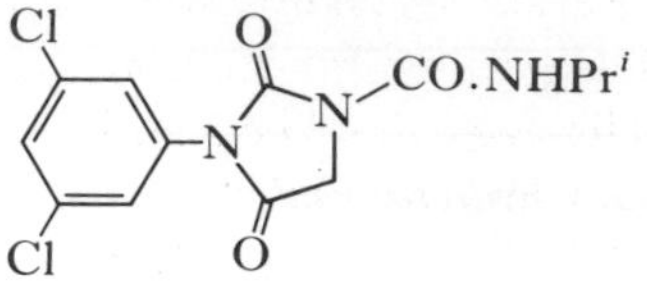

Procymidone

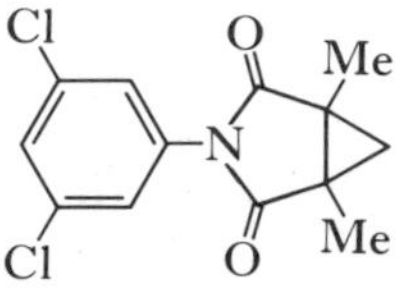

Vinclozolin

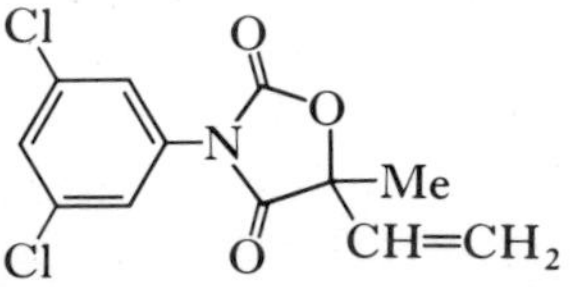

An *Aspergillus* strain producing green conidia was grown *in vitro* (conidia are spores produced vegetatively so that they contain the same genetic information as the parent mycelium). Each *Aspergillus* cell also contained genes coding for white and yellow conidia, but these were normally masked. When the compounds of this class were included in the medium, their effect was to inhibit growth and increase the frequency of production of yellow and white conidia rather than green ones. This must have occurred because the gene coding for green conidia had been lost, or rendered inoperative, and it is taken as a particular example of the interference of the compounds with the passage of information to daughter cells at cell division.

The dicarboximides were much more effective than aromatic hydrocarbons and a good correlation was obtained between growth inhibition and the number of different coloured colonies induced. The effects were observed at micromolar concentrations in the case of the dicarboximides, and Georgopoulos *et al.* (1979) consider that interference with mitosis is therefore the likely primary cause of fungitoxicity in this class of compound.

Two suggestions as to the possible mechanism have been made. Azevedo *et al.* (1977) have suggested that the compounds might prevent the correct separation of the chromosomes to opposite ends of their cells (by, for instance, interfering with the mitotic spindle) whereas evidence obtained by Kappas (1978) has raised the possibility that the chromosome itself could be the site of action of the fungicides.

The biological effects are consistent with the proposed mode of action. Fungal cells typically swell and burst and this could be caused by continuous cellular biosynthesis in the absence of cell division (Georgopoulos *et al.*, 1979). Earlier suggestions of effects of particular compounds on chitin synthesis and cell membranes have now been refuted, but interference at other sites is still a possibility (Georgopoulos *et al.*, 1979). General effects of procymidone on *Botrytis* have been described by Fritz *et al.* (1977) and Hisada *et al.* (1978).

We should note that chloroneb was included amongst the compounds in part *A* of Table 5.6 on the basis of some structural similarity to other compounds in the group and cross-resistance tests (Georgopoulos *et al.*, 1979). An action on cell division would, however, tie in with inhibitory effects of micromolar concentrations of chloroneb on DNA synthesis in *Rhizoctonia solani* (Hock and Sisler, 1969; Kataria and Grover, 1975). Indeed Tillman and Sisler (1973) found that chloroneb led to an eventual cessation of DNA synthesis in sporidia of *Ustilago maydis*, but concluded from the timing of the effect that this was a consequence of the failure of normal cell division that they observed.

Turning now to other effects, Kataria and Grover (1975), in the study mentioned above, found that chloroneb at 4 µM caused 38% inhibition of cellular respiration after 10 minutes and 43% inhibition of cytochrome oxidase *in vitro*. Under their conditions respiration was more rapidly affected than incorporation of thymidine into DNA. This fits with the observation that oxygen consumption by mitochondria from a sensitive *Mucor* strain was half-inhibited by 70 µM chloroneb whereas much higher concentrations were needed to half-inhibit mitochondria from rat liver or insensitive fungi (Lyr and Werner, 1982). In this study, damage to mitochondrial and nuclear membranes was observed by electron microscopy.

Inhibitory effects on respiration have also been reported for dicloran. Rat liver mitochondria were slightly uncoupled at 10 µM and ADP phosphorylation was severely inhibited by a 100 µM concentration (Gallo *et al.*, 1976). However, Fritz *et al.* (1977) found that dicloran at about 5 µM did not inhibit respiration in *Botrytis*.

It is not yet clear how the effects of chloroneb and dicloran on respiration relate to those on cell division.

III. COMPOUNDS INTERFERING WITH INSECT HORMONE ACTION
A. Introduction

No compound in the *Pesticide Manual* (Worthing, 1979) is known to fall into this class, but there has been considerable interest in this area and we will therefore review it briefly.

B. Juvenile hormones and their analogues
1. *The natural hormones*

The juvenile hormones are secreted by the insect corpora allata, two glands at the base of the brain. The hormones' major function is to maintain insect larvae in a juvenile stage at a moult, so that they prevent a progressive change from one developmental stage to the next. Accounts of their discovery, chemistry and function are given by Riddiford and Truman (1978) and Menn and Henrick (1981) and the proceedings of a recent symposium have been published (Pratt and Brooks, 1981). The structures of the hormones, which are found in different amounts in different tissues, are given below.

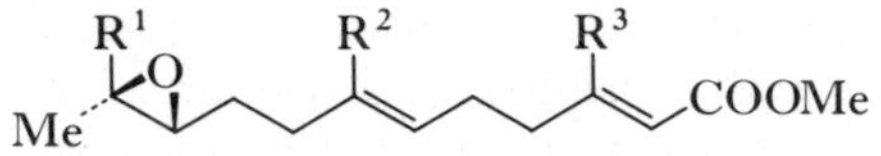

$$Me \overset{R^1}{\underset{O}{\bigtriangleup}} \diagdown \diagup \diagdown \overset{R^2}{=} \diagup \diagdown \overset{R^3}{=} \diagdown COOMe$$

	R^1	R^2	R^3
JH 0	Et	Et	Et
JH I	Et	Et	Me
JH II	Et	Me	Me
JH III	Me	Me	Me

The molecular mode of action of the juvenile hormones is not established but current ideas are reviewed briefly by Riddiford and Truman (1978). A plausible working hypothesis is that juvenile hormone enters a cell, combines with a cytoplasmic receptor protein, and is admitted to the nucleus where it influences the pattern of messengers being transcribed from the DNA. This would involve, in conjunction with other hormones, maintaining the expression of genes required in the juvenile phase, while ensuring that the adult genes are not expressed.

2. *Synthetic analogues*

The discovery of the juvenile hormones opened the way for the synthesis of analogues intended to control insect species by interfering with their normal development. Many such compounds have now been prepared and the work has been reviewed, for instance, by Menn and Henrick (1981) and Staal (1982). The presence of such compounds when the natural juvenile hormone levels are low disturbs normal morphogensis to the adult.

The resultant insects combine juvenile and adult characteristics, cannot feed or reproduce, and soon die.

The compound in this class which has been most successful commercially is methoprene, whose conceptual development has been traced by Bowers (1982). Methoprene was the most potent of a series of juvenile hormone

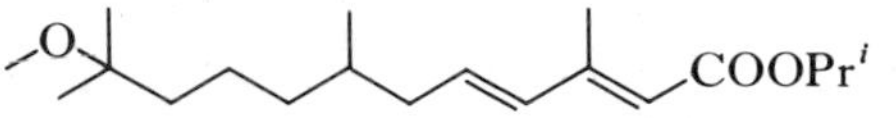

methoprene

analogues synthesised by Henrick *et al.* (1973). Activity was measured in synchronised sensitive stages of *Galleria mellonella* (greater wax moth), *Tenebrio molitor* (yellow mealworm) and *Aedes aegypti* (yellow fever mosquito), being respectively 0.055×, 130× and 1070× as active than JHI (as judged by the dose or concentration required to give 50% inhibition of growth). The compound is used as a larvicide against flood water mosquitoes, being effective in a microencapsulated form at less than 1 ppb (Riddiford and Truman, 1978). The fact that methoprene is not rapidly degraded by an esterase preparation which metabolises juvenile hormones (Terriere and Yu, 1977) presumably contributes to its effectiveness.

Continued interest in this type of compound is illustrated by the recent introduction of the juvenile hormone analogue MV-678. This material can be used to exterminate fire ant colonies by incorporation in a bait which is

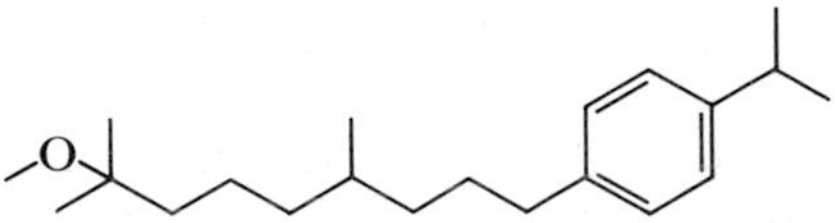

MV-678

taken back to the nest where it disrupts the development of new workers (Anon., 1982). The activity of the compound as an insect growth regulator has been known for some time (see, for example, Schwarz *et al.*, 1974).

C. Precocenes
1. *Discovery*

The background to the discovery of the precocenes by Bowers and his group has been reviewed (Bowers *et al.*, 1976; Menn and Henrick, 1981). Recognising that an excess of juvenile hormone or a mimic will only disrupt development at limited stages in the insect life cycle Bowers *et al.* (1976) reasoned that, since juvenile hormone levels are high for most of the life cycle, a compound interfering with the synthesis, movement or action of juvenile hormone might be more generally disruptive.

This was followed by a screening programme to find plant extracts with 'anti-juvenile hormone' activity culminating in the discovery that extracts of *Ageratum houstonianum* induced precocious (hence the name) meta-morphosis in immature hemipterans and prevented ovarian development in normal adult insects. Subsequent purification yielded the precocenes, 1 and 2.

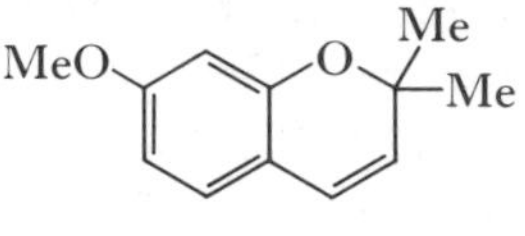

precocene 1 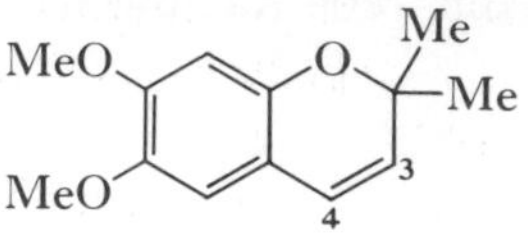   precocene 2

2. *Activity*

The most evident effect of the precocenes is to cause early instar nymphs of, for example, the milkweed bug *Oncopeltus fasciatus*, to moult to miniature adults. A treated nymphal stage first gives rise to an additional nymphal instar, which then moults to give the precocious adult (Bowers *et al.*, 1976).

The ovaries of precocious females of *Oncopeltus* did not develop, and normal adults did not develop ovaries if they were treated with precocene (especially the more active precocene 2) soon after emergence. This indicated inactivation of the corpora allata. A further effect was the induction of diapause (a period of inactivity during which insects do not feed, mate or reproduce), a condition shown to depend on the cessation of juvenile hormone secretion (Bowers *et al.*, 1976).

3. *Mode of action*

The precocenes thus had the properties expected of compounds interfering with juvenile hormone action. It was then shown that precocious maturation could be prevented by administering juvenile hormone III along with precocene 2 (Bowers *et al.*, 1976), and that incubation of precocene 2 with isolated corpora allata from cockroaches (*Periplaneta americana*) caused an inhibition of juvenile hormone synthesis (Pratt and Bowers, 1977). Subsequent to this and other work Brooks *et al.* (1979) obtained evidence that the action of the precocenes on immature *Oncopeltus* and *Locusta* required an oxidative activation within the corpora allata, and also showed that the 3,4-double bond was required for activity. This led to a study of the metabolism by *Locusta* corpora allata of radiolabelled precocene 1 (Pratt *et al.*, 1980). Radioactivity appeared in thin-layer chromatograms in the position of the *cis*- and *trans*-3,4-dihydro-diols when corpora allata tissue was used but not when other tissues were used as a control. Dihydro-diol formation was prevented by a mixed function oxidase inhibitor. Some radioactivity remained bound (probably covalently) to the corpora allata.

The stereochemistry of the diols produced led to the conclusion that the double-bond was probably enzymically epoxidised (though the subsequent hydration to the diol was likely to be chemical) (Pratt *et al.*, 1980). Since terminal epoxidation occurs in the synthesis of the juvenile hormones, it was suggested that the enzyme catalysing this conversion epoxidised the precocene. The epoxide could then be attacked by a nucleophile at the active site of the enzyme, in which case the precocene would be a classical k_{cat} or 'suicide substrate' inhibitor (Rando, 1974; Walsh, 1977; Bowers *et al.*, 1982). Such an enzyme inhibitor is chemically inert initially, but is converted to a reactive species by the enzyme, which then reacts irreversibly with the derivative to bring about its own inactivation. Alternatively the epoxide could move away from the active site and react with some other important component of the corpora allata cells, or with water to form the 3,4-diol mentioned above (see also Soderlund *et al.*, 1980).

Fridman-Cohen and Pener (1980) have obtained evidence that precocene 2 can temporarily activate the corpora allata of last instar *Locusta* nymphs so that an outburst of JH biosynthesis precedes the destructive effect described above. Activation of the enzymes of JH biosynthesis could conceivably be a prerequisite for conversion of the precocene to the epoxide, which then inhibits the whole process.

REFERENCES

Abel, A. L. (1962). *Rep. Prog. App. Chem.* **47**, 552–558
Abeles, F. B. (1968). *Weed Sci.* **16**, 498–500
Addicott, F. T. (1976). In "Herbicides. Physiology, Biochemistry, Ecology" (L. J. Audus, ed.) Vol. 1, pp. 191–217. Academic Press, London and New York
Amrhein, N. (1979). *Prog. Bot.* **41**, 108–134
Anonymous (1982). *Chem. and Eng. News* **60**, 27–29
Ashton, F. M. and Crafts, A. S. (1973). "Mode of Action of Herbicides". John Wiley and Sons, New York
Ashton, F. M. and Crafts, A. S. (1981). "Mode of Action of Herbicides", 2nd edn. John Wiley and Sons, New York.
Azevedo, J. L., Santana, E. P. and Bonatelli, R. (1977). *Mutat. Res.* **48**, 163–172.
Bartels, P. G. and Hilton, J. L. (1973). *Pestic. Biochem. Physiol.* **3**, 462–472
Bearder, J. R. (1980). In "Hormonal Regulation of Development I" (J. MacMillan, ed.) Encycl. Plant Physiol., New Series, Vol. 9, pp. 9–112. Springer-Verlag, Berlin
Ben-Aziz, A. and Aharonson, N. (1974). *Pestic. Biochem. Physiol.* **4**, 120–126
Bentley, J. A. (1950). *Nature (Lond.)* **165**, 449
Block, A. M. and Clements, R. G. (1975). *Int. J. Quantum Chem. Quantum Biol. Symp.* **2**, 197–202
Bonaly, J. (1971). *C. R. Acad. Sci. Ser. D* **273**, 150–153
Borchert, R., McChesney, J. D. and Watson, D. (1974). *Plant Physiol.* **53**, 187–191
Borgers, M. and De Brabander, M. (eds) (1975). "Microtubules and Microtubule Inhibitors". Elsevier, Amsterdam
Bowers, W. S. (1982). *Ent. Exp. and Appl.* **31**, 3–14

Bowers, W. S., Ohta, T., Cleere, J. S. and Marsella, P. A. (1976). *Science* **193**, 542–547

Bowers, W. S., Evans, P. H., Marsella, P. A., Soderlund, D. M. and Bettarini, F. (1982). *Science* **217**, 647–648

Brooks, G. T., Pratt, G. E. and Jennings, R. C. (1979). *Nature (Lond.)* **281**, 570–572

Brookes, R. F. and Leafe, E. L. (1963). *Nature (Lond.)* **198**, 589–590

Brower, D. L. and Hepler, P. K. (1976). *Protoplasma* **87**, 91–111

Brown, D. L. and Bouck, G. B. (1974). *J. Cell Biol.* **61**, 514–536

Buchenauer, H., Edgington, L. V. and Grossmann, F. (1973). *Pestic. Sci.* **4**, 343–348

Canvin, D. T. and Friesen, G. (1959). *Weeds* **7**, 153–156

Carlson, W. C., Lignowski, E. M. and Hopen, H. J. (1975). *Weed Sci.* **23**, 155–161

Chang, C. T. and Smith, D. (1972). *Weed Sci.* **20**, 220–225

Clemons, G. P. and Sisler, H. D. (1969). *Phytopathol.* **59**, 705–706

Clemons, G. P. and Sisler, H. D. (1971). *Pestic. Biochem. Physiol.* **1**, 32–43

Collis, P. S. and Weeks, D. P. (1978). *Science* **202**, 440–442

Coolbaugh, R. C. and Hamilton, R. (1976). *Plant Physiol.* **57**, 245–248

Coolbaugh, R. C., Hirano, S. S. and West, C. A. (1978). *Plant Physiol.* **62**, 571–576

Coolbaugh, R. C., Swanson, D. I. and West, C. A. (1982). *Plant Physiol.* **69**, 707–711

Coss, R. A. and Pickett-Heaps, J. D. (1974). *J. Cell Biol.* **63**, 84–98

Coss, R. A., Bloodgood, R. A., Browers, D. L., Pickett-Heaps, J. D. and McIntosh, J. R. (1975). *Exp. Cell Res.* **92**, 394–398

Coupland, D. and Peel, A. J. (1971). *Physiol. Plant.* **25**, 141–144

Cradwick, P. D. (1975). *Nature (Lond.)* **258**, 774

Davidse, L. C. (1973). *Pestic. Biochem. Physiol.* **3**, 317–325

Davidse, L. C. (1975). In "Systemic Fungicides" (H. Lyr and C. Polter, eds) pp. 137–143. Akademie-Verlag, Berlin

Davidse, L. C. (1976). *Pestic. Biochem. Physiol.* **6**, 538–546

Davidse, L. C. (1979). In "Systemic Fungicides" (H. Lyr and C. Polter, eds), pp. 277–286. Akademie-Verlag, Berlin

Davidse, L. C. and Flach, W. (1977). *J. Cell. Biol.* **72**, 174–193

Davidse, L. C. and Flach, W. (1978). *Biochim. Biophys. Acta* **543**, 82–90

Davis, F. S., Villareal, A., Bauer, J. R. and Goldstein, I. S. (1972). *Weed Sci.* **20**, 185–188

Dennis, F. G. Jr., Wilczynski, H., de la Guardia, M. and Robinson, R. W. (1970). *Hort. Sci.* **5**, 168-170

Dicks, J. W. (1980). In "Recent Developments in the Use of Plant Growth Retardants" (D. R. Clifford and J. R. Lenton, eds) pp. 1–14. British Plant Growth Regulator Group, A.R.C. Letcombe Laboratory, Wantage, U.K.

Dodds, J. H. and Hall, M. A. (1980). *Science Progress (Oxford)* **66**, 513–535

Dodel, J. B. and Carraro, G. A. (1981). Compte-rendu 11eme Conf., COLUMA Versailles, **3**, 764–771

Douglas, T. J. and Paleg, L. G. (1974). *Plant Physiol.* **54**, 238–245

Douglas, T. J. and Paleg, L. G. (1978). *Phytochem.* **17**, 713–718

Dubrovin, K. P. (1959). Proc. N. Cent. Weed Control Conf., U.S.A., p. 15

Duncan, J. D. and West, C. A. (1981). *Plant Physiol.* **68**, 1128–1134

Dustin, P. (1978). "Microtubules". Springer-Verlag, Berlin

Eckert, J. W. and Rahm, M. L. (1979). *Pestic. Sci.* **10**, 473–477

Edgington, L. V. (1972). In "Herbicides, Fungicides, Formulation Chemistry" (A. S. Tahori, ed.) pp. 421–423. Gordon and Breach, New York

Eisinger, W. R. and Morré, D. J. (1971). *Can. J. Bot.* **49**, 889–897

Farrimond, J. A., Elliott, M. C., and Clack, D. W. (1978). *Nature (Lond.)* **274**, 401–402

Farrimond, J. A., Elliott, M. C. and Clack, D. W. (1981). *Phytochem.* **20**, 1185–1190

Fawcett, C. H., Wain, R. L. and Wightman, F. (1958). *Nature (Lond.)* **181**, 1387–1389

Filner, P. and Varner, J. E. (1967). *Proc. Natl. Acad. Sci. U.S.A.* **58**, 1520–1526

Flavin, M. and Slaughter, C. (1974). *J. Bacteriol.* **118**, 59–69

Frear, D. S. (1975). In "Herbicides. Chemistry, Degradation and Mode of Action" (P. C. Kearney and D. D. Kaufman, eds) Vol. 2, pp. 541–607. Dekker, New York

Fridman-Cohen, S. and Pener, M. P. (1980). *Nature (Lond.)* **286**, 711–713

Fritz, R., Leroux, P. and Gredt, M. (1977). *Phytopathol. Z.* **90**, 152–163

Frost, R. G. and West, C. A. (1977). *Plant Physiol.* **59**, 22–29

Fuchs, A., van den Berg, G. A. and Davidse, L. C. (1972). *Pestic. Biochem. Physiol.* **2**, 191–205

Gallo, M. A., Bachmann, E. and Goldberg, L. (1976). *Toxicol. Appl. Pharmacol.* **35**, 51–61

Garraway, J. L. and Wain, R. L. (1976). *Medicinal Chem.* **11** (7, Drug Res.) 115–164

Georgopoulos, S. G. and Zaracovitis, C. (1967). *Ann. Rev. Phytopathol.* **5**, 109–130

Georgopoulos, S. G., Sarris, M. and Ziogas, B. N. (1979). *Pestic. Sci.* **10**, 389–392

Goldsmith, M. H. M. (1977). *Ann. Rev. Plant Physiol.* **28**, 439–478

Graebe, J. E. and Ropers, H. J. (1978). In "Phytohormones and Related Compounds – A Comprehensive Treatise" (D. S. Letham, P. B. Goodwin and T. J. V. Higgins, eds) Vol. 1, pp. 107–204. Elsevier, Amsterdam

Gunning, B. E. S. and Hardham, A. R. (1979). *Endeavour* (New Series) **3**, 112–117

Gunning, B. E. S. and Hardham, A. R. (1982). *Ann. Rev. Plant Physiol.* **33**, 651–698

Hammerschlag, R. S. and Sisler, H. D. (1973). *Pestic. Biochem. Physiol.* **3**, 42–54

Hampel, H. and Löcher, F. (1973). Proc. British Insectic. Fungic. Conf., Vol. 1, pp. 127–134

Hanson, J. B. and Slife, F. W. (1969). *Residue Reviews* **25**, 59–67

Hanson, J. B. and Trewavas, A. J. (1982). *New Phytol.* **90**, 1–18

Hatzios, K. K. and Howe, C. M. (1982). *Pestic. Biochem. Physiol.* **17**, 207–214

Hedden, P., MacMillan, J. and Phinney, B. O. (1978). *Ann. Rev. Plant Physiol.* **29**, 149–192

Henrick, C. A., Staal, G. B. and Siddall, J. B. (1973). *J. Agric. Food Chem.* **21**, 354–359

Hepler, P. K. and Jackson, W. T. (1969). *J. Cell Sci.* **5**, 727–743

Hertel, C., Quader, H., Robinson, D. G. and Marme, D. (1980). *Planta* **149**, 336–340

Hertel, C., Quader, H., Robinson, D. G., Roos, I., Carafoli, E. and Marme, D. (1981). *FEBS Lett.* **127**, 37–39

Hess, F. D. (1979). *Exp. Cell Res.* **119**, 99 109

Hess, F. D. and Bayer, D. (1974). *J. Cell Sci.* **15**, 429–441

Hess, F. D. and Bayer, D. (1977). *J. Cell Sci.* **24**, 351–360

Hisada, Y., Kato, T. and Kawase, Y. (1978). *Ann. Phytopathol. Soc. Japan* **44**, 509–518

Hock, W. K. and Sisler, H. D. (1969). *Phytopathol.* **59**, 627–632

Hoffman, T. and Parups, E. V. (1964). *Residue Reviews* **7**, 96–113

Jackson, W. T. and Stetler, D. A. (1973). *Can. J. Bot.* **51**, 1513–1518

Jacobs, M. and Hertel, R. (1978). *Planta* **142**, 1–10

Jacobs, W. P. (1979). "Plant Hormones and Plant Development". Cambridge University Press

Jacobsen, J. V. (1977). *Ann. Rev. Plant Physiol.* **28**, 537–564

Jones, R. L., Metcalfe, T. P. and Sexton, W. A. (1954). *J. Sci. Food Agric.* **5**, 32–38

Kaethner, T. M. (1977). *Nature (Lond.)* **267**, 19–23

Kamuro, Y. (1981). *Japan Pesticide Information* No. 39, 17–19

Kappas, A. (1978). *Mutation Res.* **51**, 189–197

Kataria, H. R. and Grover, R. K. (1975). *Indian J. Exp. Biol.* **13**, 281–285

Katekar, G. F. (1976). *Phytochem.* **15**, 1421–1424

Katekar, G. F. (1979). *Phytochem.* **18**, 223–233

Katekar, G. F. and Geissler, A. E. (1977). *Plant Physiol.* **60**, 826–829

Katekar, G. F. and Geissler, A. E. (1980). *Plant Physiol.* **66**, 1190–1195

Katekar, G. F. and Geissler, A. E. (1981). *Phytochem.* **20**, 2465–2469

Katekar, G. F. and Geissler, A. E. (1982). *Phytochem.* **21**, 257–260

Katekar, G. F., Nave, J.-F. and Geissler, A. E. (1981). *Plant Physiol.* **68**, 1460–1464

Kefford, N. P. and Caso, O. H. (1966). *Bot. Gaz.* **127**, 159–163

Keitt, G. W. (1967). *Physiol. Plant.* **20**, 1076–1082

Keitt, G. W. and Baker, R. A. (1966). *Plant Physiol.* **41**, 1561–1569

Kende, H. and Gardner, G. (1976). *Ann. Rev. Plant Physiol.* **27**, 267–290

Kidd, B. R., Stephen, N. H. and Duncan, H. J. (1982). *Plant Sci. Lett.* **26**, 211–217

Kiermayer, O. and Fedtke, C. (1977). *Protoplasma* **92**, 163–166

Killmer, J. L., Widholm, J. M. and Slife, F. W. (1980). *Plant Sci. Lett.* **19**, 203–208

Kirby, C. (1980). "The Hormone Weedkillers". British Crop Protection Council, London

Knypl, J. S. (1977). *Experientia* **33**, 725–726

Kumari, L., Decallonne, J. R., Meyer, J. A. and Talpaert, M. (1977). *Pestic. Biochem. Physiol.* **7**, 273–282

Laclette, J. P., Guerra, G. and Zetina, C. (1980). *Biochem. Biophys. Res. Commun.* **92**, 417–423

Lang, A. (1970). *Ann. Rev. Plant Physiol.* **21**, 537–570

Lehmann, P. A. F. (1978). *Chem.-Biol. Interact.* **20**, 239–249

Lembi, C. A., Morré, D. J., Thomson, K. and Hertel, R. (1971). *Planta* **99**, 37–45

Lendzian, K. J., Ziegler, H. and Sankhla, N. (1978). *Planta* **141**, 199–204

Leopold, A. C. (1971). *Plant Physiol.* **48**, 537–540

Letham, D. S., Goodwin, P. B. and Higgins, T. J. V. (eds) (1978). "Phytohormones and Related Compounds: A Comprehensive Treatise". Vol. 1: "The Biochemistry of Phytohormones and Related Compounds". Elsevier, Amsterdam

Lieberman, M. (1979). *Ann. Rev. Plant Physiol.* **30**, 533–591

Lignowski, E. M. and Scott, E. G. (1971). *Plant Cell Physiol.* **12**, 701–708

Lignowski, E. M. and Scott, E. G. (1972). *Weed Sci.* **20**, 267–270

Linscott, D. L. and Hagin, R. D. (1970). *Weed Sci.* **18**, 197–198

Loos, M. A. (1975). In "Herbicides. Chemistry, Degradation and Mode of Action" (P. C. Kearney and D. D. Kaufman, eds) Vol. 1, pp. 1–128. Dekker, New York

Lyr, H. and Werner, P. (1982). *Pestic. Biochem. Physiol.* **18**, 69–76

MacMillan, J. (1971). In "Aspects of Terpenoid Chemistry and Biochemistry" (T. W. Goodwin, ed.) pp. 153–180. Academic Press, London and New York

Mann, J. D., Cota-Robles, E., Yung, K.-H., Pu, M. and Haid, H. (1967). *Biochim. Biophys. Acta* **138**, 133–139

Mathur, S. N. and Yadav, S. R. (1975). *Indian J. Plant Physiol* **18**, 8–11

McIlrath, W. J. and Ergle, D. R. (1953). *Plant Physiol.* **28**, 693–702

McLeish, J. (1952). *Heredity* **6** (suppl.), 125–147

Menhenett, R. (1978). *Monogr. Br. Crop Prot. Counc.* **21** (Oppor. Chem. Plant Growth Regul.), pp. 187–194

Menhenett, R. (1979). *Ann. Bot.* **43**, 305–318

Menn, J. J. and Henrick, C. A. (1981). *Phil. Trans. Roy. Soc. B.* **295**, 57–71

Mitchell, J. W. and Livingston, G. A. (1968). "Methods of Studying Plant Hormones and Growth-regulating Substances", Agriculture Handbook No. 336. Agricultural Research Service, United States Department of Agriculture

Mizuno, K., Koyama, M. and Shibaoka, H. (1981). *J. Biochem. (Tokyo)* **89**, 329–332

Montague, M. J. (1975). *Plant Physiol.* **56**, 167–170

Moreland, D. E. and Huber, S. C. (1979). *Pestic. Biochem. Physiol.* **11**, 247–257

Moreland, D. E., Farmer, F. S. and Hussey, G. G. (1972a). *Pestic. Biochem. Physiol.* **2**, 342–353

Moreland, D. E., Farmer, F. S. and Hussey, G. G. (1972b). *Pestic. Biochem. Physiol.* **2**, 354–363

Morgan, P. W. (1976). In "Herbicides. Physiology, Biochemistry, Ecology" (L. J. Audus, ed.) Vol. 1, pp. 255–280. Academic Press, London and New York

Noguchi, T., Ohkuma, K. and Kosaka, S. (1971). In "Pesticide Terminal Residues" (A. S. Tahori, ed.) pp. 257–270. Butterworths, London

Noodén, L. D. (1969). *Physiol. Plant* **22**, 260–270

Noodén, L. D. (1972). *Plant Cell Physiol.* **13**, 609–621

Normand, G., Hartmann, M. A., Schuber, F. and Benveniste, P. (1975). *Physiol. Veg.* **13**, 743–761

Palm, H. L., Riggleman, J. D. and Allison, D. A. (1980). Proc. British Crop Protection Conf. "Weeds", Vol. 1 pp. 1–6

Parka, S. J. and Soper, O. F. (1977). *Weed Sci.* **25**, 79–87

Pillmoor, J. B. and Gaunt, J. K. (1981). In "Progress in Pesticide Biochemistry" (D. H. Hutson and T. R. Roberts, eds) Vol. 1, pp. 147–218. John Wiley and Sons, Chichester

Porter, W. L. and Thimann, K. V. (1965). *Phytochem.* **4**, 229–243

Pratt, G. E. and Bowers, W. S. (1977). *Nature (Lond.)* **265**, 548–550

Pratt, G. E. and Brooks, G. T. (eds) (1981). "Juvenile Hormone Biochemistry. Action, Agonism and Antagonism". Elsevier/North-Holland, Amsterdam.

Pratt, G. E., Jennings, R. C., Hamnett, A. F. and Brooks, G. T. (1980). *Nature (Lond.)* **284**, 320–323

Pratt, H. K. and Goeschl, J. D. (1969). *Ann. Rev. Plant Physiol.* **20**, 541–548

Quader, H. and Filner, P. (1980). *Eur. J. Cell Biol.* **21**, 301–304

Rakhaminova, A. B., Khavin, E. E. and Yaguzhinskii, L. S. (1978). *Biokhimya* **43**, 639–653 (English version)

Rando, R. R. (1974). *Science* **185**, 320–324

Ray, P. M. (1977). *Plant Physiol.* **59**, 594–599

Ray, T. B. (1982). *Pestic. Biochem. Physiol.* **17**, 10–17

Reid, D. M. and Crozier, A. (1972). In "Plant Growth Substances, 1970" (D. J. Carr, ed.) pp. 420–427. Springer-Verlag, Berlin

Richmond, D. V. and Phillips, A. (1975). *Pestic. Biochem. Physiol.* **5**, 367–379

Riddiford, L. M. and Truman, J. W. (1978). In "Biochemistry of Insects" (M. Rockstein, ed.) pp. 307–357. Academic Press, London and New York

Rikin, A., Atsmon, D. and Gitler, C. (1982). *Planta* **154**, 402–406
Roberts, K. and Hyams, J. (1980) (eds). "Microtubules". Academic Press, London and New York
Robinson, D. G. and Herzog, W. (1977). *Cytobiologie* **15**, 463–474
Robinson, S. J., Yocum, D. F., Ikuma, H. and Hayashi, F. (1977). *Plant Physiol.* **60**, 840–844
Rubery, P. H. (1981). *Ann. Rev. Plant Physiol.* **32**, 569–596
Rusness, D. G. and Still, G. G. (1974). *Pestic. Biochem. Physiol.* **4**, 109–119
Sabnis, D. D. and Hart, J. W. (1982). In "Nucleic Acids and Proteins in Plants I" (D. Boulter and B. Parthier, eds) Encycl. Plant Physiol., New Series, Vol. 14A, pp. 401–437. Springer-Verlag, Berlin
Schneider, G. (1970). *Ann. Rev. Plant Physiol.* **21**, 499–536
Schwarz, M., Miller, R. W., Wright, J. E., Chamberlain, W. F. and Hopkins, D. E. (1974). *J. Econ. Entomol.* **67**, 598–601
Selling, H. A., Vonk, J. W. and Kaars-Sijpesteijn, A. (1970). *Chem. Ind.* 1625–1626
Sheir-Neiss, G., Lai, M. H. and Morris, N. R. (1978). *Cell* **15**, 639–647
Shive, J. B. and Sisler, H. D. (1976). *Plant Physiol.* **57**, 640–644
Sims, J. J., Mee, H. and Erwin, D. C. (1969). *Phytopathol.* **59**, 1775–1776
Soderlund, D. M., Messeguer, A. and Bowers, W. S. (1980). *J. Agric. Food Chem.* **28**, 724–731
Solel, Z., Schooley, J. M. and Edgington, L. V. (1973). *Pestic. Sci.* **4**, 713–718
Sommer, L. H., Bailey, D. L. and Whitmore, F. C. (1948). *J. Am. Chem. Soc.* **70**, 2869–2872
Staal, G. B. (1982). *Ent. Exp. and Appl.* **31**, 15–23
Stephen, N. H., Cook, G. T. and Duncan, H. J. (1980). *Ann. Appl. Biol.* **96**, 227–234
Sterrett, R. B. and Fretz, T. A. (1975). *Hort. Science* **10**, 161–162
Strachan, S. D. and Hess. F. D. (1982). *J. Agric. Food Chem.* **30**, 389–391
Sumida, S. and Ueda, M. (1974). In "Mechanism of Pesticide Action" (G. K. Kohn, ed.) pp. 156–168. American Chemical Society, Washington, D.C.
Sumida, S. and Ueda, M. (1976). *Plant Cell Physiol.* **17**, 1351–1354
Sumida, S. and Yoshida, R. (1982). In "Biochemical Responses Induced by Herbicides" (D. E. Moreland, J. B. St. John and F. D. Hess, eds) pp. 251–260. American Chemical Society, Washington, D.C.
Sussman, M. R. and Gardner, G. (1980). *Plant Physiol.* **66**, 1074–1078
Sussman, M. R. and Goldsmith, M. H. M. (1981). *Planta* **151**, 15–25
Swanson, C. R. (1972.) In "Herbicides, Fungicides, Formulation Chemistry" (A. S. Tahori, ed.) pp. 87–112. Gordon and Breach, New York
Synerholm, M. E. and Zimmerman, P. W. (1947). *Contrib. Boyce Thompson Inst.* **14**, 369–382
Templeman, W. G. and Sexton, W. A. (1945). *Nature (Lond.)* **156**, 630
Terriere, L. C. and Yu, S. J. (1977). *Pestic. Biochem. Physiol* **7**, 161–168
Thimann, K. V. (1969). In "The Physiology of Plant Growth and Development" (M. B. Wilkins, ed.) pp. 1–45, McGraw-Hill, London
Thomson, K.-St. (1972). In "Hormonal Regulation of Plant Growth and Development" (N. Kaldeway and Y. Vardar, eds) pp. 83–88. Verlag-Chemie, Weinheim
Thomson, K.-St. and Leopold, A. C. (1974). *Planta* **115**, 259–270
Tillman, R. W. and Sisler, H. D. (1973). *Phytopathol.* **63**, 219–225.
Tolbert, N. E. (1960). *J. Biol. Chem.* **235**, 475–479
Torrey, J. G. (1976). *Ann. Rev. Plant Physiol.* **27**, 435–459

Travis, R. L. and Woods, W. G. (1977). *Plant Physiol.* **60**, 54–57

Trewavas, A. J. (1982). *Physiol Plant.* **55**, 60–72

Upadhyaya, M. K. (1978). "Mode of Dinitroaniline Herbicide Action". Ph.D. Thesis, Univesity of Michigan, U.S.A. (Univ. Microfilms Int. No. 78/7188)

Upadhyaya, K. K. and Noodén, L. D. (1980). *Plant Physiol.* **66**, 1048–1052

Veerasekaran, P., Kirkwood, R. C. and Parnell, E. W. (1981). *Pestic. Sci.* **12**, 330–338

Venis, M. A. (1978). *Biochem. Soc. Trans* **6**, 325–333

Venis, M. A. (1979). In "Advances in Pesticide Science" (H. Geissbühler, ed.) Pt 3, pp. 487–493. Pergamon, Oxford

Vonk, J. W. and Kaars Sijpesteijn, A. (1971). *Pestic. Sci.* **2**, 160–164

Vonk, J. W., Mihanovic, B. and Kaars Sijpesteijn, A. (1977). *Neth. J. Plant Physiol.* **83**, (suppl. l), 269–276

Wain, R. L. and Fawcett, C. H. (1969). In "Plant Physiology" (F. C. Steward, ed.) pp. 231–296. Academic Press, London and New York

Wain, R. L. and Smith, M. S. (1976). In "Herbicides. Physiology, Biochemistry, Ecology" (L. J. Audus, ed.) Vol. 2, pp. 279–302. Academic Press, London and New York

Walsh, C. T. (1977), *Horiz. Biochem. Biophys.* **3**, 36–81

Warner, H. L. and Leopold, A. C. (1969). *Plant Physiol.* **44**, 156–158

Webb, J. L. (1966). "Enzyme and Metabolic Inhibitors", Vol. III. Academic Press, London and New York

West, C. A. (1973). In "Biosynthesis and its Control in Plants" (B. V. Milborrow, ed.) pp. 143–169. Academic Press, London and New York

West, C. A. and Fall, R. R. (1972). In "Plant Growth Substances 1970" (D. J. Carr, ed.) pp. 133–142. Springer-Verlag, Berlin

Worthing, C. R. (1979) (ed.). "The Pesticide Manual", 6th edn. British Crop Protection Council, London

W.S.A. Handbook (1979). "Herbicide Handbook of the Weed Science Society of America". Illinois, U.S.A.

Ziegler, H. J. (1970). *Endeavour* **XXIX**, 112–116.

Zimmermann, P. W. and Hitchcock, A. E. (1951). *Contrib. Boyce Thompson Inst.* **16**, 209–213

6 | Pesticides Thought to Inhibit Biosynthetic Reactions

This chapter is concerned with pesticides which may act by interfering with biosynthetic reactions, though chemicals affecting the biosynthesis of compounds involved in the control of growth are excluded since they have already been considered.

I. FUNGICIDES AND HERBICIDES INHIBITING NUCLEIC ACID AND PROTEIN SYNTHESIS
A. Biochemical background

The information for all cellular activities is carried in the DNA of the cells and is encoded in the precise sequences of the bases adenine, guanine, cytidine and thymidine, which are linked to the sugar-phosphate backbone of the linear polymer. All the cells of a multicellular organism contain the same DNA complement, arranged in genes, each of which codes for a particular polypeptide chain such as an enzyme.

The central dogma of molecular biology states that information flows from nucleic acid to protein, and not vice versa. Reference to Fig. 6.1, which outlines the normal sequence of events during protein synthesis from the DNA template, shows the initial transfer (transcription) of information from the DNA of the gene to messenger ribonucleic acid (mRNA), which may be regarded as convenient working copy of the information. In the course of transcription the mRNA obtains a base sequence complementary to that of the DNA by a process of base-pairing.

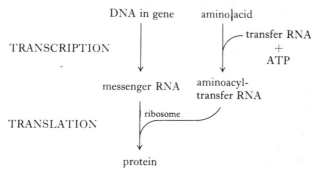

Fig. 6.1 A schematic summary of protein synthesis

The next transfer of information, from mRNA to protein, is referred to as translation and involves small specialised sub-cellular organelles, the ribosomes, each of which is composed of one large and one small subunit. The function of the ribosome is the construction of the linear amino acid sequence which constitutes the protein. The sequence is dictated by the mRNA molecules, which become attached to the ribosomes. As shown in Fig. 6.1, there are three steps in the actual synthesis of the protein from free amino acids in the cytoplasm. Firstly, the free amino acid reacts with ATP and a specific transfer ribonucleic acid (tRNA) molecule to form an aminoacyl-tRNA. By virtue of the specificity of the enzymes involved each tRNA will only become coupled to a particular amino acid. Secondly, by a process of base pairing, the aminoacyl-tRNA becomes attached to the mRNA through a specific binding site on the transfer molecule (Fig. 6.2). Thus, the tRNA acts as a molecular adaptor into which the amino acid is plugged so that it can be adapted to the base sequence language of the mRNA. The third step is the actual synthesis of protein by the ribosome, in which the amino acid supplied by the tRNA is joined onto the end of a growing protein chain by the formation of a peptide bond. In this way, the mRNA is 'read' by the passage of the ribosome along the molecule. Many ribosomes may read the RNA message simultaneously and the resultant assembly of ribosomes on a message is called a polysome.

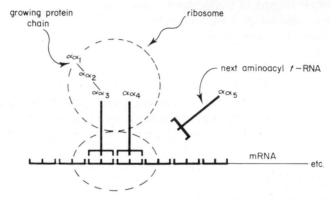

FIG. 6.2 Diagrammatic representation of some of the elements of protein synthesis, in which the information coded in the mRNA is translated into a protein sequence. αα, amino acid

B. Nucleic acid synthesis
1. *Furalaxyl and metalaxyl*

Kerkenaar (1981) has recently studied the effect of metalaxyl on various biochemical functions of the fungus *Pythium splendens*. At a concentration in the medium of about 0.3 μM very little metalaxyl was taken up by the fungus, but even so nucleic acid synthesis was noticeably depressed within 30 minutes and became half-inhibited in 1–2 hours. Other systems examined were much less sensitive.

metalaxyl   furalaxyl

Fisher and Hayes (1982) conducted a very similar study of the effects of metalaxyl on (*inter alia*) *Phytophthora palmivora*, finding that RNA synthesis was the most sensitive process. They extended the conclusion to furalaxyl, which half-inhibited RNA synthesis in cultures exposed to an LC_{95} concentration of the order of micromolar,

The mode of action has not yet been worked out in molecular terms. Kerkenaar (1981) points out the structural resemblance between these fungicidal acylalanines and the herbicidal chloroacetanilides (Chapter 7) and speculates that the modes of action of these two classes of compound may be similar. However, present evidence seems to indicate that the chloroacetanilides act in a more general way (p. 295).

2. Hymexazol

Kamimura *et al.* (1976) studied the effect of the soil fungicide hymexazol on various functions of a *Fusarium* species growing in culture. Under these conditions, the compound was fungistatic rather than fungicidal, and about 3 mM compound was needed to suppress growth by half. The

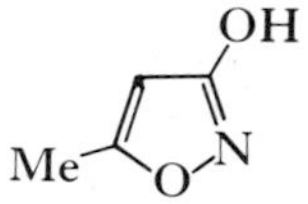

hymexazol

chemical had no effect on respiration or cell membrane integrity but, at 3 mM, inhibited protein, RNA and DNA synthesis by 20%, 30% and 50% respectively. The concentrations used were high, but DNA synthesis displayed the same sensitivity as fungal growth. On this basis Kamimura *et al.* (1976) suggested that hymexazol may exert its antifungal activity by interfering with DNA metabolism.

C. Protein synthesis

The agricultural use of antibiotics, together with information on their mode of action has been reviewed by Siegel (1977), and Vázquez (1979) has collated the evidence for the mode of action of most inhibitors of protein synthesis.

1. Fungicidal antibiotics

(a) Cycloheximide

This agricultural fungicide (Table 6.1) caused 50% inhibition of amino acid incorporation into protein in cell free extracts of the yeast *Saccharomyces pastorianus* at approximately 0.7 μM, which is about the concentration required to inhibit the growth of the intact cells (Siegel and Sisler, 1963). It is likely that this inhibition of protein synthesis explains the antifungal action of the compound since Kerridge (1958) showed that approximately 2 to 4 μM cycloheximide, which was the minimum concentration required to inhibit growth of the yeast *Saccharomyces carlsbergensis* completely inhibited the synthesis of protein and also DNA in the intact cells. The inhibition of DNA synthesis could be a result of the initial interference with protein synthesis but more probably represents a secondary effect of the chemical (Gale et al., 1981).

Table 6.1   Antibiotics which inhibit protein synthesis

Name	Structure
Blasticidin S	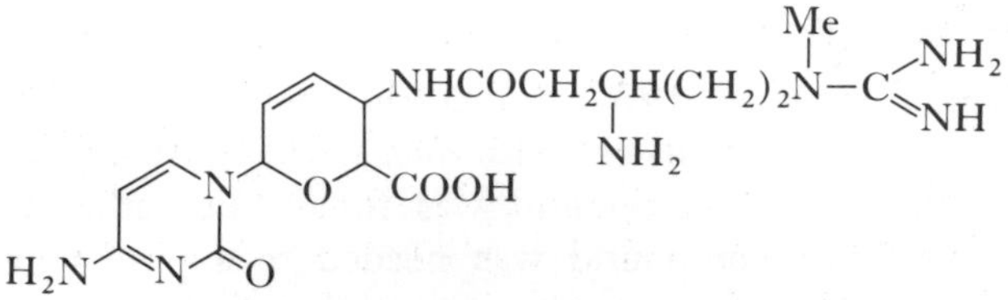
Cycloheximide	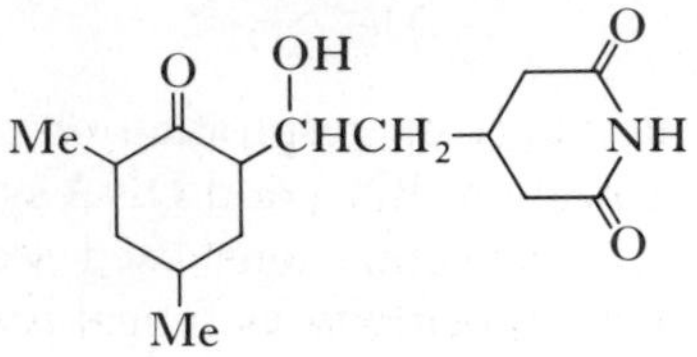
Kasugamycin	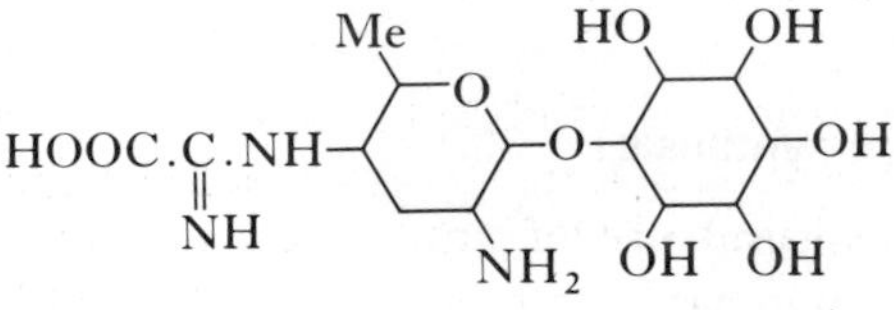

The use of fungal mutants has shown that cycloheximide works by interaction with the large subunit of the ribosome (see Gale et al., 1981; Georgopoulos and Vomvoyanni, 1972), although the exact mechanism is uncertain. Cycloheximide certainly prevents movement of ribosomes along the mRNA, but this might perhaps be due to inhibition of specific steps involved in the initiation of the translation process (Gale et al., 1981).

Cycloheximide is not only toxic to fungi, but will also kill plants and animals (Worthing, 1979), and Ennis and Lubin (1964) showed that cyclo-heximide inhibited protein synthesis in rat liver extracts. However, it did not inhibit protein synthesis in extracts from the bacterium *Escherichia coli*, and has little effect on bacterial growth, a selective effect due to the difference in ribosome structure found between bacteria and the majority of ribosomes found in the cells of plants and animals. Although all ribosomes are made of RNA and various protein molecules on an essentially similar ground plan, those from bacteria are somewhat smaller and contain a higher proportion of RNA than the majority of those found elsewhere.

(b) Blasticidin S

This antibiotic (Table 6.1) is used in Japan to control rice blast (*Pyricularia oryzae*). The minimum concentration required to inhibit growth of the mycelium of this fungus is approximately 2 μM, at which level blasticidin S also inhibits the incorporation of radioactive glutamic acid into protein in the fungus (see review by Misato, 1969).

There is evidence that it binds to a single site on the larger ribosomal subunit, as a result of which the incoming aminoacyl-tRNA molecule cannot gain access to its binding site and the elongation of the protein chain is prevented (Siegel, 1977 and references therein).

(c) Kasugamycin

This antibiotic (Table 6.1), like blasticidin S, is only used in Japan against rice blast, and it is effective at similar concentrations (Misato, 1969). Most mode of action work has been done with bacteria (whose smaller ribosomes resemble only a proportion of those found in fungal cells; see above) and it indicates that the pesticide binds to the smaller of the two ribosomal subunits and inhibits protein synthesis initiation at lower concentrations than those at which it inhibits chain elongation (Siegel, 1977). Initiation is inhibited because the complex between the small ribosomal subunit and the special initiator aminoacyl-tRNA is destabilised (Poldermans *et al.*, 1979).

2. Miscellaneous herbicides

(a) 2-(4-Methyl-2,6-dinitroanilino)-N-methylpropionamide

This compound (referred to as MDMP) is chiral and only the *R*-isomer is herbicidal (Kerr and Avery, 1972), though it is not in commercial use.

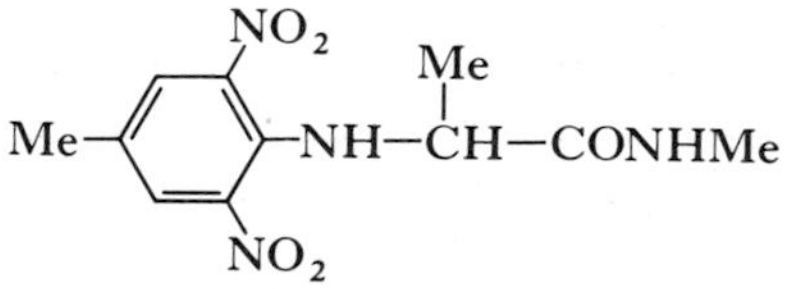

Using *in vitro* translation systems Weeks and Baxter (1972) reported that protein synthesis was inhibited at the initiation step, and later work suggested that this could be due to the ability of the compound to interfere with the attachment of ribosomes to messenger RNA (Baxter and McGowan, 1976). Like cycloheximide, MDMP is active against the larger ribosomes of plants and animals (Baxter *et al.*, 1973) but not against the smaller ribosomes found in bacteria and chloroplasts, and this selective action can be of use in distinguishing cytoplasmic and chloroplast protein synthesis (e.g. see Ellis, 1975).

The compound has a certain resemblance to the *bis-N*-substituted alanine esters, such as benzoylprop-ethyl, which are collated in Chapter 8 (p. 315). As will be described, these esters are active after hydrolysis to the acid, but there seems to be no evidence for cleavage of the amide bond of, e.g., benzoylprop-ethyl which would lead to compounds of a similar structure to MDMP.

(b) Endothal

Very little work appears to have been done on the mode of action of this herbicide, but it might act by inhibiting protein synthesis since an 11 μM

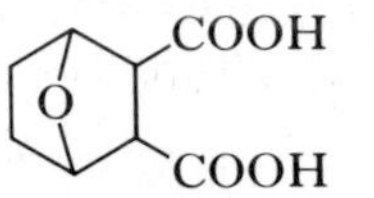

endothal

solution caused 63% inhibition of leucine incorporation into protein by segments of the legume *Sesbania exaltata*, though the effect on barley segments was much less (Mann *et al.*, 1965). However, an effect on lipid synthesis is possible since a concentration of 27 μM led to a 45% inhibition of lipid synthesis in *Sesbania* segments (Mann and Pu, 1968).

II. HERBICIDES INTERFERING WITH CAROTENOID ACCUMULATION
A. Introduction

The compounds given in Table 6.2 are inhibitors of carotenoid accumulation and this is probably their primary effect. Other biochemical responses which have been investigated will be mentioned later.

Most of the work has been done with the pyridazinones, so we shall consider them first. Sandoz 6706 is not a commercial herbicide, but has been widely used in mode of action work, and is therefore included. San 6706 is converted to norflurazon *in vivo* (see Devlin *et al.*, 1976) so that their effects on carotenoid accumulation are probably equivalent. Several other pyridazinones have been used in mode of action studies, and these will be found in the literature survey by Eder (1979).

B. Biochemical background

Carotenoids are yellow to red pigments found throughout the plant kingdom, and they are named after the yellow pigment of carrot, β-carotene (p. 241). Their biochemistry has been reviewed in detail by Goodwin

(1981). They function as accessory pigments to chlorophyll in the light-harvesting process (Junge, 1977) and also to protect the cell against photosensitised oxidations which would otherwise prove lethal to the plant (Krinsky, 1966). For instance, an albino mutant of corn lacking carotenoids produced chlorophyll in dim light, but not in bright light, which caused total destruction of the chlorophyll in twenty minutes (Anderson and Robertson, 1960). A plant with chlorophyll but no coloured carotenoids is never found in nature since it represents a lethal mutation (Goodwin, 1971).

Carotenoid biosynthesis (see Britton, 1979) occurs via the isoprenoid pathway which we have met earlier in relation to gibberellin biosynthesis (pp. 195, 197). Geranylgeranyl pyrophosphate gives rise to phytoene, the first unique member of the carotenoid synthesis pathway, and the process thereafter is one of progressive desaturation through phytofluene, ζ-carotene, neurosporene and lycopene, which gives rise to β-carotene by cyclisation. Phytoene itself is colourless, and the colour develops through the sequence with increasing degrees of conjugation of the double-bond system.

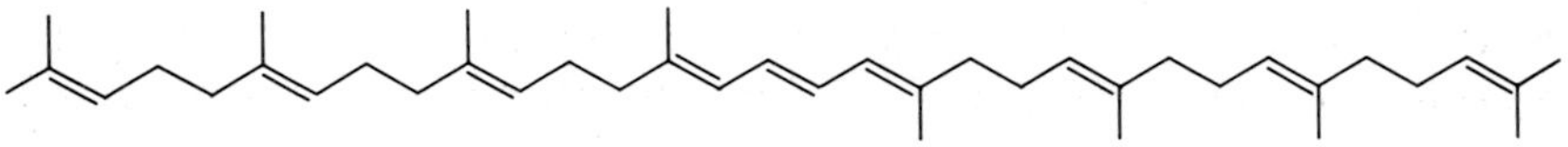

phytoene

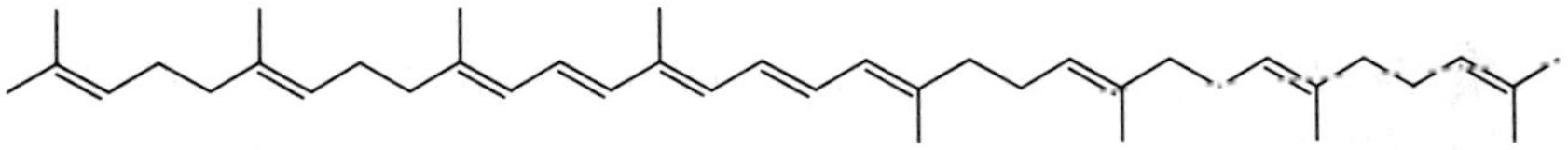

phytofluene

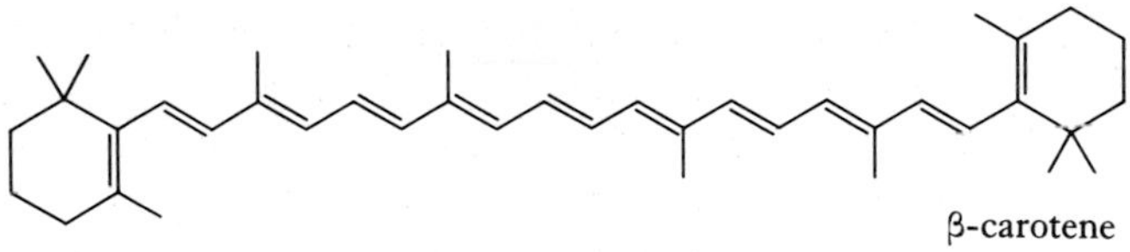

β-carotene

C. Pyridazinones (Table 6.2)

Norflurazon (Table 6.2) is the pyridazinone of practical importance, being a bleaching herbicide used for selective weed control in, e.g., cotton. Information pertinent to the mode of action of the pyridazinones has been surveyed by Eder (1979). We will first describe their effect on carotenoid synthesis in intact plants and algae, and then consider work on *in vitro* systems. This will be followed by an account of other effects caused by these compounds, and some general conclusions.

Table 6.2 Compounds interfering with carotenoid accumulation

Name	Structure
Pyridazinones Norflurazon	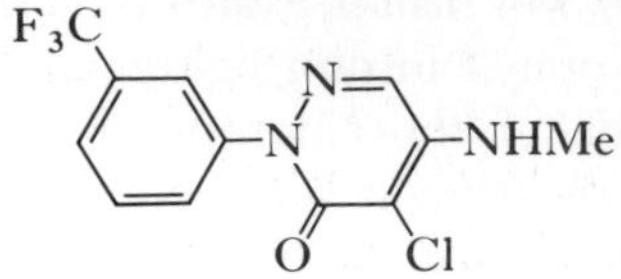
San 6706	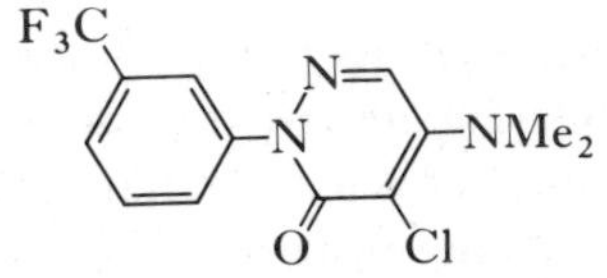
Others Aminotriazole	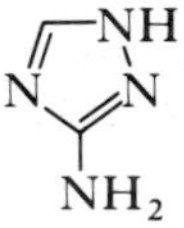
Fluridone	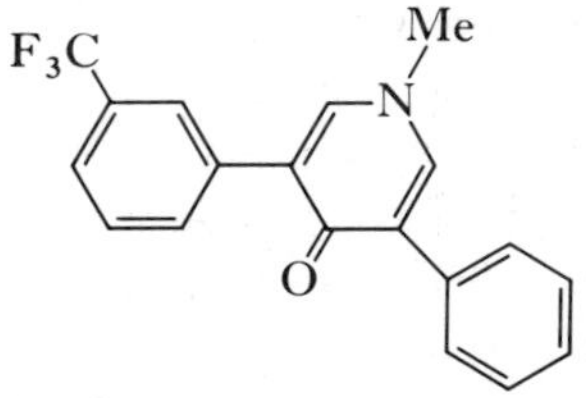
Methoxyphenone	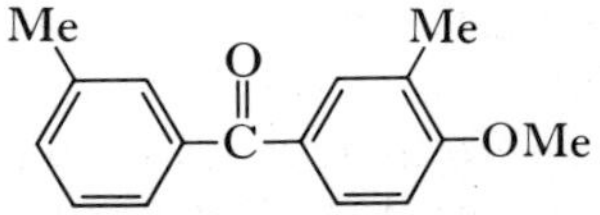

1. *Studies on inhibition of carotenoid synthesis in intact plants and algae*

In plants treated with pyridazinone herbicides, β-carotene levels are lowered and various precursors accumulate, usually phytoene (Bartels and McCullough, 1972; Vaisberg and Schiff, 1976; Pardo and Schiff, 1980) and also phytofluene (Ben-Aziz and Koren, 1974; Kümmel and Grimme, 1975; Ridley and Ridley, 1979). As discussed already, one of the functions of the carotenoids is to protect the photosynthetic apparatus against photo-destruction, so that the decrease in chlorophyll content which accompanies the effect on carotenoid levels at high light intensities (Hilton *et al.*, 1969; Devlin *et al.*, 1976; Lichtenthaler and Kleudgen, 1977) is easily explained; it does not occur at low light intensities (Ridley and Ridley, 1979), or when illumination is with light of a wavelength not absorbed by chlorophyll (Frosch *et al.*, 1979).

Frosch *et al.* (1979) examined the effects of pyridazinones on lipid, chlorophyll and carotenoid content in mustard seedlings exposed to white light, red light and far-red light, the last of which is not absorbed by chlorophyll. This approach enabled them to separate the effect of the herbicide from the photo-oxidising effect of light, and they concluded that, in the case of norflurazon, a primary effect on carotenoid synthesis is sufficient to explain the herbicidal activity. Photodestructive reactions may proceed with the involvement of superoxide (p. 90; Feierabend and Winkelhüsener, 1982).

Ridley and Ridley (1979) have proposed that the site of action of San 6706 is located in the cytoplasm. The biosynthesis of carotenoids from mevalonic acid is believed to take place entirely within the chloroplast (Goodwin, 1981) and a recent study has shown that enzymes catalysing the formation and desaturation of phytoene are in fact associated with the chloroplast outer membrane (Lütke-Brinkhaus *et al.*, 1982). Nevertheless, the biosynthesis is dependent on nuclear DNA, so, for instance, enzymes might be synthesised in the cytoplasm and then moved to the chloroplast (see Ridley and Ridley, 1979 for original references). These workers showed that photosynthesis was not inhibited in San 6706-treated plants if low-intensity light was used. However, San 6706 was a good inhibitor of the Hill reaction *in vitro* ($I_{50} = 1.6$ μM). Ridley and Ridley (1979) reasoned that, if San 6706 had been able to gain access to the chloroplast, it would have inhibited photosynthesis, and they concluded therefore that it must have been confined to the cytoplasm. Such a view is supported by the work of Grumbach and Drollinger (1980) who examined the effect of protein synthesis inhibitors on San 6706-treated greening radish plants, and deduced that protein synthesis in the cytoplasm was required for the herbicidal effect.

Radish was also used by Grumbach and Bach (1979), who found that, although the pigment content of San 6706-treated plants was very low, the incorporation of radioactivity from acetate and mevalonate into, for instance, chlorophyll and β-carotene, was, surprisingly, much higher than in the controls, and this was associated with increased activity of the enzyme 3-hydroxy-3-methylglutaryl CoA reductase, which catalyses the formation of mevalonic acid (see Fig. 5.1 p. 197). This indicated that their San 6706-

treated plants could synthesise, but not accumulate chlorophylls and carotenoids, and they suggested that this could have been because other structural components of the thylakoid were missing.

The effect of pyridazinones on intact algae has also been studied. Sandmann *et al.* (1981) analysed the relationship between the structure of a range of pyridazinones and their ability to cause bleaching effects in *Scenedesmus*. Three substitution positions were examined, and optimal activity required electron-donating and withdrawing groups leading to the charge distribution as shown:

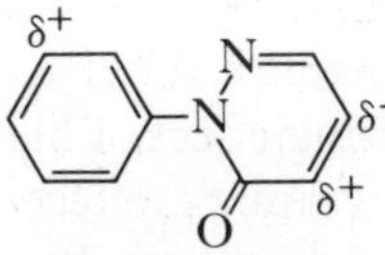

Subsequent work has largely confirmed this conclusion and put it on a quantitative basis by Hansch analysis (Sandmann and Böger, 1983).

2. Effects on carotenoid synthesis in in vitro systems

Sandmann *et al.* (1980) studied the effect of norflurazon on a carotenoid-synthesising system from the fungus *Phycomyces blakesleeanus* though here they pinpointed a site of action earlier in the pathway than phytoene. If this were true also in plants it would be irreconcilable with the body of data indicating phytoene and phytofluene accumulation in treated plants. More reconcilable results were obtained by Beyer *et al.* (1980) in a study of the effects of San-6706 on β-carotene synthesis from [^{14}C]-isopentenyl pyrophosphate by a cell-free preparation from daffodil flowers. Inclusion of 50 μM compound led to about 80% less label in β-carotene, and a build-up of geranylgeraniol and phytoene. Finally it has recently been shown that a cell-free system from the blue-green alga *Aphanocapsa* could convert radioactive geranylgeranyl pyrophosphate into β-carotene. When norflurazon was added the ratio of radiolabelled phytoene to β-carotene changed from 0.16 (no additions) to 5.40 (1 μM norflurazon), consistent with a block in the pathway between phytoene and β-carotene (Clarke *et al.*, 1982).

As mentioned above, enzymes catalysing the formation of lycopene from geranylgeranyl pyrophosphate have recently been shown to be associated with the outer membranes of spinach chloroplasts (Lütke-Brinkhaus *et al.*, 1982). Clearly it will be of interest to examine the effect of pyridazinone herbicides on these enzymes.

The results obtained in the *in vitro* systems lend weight to the view that pyridazinones act by inhibiting carotenoid biosynthesis, though they do not explain some of the results obtained with intact plants, as described above.

3. Other effects

Chloridazon (formerly pyrazon) is a pyridazinone which is an inhibitor of the Hill reaction (p. 80) with very little effect on carotenoid biosynthesis or fatty acid desaturation (St. John and Hilton, 1976). As already mentioned, San 6706 is quite a good inhibitor of the Hill reaction *in vitro* (Hilton *et al.*, 1969; Ridley and Ridley, 1979; Eder, 1979) though there is evidence that this does not happen *in vivo* since it may not gain entry to the chloroplast (Ridley and Ridley, 1979). Norflurazon, however, is only a poor inhibitor of the Hill reaction, being reported as much less active than chloridazon, which is itself ten times less active than atrazine (Eder, 1979).

Feierabend and his colleagues have conducted a detailed investigation of the effects of the bleaching herbicides on enzymes of peroxisomes (cellular organelles containing many oxidase enzymes) and chloroplasts (Feierabend and Schubert, 1978; Feierabend *et al.*, 1979). The effects were complicated and various, but since they did not occur in plants maintained in darkness, the authors concluded that they were secondary consequences of photo-destruction.

A number of studies have provided evidence that the pyridazinones also interfere with plant lipids. Galactolipids (glycerol esterified on the 1- and 2-positions by fatty acids and substituted on the 3-position by glycosidically-linked galactose) are major chloroplast lipids whose fatty acid composition is altered by pyridazinones. The effects differ in detail according to plant source and compound tested, but a typical effect is to raise the proportion of linoleic acid (18 carbon atoms, 2 double-bonds) to linolenic acid (18 carbon atoms, 3 double-bonds) (St. John and Hilton, 1976; St. John *et al.*, 1979; Eder, 1979; Khan *et al.*, 1979). Since certain structural analogues of the pyridazinones considered above can bring about this effect on lipids without inhibiting pigment formation, it is very likely that the influence on lipid constitution is a separate effect and not a consequence of carotenoid loss; this view is further supported by the expectation that photo-oxidative reactions would not give the rather specific changes caused by some pyridazinones (St. John, 1982). Nevertheless, the inhibition of lipid desaturation might have some parallel with the inhibition of desaturation of carotenoids already described (Britton, 1979).

4. Conclusions

Two attempts have been made at unifying hypotheses which might explain all the results.

Lichtenthaler and Kleudgen (1977) proposed that the observed loss of chloroplast ribosomes might account for the activity of the compounds. However, since the ribosomes are retained in the dark (Bartels and Watson, 1978) and since, as we have seen, light is required for pigment breakdown to occur, the ribosomes are probably lost as a consequence of photo-oxidations taking place in the chloroplast (Feierabend and Schubert, 1978). Nevertheless, some more subtle interaction at the ribosome level may still be a possibility (see Kleudgen, 1979).

Vaisberg and Schiff (1976) examined the effect of norflurazon on *Euglena*, and their results suggested that a primary effect on carotenoid synthesis could have consequences for the synthesis and assembly of other components of chloroplast membranes via normal metabolic control mechanisms. Subsequent studies have shown that such control mechanisms appear not to operate in higher plants (Pardo and Schiff, 1980; Frosch et al., 1979), i.e., the inhibition of carotenoid synthesis does not, per se, prevent synthesis of other chloroplast membrane components or halt membrane assembly. This contrast provides a reminder that, in studies on herbicides, higher plant tissue should be used as a first choice unless the reasons for using a model system are compelling.

In conclusion, a common denominator in the reported work is the effect of pyridazinones on carotenoid levels and it therefore seems likely that this lies at the heart of their herbicidal action. The lack of the protective carotenoids leads to photodestructive reactions which kill the plant. It is possible that treated plants do not lose their ability to make coloured carotenoids, but that, for some reason, these cannot be retained. The site of action could be in the cytoplasm in green plant tissue, but inhibition of carotenoid synthesis has so far been demonstrated only in cell-free preparations from other sources.

D. Aminotriazole

There is an extensive literature on the mode of action of aminotriazole (amitrole; Table 6.2) which was introduced as a herbicide in 1954. A wide range of biochemical effects has been documented, and these are reviewed by Hilton (1969) and Carter (1975). Aminotriazole is, like the pyridazinones, a bleaching herbicide, and it may also interfere with carotenoid biosynthesis.

When applied to whole plants aminotriazole has the characteristic effect of preventing the formation of chlorophyll in newly expanded leaves (Hilton, 1969), and it has been shown that the herbicide only affects developing tissue (Bartels et al., 1967). Thus, treated plants grown in the dark did not become green on exposure to light if the herbicide had been applied at germination, but chlorophyll did form if the leaves were first allowed to mature in the dark, and were then treated with aminotriazole and exposed to the light. An electron microscope and ultracentrifuge study showed that the chloroplasts of aminotriazole-treated seedlings that had been grown in the light had considerable abnormalities, but these were not present in treated seedlings kept in the dark (Bartels *et al.*, 1967).

This evidence suggests that aminotriazole prevents the synthesis of a component that is required for the accumulation of chlorophyll. However, the herbicide does not affect the synthesis of chlorophyll itself since it does not prevent the accumulation of protochlorophyllide by seedlings grown in the dark, nor the further conversion to chlorophyllide a, an immediate precursor of normal chlorophyll (Burns *et al.*, 1971).

Burns *et al.* (1971) pointed out that wheat seedlings grown in the presence of aminotriazole closely resembled seedlings of a wheat mutant lacking carotenoids, and they found that spectral and chromatographic evidence from extracts of dark-grown seedlings from wheat seeds shaken with 100 μM aminotriazole for 24 hours indicated that β-carotene levels were lower while more saturated precursors accumulated. Since, as we have seen, carotenoids such as β-carotene are necessary to prevent the photo-oxidation of chlorophyll and the disruption of chloroplast structure it is possible that the inhibition of carotenoid biosynthesis is the primary site of action of aminotriazole. Even if this is so, the effect is not identical to that of the pyridazinones, since bleaching of existing carotenoids was much slower when aminotriazole-treated seedlings were first exposed to bright light than when the seedlings had been treated with San 6706 (Feierabend *et al.*, 1979).

Recently, Rüdiger and colleagues (see Rüdiger and Benz, 1979) found that treatment of wheat seedlings with aminotriazole (1 mM during the first 24 h of a 96 h growth period) resulted in a change in the structure of their chlorophyll so that it carried, instead of the normal phytol side-chain, a chain consisting of geranylgeraniol and dihydrogeranylgeraniol; chlorophyll molecules with these side-chains may be normal chlorophyll precursors. On this basis Rüdiger and Benz (1979) propose that a major effect of aminotriazole is to inhibit the reactions involved in reducing the geranylgeraniol side chain to phytol. However, it is difficult to predict the consequences of this effect, and to integrate it with the other work described, though it may be related to the discovery that the content of plasto-quinone, which also has an isoprenoid side-chain, is affected by amino-triazole (Vivekanadan and Gnanam, 1975).

geranylgeraniol

phytol

E. Fluridone

Fluridone (Table 6.2) is a pre-emergence herbicide, structurally related to the pyridazinones and of particular use for selective weed control in cotton. Similar studies to the ones described above have also been conducted with this compound. Bartels and Watson (1978) found that 50 and 100 μM fluridone caused β-carotene depletion and phytoene accumulation in wheat seedlings as did norflurazon. Lower concentrations were not used. In another study carotenoid levels were severely depleted in 6-day-old wheat and maize seedlings treated with low concentrations (< 10 μM) of fluridone, both in high and low light intensities, and chlorophyll was affected secondarily (Devlin *et al.*, 1978). Effects on other metabolic processes have been reported in isolated mesophyll cells of cotton and kidney bean but much higher concentrations were needed (Rafii *et al.*, 1979).

F. Methoxyphenone

Apparently the only work on the mode of action of this pre-emergence herbicide (Table 6.2) is reported by Fujii *et al.* (1977). Accumulation of β-carotene in dark-grown barnyard millet (*Echinochloa frumentacea*) was strongly inhibited by the herbicide, and chlorophyll accumulation was also prevented in light. β-Carotene depletion was accompanied by the accumulation of phytofluene and ζ-carotene, so the evidence that the compound inhibits carotenoid synthesis is good.

III. HERBICIDES AND FUNGICIDES INHIBITING LIPID SYNTHESIS
A. Introduction

We shall be concerned in this section with pesticides that probably act by interfering with the synthesis of one of three types of lipid: *fatty acids*, which possess a long hydrocarbon chain and a terminal carboxyl group; *sterols* which are compounds based on the structure

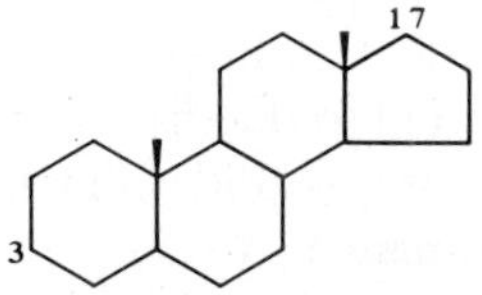

and carry a hydroxyl group at position 3 and a branched aliphatic chain of 8 or more carbon atoms at position 17; and *phosphoglycerides* which comprise a molecule of glycerol, esterified on two of its hydroxyl groups with fatty acids, and on the third with a phosphoryated alcohol.

B. Synthesis of fatty acids and their products
1. *Biosynthesis and function of the cuticle*

The cuticle of plants consists of a polymer of interesterified hydroxy-carboxylic acids of C_{16} or C_{18} chain length, embedded in wax, itself a complex mixture of very long-chain non-polar lipids. Hydrocarbons are a major component of these non-polar lipids, and they usually contain an odd number of carbon atoms between 25 and 35, although branched, unsaturated and cyclic hydrocarbons are known. Waxes also contain ketones and secondary alcohols of similar chain length, with the functional group near the centre of the molecule (Kolattukudy, 1980).

Other major components of plant cuticular lipids are esters, usually of C_{20}-C_{24} *n*-alkanoic acids with C_{24}-C_{28} *n*-alkan-1-ols (the so-called wax esters), and the free acids and alcohols also occur. Minor components are also known, and the balance of the constituents varies between plants (Kolattukudy, 1980).

All these components are thought to derive from fatty acids, which are built up by the successive addition of 2-carbon fragments from malonic

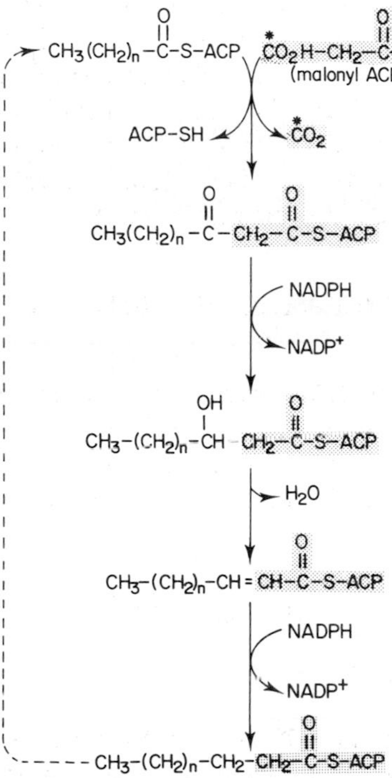

Fig. 6.3 An outline of fatty acid biosynthesis. n is an even number from 0 to 12, ACP-SH, acyl carrier protein

acid (which becomes decarboxylated) to a growing hydrocarbon chain, both components being esterified to the thiol group of a special acyl carrier protein, ACP (Stumpf, 1980). The carbonyl group in the growing chain is removed by reduction, dehydration and another reduction to give the hydrocarbon chain (for outline see Fig. 6.3). When the chain length reaches 16 carbon atoms the acid (palmitic acid) is released by cleavage of the thioester.

The chain is extended further by separate elongation systems, one to convert palmitate to stearate (C_{18}) and one for further elongation of stearate (Harwood, 1975). Double bonds are introduced at this stage by desaturase enzymes operating on the C_{16} or C_{18} saturated fatty acid as appropriate. Alcohols derive from the acids, and the acids probably give rise to hydrocarbons by decarboxylation so that the chain contains an odd number of carbon atoms (reviewed by Kolattukudy, 1980).

The cuticle is considered to give essential protection to the cells against injury caused by wind, physical abrasion, frost, radiation, pathogens and chemicals (Martin and Juniper, 1970). It is clear, therefore, that inhibition of its formation will be detrimental to the plant.

2. Thiocarbamates

(a) Gross effects

Thiocarbamate herbicides (Table 6.3) are volatile liquids that are normally incorporated into the soil and which exert their herbicidal action at the early stages of seedling growth. Their metabolism and chemical properties have been reviewed by Fang (1975).

Table 6.3   Thiocarbamate herbicides

Name	Structure
A. Alkyl and aryl thiocarbamates	
Butylate	$^{i}Bu_2N.\overset{\displaystyle O}{\overset{\displaystyle \|}{C}}.SEt$
Cycloate	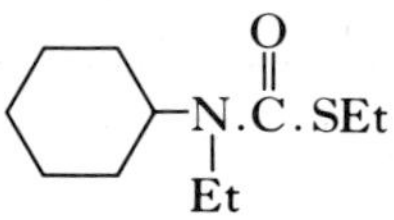 $\langle \hspace{-0.3em} \bigcirc \hspace{-0.3em} \rangle$—$\overset{\displaystyle O}{\overset{\displaystyle \|}{\underset{\displaystyle \underset{Et}{\|}}{N}}}.\overset{\displaystyle }{C}.SEt$
EPTC	$Pr_2N.\overset{\displaystyle O}{\overset{\displaystyle \|}{C}}.SEt$

Name	Structure
Molinate	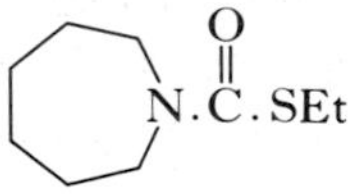
Pebulate	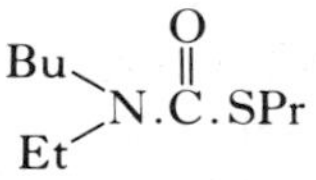 Bu $\diagdown$ $N.C.SPr$, Et, with O double bonded to C
Thiobencarb	$Et_2N.\overset{\overset{O}{\|\|}}{C}.SCH_2$—⟨benzene ring⟩—$Cl$
Tiocarbazil	$\left(\overset{Me\diagdown}{\underset{Et\diagup}{C}}H-\right)_2 N.\overset{\overset{O}{\|\|}}{C}.SCH_2$—⟨benzene ring⟩
Vernolate	$Pr_2N.\overset{\overset{O}{\|\|}}{C}.SPr$
B. *S-Chloroallyl thiocarbamates** Di-allate	$^{i}Pr_2N.\overset{\overset{O}{\|\|}}{C}.SCH_2\overset{\overset{Cl}{\|}}{C}=CHCl$
Sulfallate	$Et_2N.\overset{\overset{S}{\|\|}}{C}.SCH_2\overset{\overset{Cl}{\|}}{C}=CH_2$
Tri-allate	$^{i}Pr_2N.\overset{\overset{O}{\|\|}}{C}.SCH_2\overset{\overset{Cl}{\|}}{C}=CCl_2$

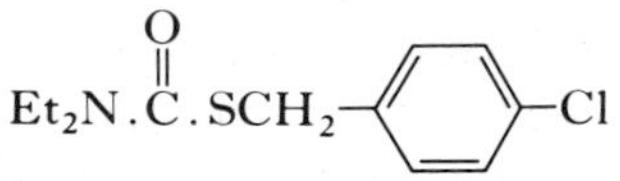

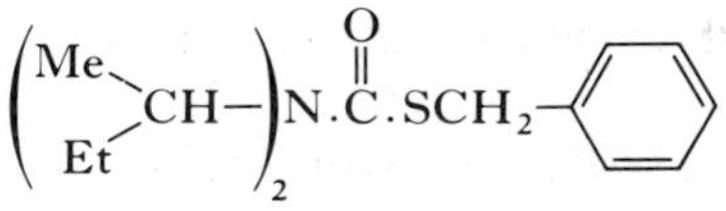

*Stereochemistry at the double bond is not defined here; for details of this see Worthing (1979).

It should be noted that sulfallate is a dithiocarbamate; it seems possible that this molecule undergoes a $>C=S$ to $>C=O$ conversion in the plant but this may not be necessary since it is active in its own right against chloroplast preparations (Wilkinson and Smith, 1975; see below).

There is no doubt that thiocarbamates inhibit the production of wax by plants. Gentner (1966) observed that post-emergence application of thiocarbamate herbicides caused obvious reductions in the amount of external wax on cabbage leaves; at herbicidal rates EPTC caused up to 90% reduction in the weight of wax produced per unit area. Wilkinson and Hardcastle (1970) found that EPTC reduced the thickness of the cuticle of sicklepod (*Cassia obtusifolia*) leaves by up to some 20% when applied at herbicidal rates. Still *et al.* (1970) grew peas in a 10 μM solution of EPTC or sulfallate, and then took electron micrographs of the leaves. They observed that EPTC caused a significant reduction in the amount of wax on both upper and lower surfaces, though sulfallate only altered the crystalline structure of the wax, and not the amount present. They also found that plants treated with di-allate vapour for seven days had 80% less surface wax per unit dry weight than untreated plants. Thus the thiocarbamates interfere with cuticle formation. We will now turn to the specific suggestions that have been made about their biochemical mode of action.

(b) *Effects on fatty acid biosynthesis*

Although the primary site of action of the thiocarbamates has not been established with certainty, effects on fatty acid biosynthesis are well documented. Wilkinson and co-workers have conducted a wide range of investigations into the effects of thiocarbamates on, principally, fatty acid levels of treated tissues (Wilkinson *et al.*, 1977; Wilkinson, 1978a). For instance, when soybean seedlings were grown in a nutrient solution containing 2.6 μM EPTC, the total leaf fatty acid content was depressed by 63% after 19 days, though fresh weight was reduced only 18%. Individual C_{16} and C_{18} fatty acid levels were examined, and the results suggested that depletion of both saturated and some unsaturated fatty acids was occurring (Wilkinson *et al.*, 1977). Effects on lipid synthesis in isolated spinach chloroplasts have also been noted; micromolar concentrations of EPTC, butylate, pebulate, sulfallate and vernolate prevented acetate incorporation into fatty acids (Wilkinson and Smith, 1975, 1976).

The more specific suggestion has been made that the thiocarbamates inhibit the biosynthesis of very long-chain fatty acids (i.e., those containing more than 18 carbon atoms), and thereby alcohols, wax esters and hydrocarbons (Fig. 6.4). For instance, in germinating peas, 10 μM EPTC and di-allate depressed incorporation of radioactivity from acetate into very long-chain fatty acids even though the incorporation into total fatty acids was slightly increased (Harwood and Stumpf, 1971). Kolattukudy and Brown (1974) went on to examine the effect of micromolar solutions of EPTC, di-allate and sulfallate on the incorporation of (1-^{14}C) acetate into cuticle components by pea-leaf slices. They found that incorporation of

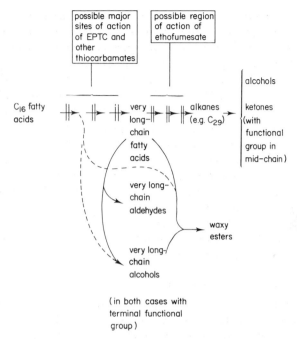

Fig. 6.4 Outline of the biosynthesis of components derived from very long-chain fatty acids; possible sites of action of thiocarbamates and ethofumesate

label into C_{31}-hydrocarbon and C_{31}-secondary alcohol was 50% inhibited at 1–2 μM di-allate, though C_{22}-fatty acid accumulated, suggesting an effect on the elongation process between these components. This was supported by the observation that incorporation of label into surface lipids was much more strongly inhibited than that into internal lipids, which do not contain such long aliphatic chains (Kolattukudy, 1970).

Effects on very long-chain fatty acid synthesis can also be studied using cut potato discs, which make these components as part of a wound response. EPTC at 10 μM caused only a slight depression of labelled acetate incorporation into total (C_{16}–C_{24}) fatty acids, but there was an almost complete absence of radioactivity in products derived from very long-chain fatty acids (C_{20}–C_{24} in this system; Bolton and Harwood, 1976). Di-allate and tri-allate were more effective inhibitors of acetate incorporation into total fatty acids, giving about 25% inhibition at 10 μM, leading to an apparent maximum of 60% inhibition at concentrations of 100 μM and above. When the individual fatty acid labelling patterns were compared with those of untreated tissue, the principal effect noted was that the proportion of label incorporated into palmitic acid (C_{16}) had risen at the expense of that into very long-chain fatty acids. Eronen and Karunen (1977) analysed directly the fatty acids of control and EPTC-treated wheat leaves, obtaining results which supported an effect of EPTC on fatty acid chain elongation. In addition, they found an increase in the ratio of saturated to unsaturated fatty acids, indicating an inhibition of the fatty acid desaturase system.

In view of the symptoms produced by EPTC it seems very likely that the effect on very long-chain fatty acid synthesis is a major effect of the thiocarbamates *in vivo*. There are no specific proposals for the manner in which a thiocarbamate might interact with a component of the elongation system, but it seems possible that the enzymes of this system may contain similar components to those in C_{16}-fatty acid synthesis and desaturation processes so that an effect on these latter systems would not be too surprising and might explain the effects seen in some experiments. Further resolution of the precise interaction of thiocarbamates with these biosynthetic processes will probably have to await progress in techniques for studying the pathways *in vitro*.

(c) Other effects

Various other effects of thiocarbamates have been reported, such as those on quinone levels in wheat after EPTC treatment (Wilkinson, 1978b), though this was considered to be a secondary effect following disruption of lipid synthesis. Weinberg and Castelfranco (1975) took a similar view of their observations on EPTC-induced changes in intracellular membrane constituents. Thiocarbamates can also stimulate peroxidase (Harvey *et al.*, 1975).

There has been recent interest in the effects of thiocarbamates on gibberellin metabolism. Gibberellin levels became lowered in treated plants (Donald *et al.*, 1979; Wilkinson and Ashley, 1979), and the growth response of sensitive species to exogenous gibberellins was reduced by thiocarbamates (Donald, 1981). Wilkinson (1983) has recently shown that, whereas EPTC did not influence total incorporation of radiolabelled mevalonic acid into isoprenoid compounds by a cell-free preparation from *Sorghum bicolor*, the distribution of label was altered so that geranylgeraniol accumulated while the levels of *ent*-kaurene fell (to about half in the presence of 1 μM EPTC). These compounds are intermediates in the pathway of gibberellin biosynthesis (see Fig. 5.1, p. 197). Wilkinson (1983) proposes that the action on this pathway is a major effect of EPTC in plants.

(d) Oxidation of thiocarbamates

Casida and his colleagues have suggested that all thiocarbamates are converted *in vivo* to their sulphoxides and that these are metabolised by

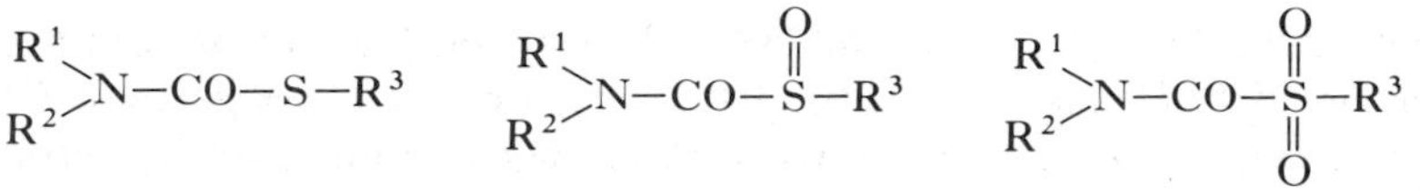

$$R^1 \diagdown N-CO-S-R^3 \qquad R^1 \diagdown N-CO-\overset{\displaystyle O}{\underset{\displaystyle }{\overset{\|}{S}}}-R^3 \qquad R^1 \diagdown N-CO-\overset{\displaystyle O}{\underset{\displaystyle O}{\overset{\|}{\underset{\|}{S}}}}-R^3$$
$$R^2 \diagup \qquad\qquad R^2 \diagup \qquad\qquad R^2 \diagup$$

| thiocarbamate | sulphoxide | sulphone |

conjugation through the thiol group of glutathione. By analogy therefore a vital thiol group involved in lipid synthesis might be the target of these herbicides (Lay and Casida, 1976). This idea is supported by the observation that the sulphoxides are herbicidal in their own right (Casida *et al.*, 1974). According to Horváth and Pulay (1980) however, the sulphoxide of EPTC was unable to react with glutathione, even in the presence of maize homogenates. They suggest that the sulphone is a better candidate for the reactive derivative, but that they did not detect it in their *in vivo* metabolism studies because it reacts with thiols or water as quickly as it is formed. Whichever is the reactive species there may, of course, be several essential thiol groups involved in lipid and wax synthesis as well as in other areas of metabolism; reaction with these might go some way to explaining the diversity of effects that have been noted above.

The compounds in Table 6.3 are divided into two groups at the suggestion of Schuphan and Casida (1979) who showed that the sulphoxides of *S*-chloroallyl thiocarbamates can rearrange to give acrolein derivatives, as shown in Fig. 6.5 for tri-allate. Di-allate will react to give 2-chloroacrolein, and both this and 2-chloroacrylyl chloride are likely to be general

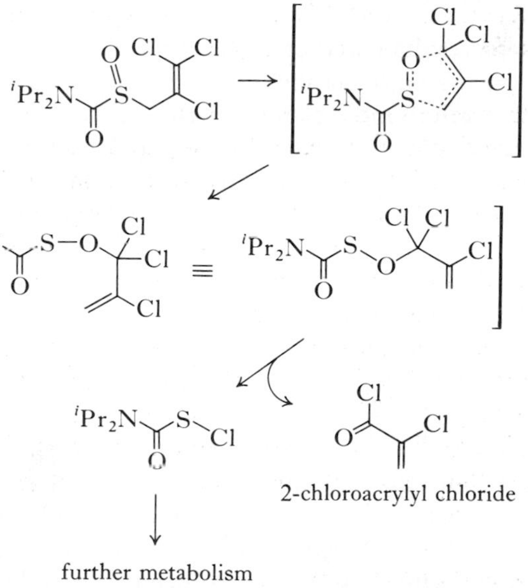

2-chloroacrylyl chloride

further metabolism

Fig. 6.5 Mechanism for the conversion of tri-allate to 2-chloroacrylyl chloride (Schuphan and Casida, 1979)

toxicants by indiscriminate reaction with cell components. Schuphan and Casida (1979) therefore suggest that in the case of these particular thio-carbamates production of acrolein derivatives contributes to the herbicidal action.

3. *Ethofumesate*

This compound is used for the control of grass and broad-leaved weeds in sugar beet.

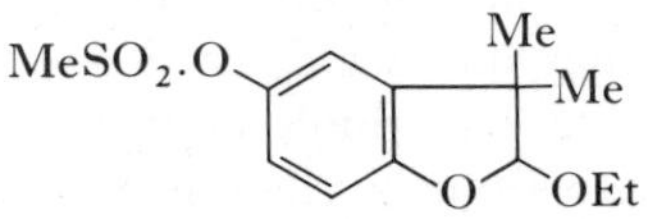

ethofumesate

Ethofumesate has been reported to inhibit epicuticular wax formation in cabbage leaves (Leavitt *et al.*, 1978). In this work wax was isolated from leaves of plants grown to the six-leaf stage in the presence of ethofumesate at 0.84 kg/ha and also from untreated leaves, and the main components were compared. In relation to the control leaves, ethofumesate caused the levels of C_{29}-alkane and C_{29}-ketone to fall by approximately 95 and 90% respectively. On the other hand the content of the long chain waxy esters containing a total of 36–44 carbon atoms (a minor cuticle component) rose by about 50%.

Putting these results together with current ideas on the biosynthesis of cuticle components outlined above, it is possible to propose as a working hypothesis that ethofumesate inhibits the conversion of very long-chain fatty acids into alkanes, as shown in Fig. 6.4. A consequence would be the increased availability of the components required for the synthesis of waxy esters. There is as yet no biochemical evidence to support this hypothesis.

C. Sterol formation
1. *Introduction*

(a) *General remarks*

Following leads in the mammalian and plant biochemistry literature, Ragsdale and Sisler (1972) showed that ergosterol biosynthesis was in-hibited in sporidia of *Ustilago maydis* exposed to the (then) new fungicide triarimol at a concentration of approximately 7 μM. The use of this com-pound is now discontinued (Worthing, 1979), but it was the first member of a rapidly expanding class of agricultural fungicides which inhibit sterol synthesis, and will therefore be included in the account which follows.

The fungicides which we shall consider are given in Table 6.4, and most are active against powdery mildew fungi with varying effectiveness towards smut and rust diseases (Buchenauer, 1977a). We have included only agricultural fungicides but not a number of drugs which are receiving attention as inhibitors of sterol biosynthesis (Buchenauer, 1978a; Marriott, 1980; van den Bossche *et al.*, 1978).

Table 6.4   Fungicidal inhibitors of sterol biosynthesis. The references (italic suffixes) are to work on the mode of action

Name	Structure
A. Imidazoles 1-(4-Chlorophenoxy)-1- (imidazol-1-yl)-3,3- dimethylbutanone	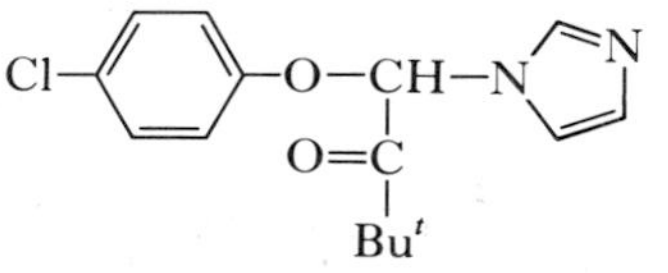 [a*]
Fenapanil	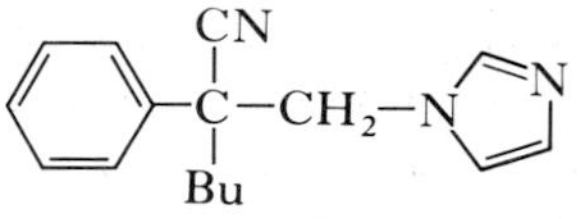 [b]
Imazalil	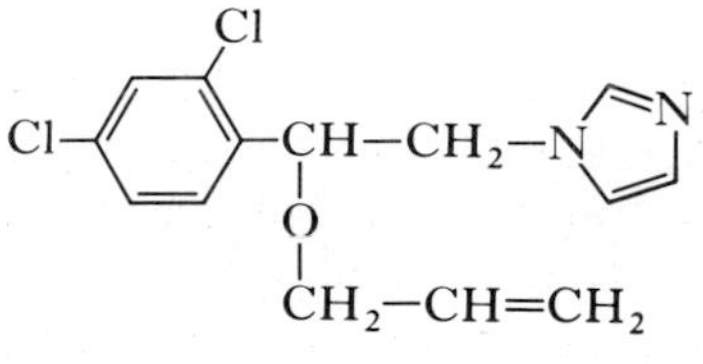 [c,d,e]
Prochloraz	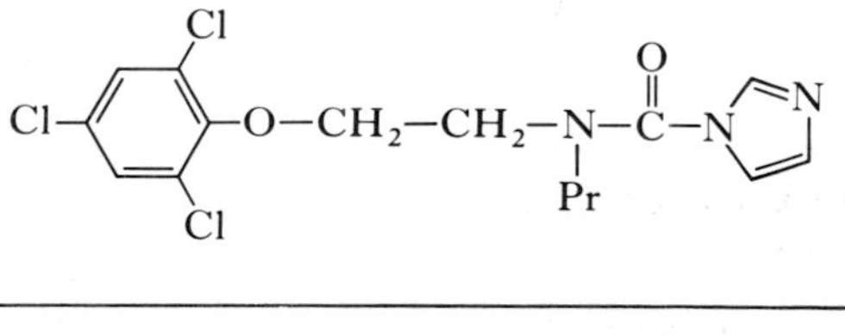 [f]
D. Triazoles Bitertanol	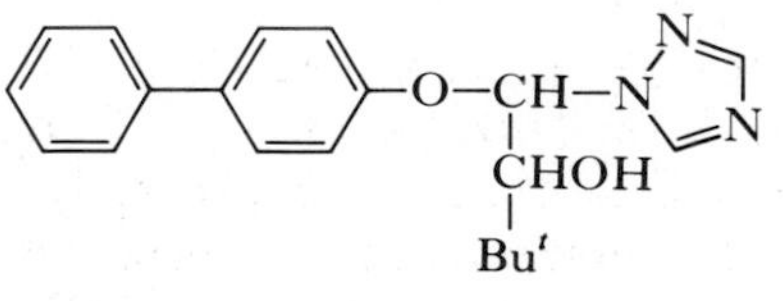 [g]

---

*No biochemical data in these publications.   †For work on non-fungal systems see, e.g., Frasinel *et al.* (1978) and Mitropoulos *et al.* (1976)

Name	Structure
Diclobutrazol	
Triadimefon	
Triadimenol	
Etaconazole	
Propiconazole	
Fluotrimazole	

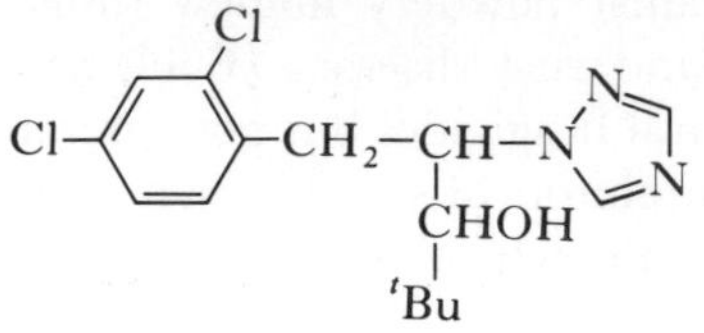

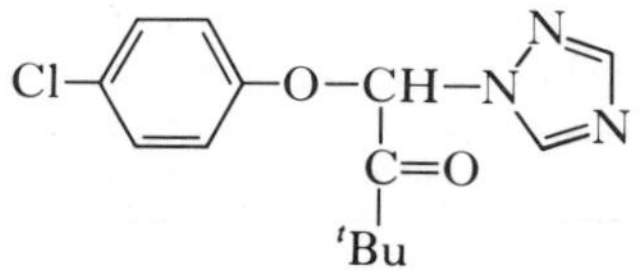

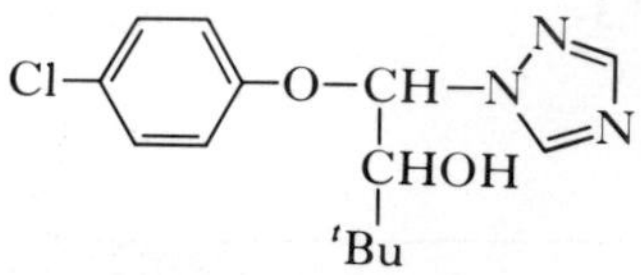

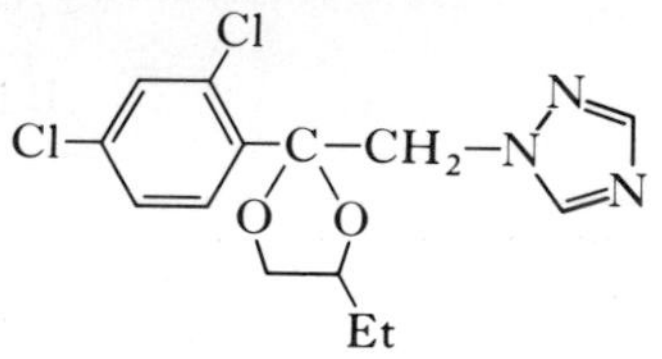

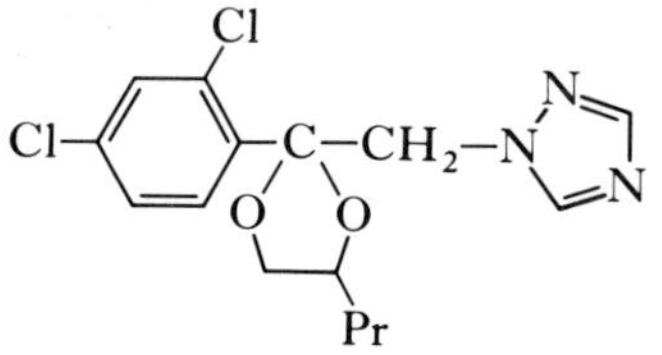

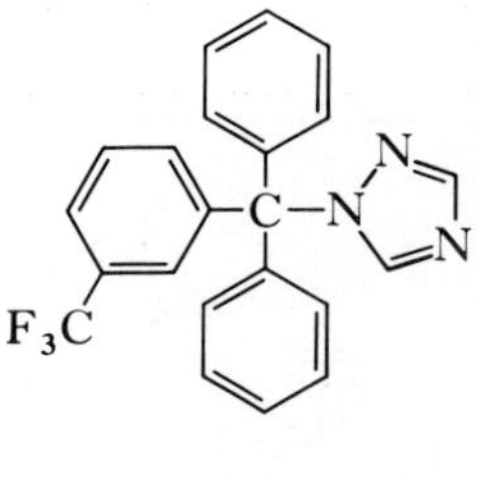

Name	Structure
C. Pyridine Buthiobate	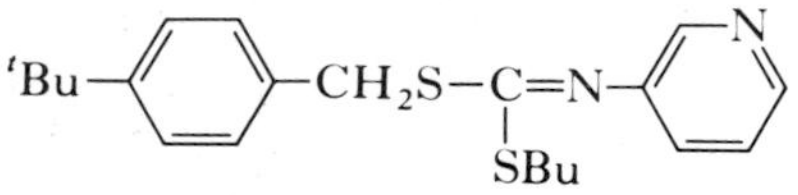 [n,o]
D. Pyrimidines Fenarimol	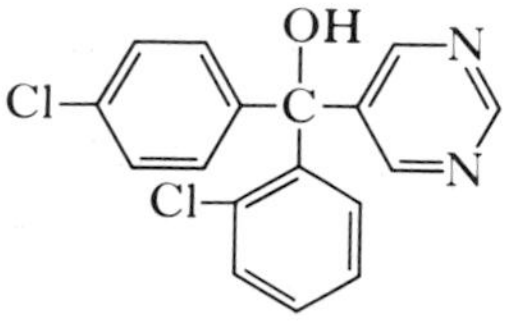 [p]
Nuarimol	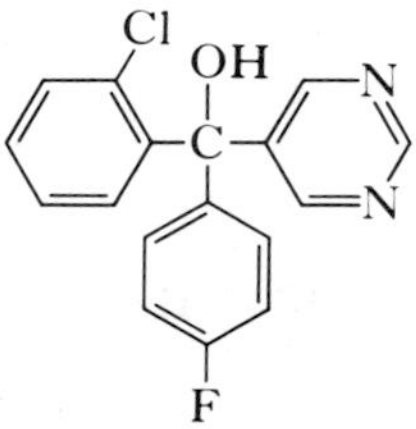 [p]
Triarimol	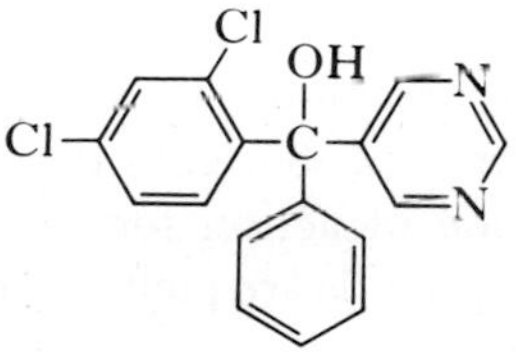 [q,r]
E. Piperazine Triforine	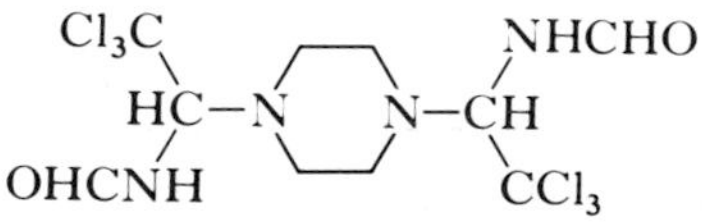 [s]
F. Morpholines Dodemorph	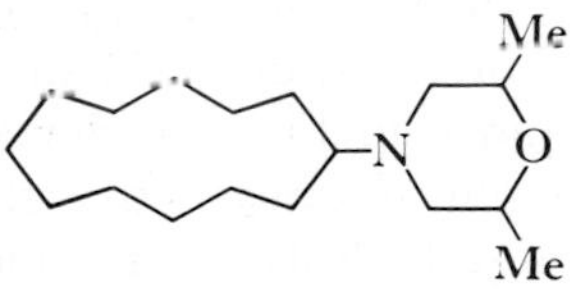

Name	Structure
Fenpropimorph	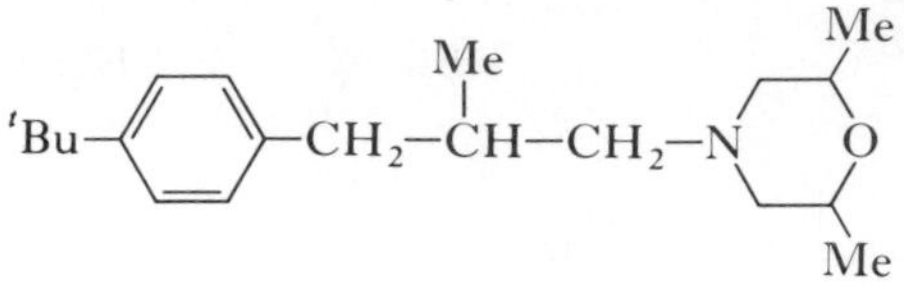 ^{t}Bu—⟨benzene⟩—CH$_2$—CH(Me)—CH$_2$—N(morpholine, 2,6-diMe) ^{t*}
Tridemorph	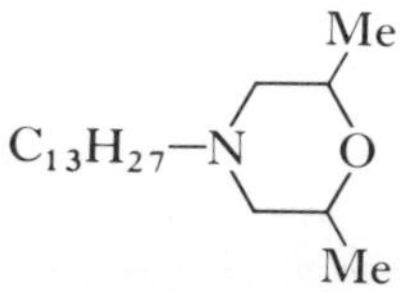 C$_{13}$H$_{27}$—N(morpholine, 2,6-diMe) ^{u,v}

[a]Worthing (1979); [b]Martin and Edgington (1982); [c]Buchenauer (1977d); [d]Leroux and Gredt (1978); [e]Siegel and Ragsdale (1978); [f]Pappas and Fisher (1979); [g]Kraus (1979); [h]Bent and Skidmore (1979); [i]Buchenauer (1977b); [j]Buchenauer (1978b); [k]Henry and Sisler (1981); [l]Buchenauer and Kemper (1981); [m]Buchenauer (1978a); [n]Kato et al. (1974); [o]Kato and Kawase (1976); [p]Buchenauer (1977c); [q]Ragsdale and Sisler (1972); [r]Ragsdale (1975); [s]Sherald and Sisler (1975); [t]Bohnen et al. (1979); [u]Kerkenaar et al. (1979); [v]Kato et al. (1980).

The methods used to identify a compound as an inhibitor of sterol synthesis are largely common to the various investigators, and these techniques will therefore be considered together. Citations to work establishing this mode of action for individual compounds are given in Table 6.4. Some compounds are included on the basis of structural analogy, in which case the reference gives biological data only, and is distinguished by an asterisk.

(b) Function and biosynthesis of ergosterol

Ergosterol is the major sterol in many fungi and it plays an important role in membrane structure and function (see Buchenauer, 1977b; Ragsdale, 1977; Weete, 1974). This function is analogous to that of the structurally related cholesterol in mammalian membranes. A likely pathway of biosynthesis is outlined in Fig. 6.6. Some of the transformations may be carried out in different orders, so that there is more than one route between 24-methylene-dihydrolanosterol and ergosterol (Sherald and Sisler, 1975; Ragsdale, 1977). The route is not necessarily the same in different organisms, so that, for instance, in *Saccharomyces cerevisiae* the introduction

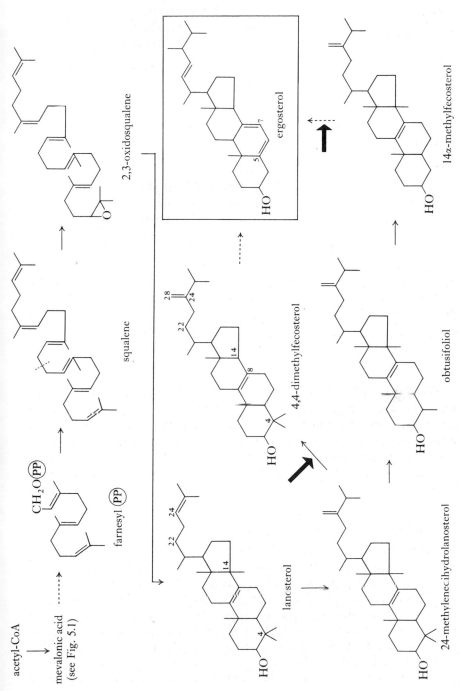

Fig. 6.6 Outline of alternative biosynthetic routes to ergosterol. Dotted arrows indicate more than one step. For further explanation see the text. Heavy arrows indicate the site of action of 14α-demethylation inhibitors. (PP) pyrophosphate

of C-28 into the side-chain is not carried out until after the completion of the demethylations at C-4 and C-14, and the introduction of the double bonds at C-5 and C-7 (Kato and Kawase, 1976). Also although the transformations tend to occur in a preferred order in a given species, this is not absolute (Weete, 1973) so that inhibition of the preferred pathway may lead to accumulation of intermediates of a parallel pathway (see Fig. 6.6). To a certain extent, elucidation of the details of the biosynthesis has occurred simultaneously with an increase in understanding of the way in which the inhibitors work.

(c) Methods used for investigation of inhibitors

Two approaches have been used to study the effects of inhibitors on fungal sterols. The first consists of growing fungi in the presence and absence of a sub-lethal concentration of fungicide and extracting, purifying and analysing the sterols, the latter using thin-layer chromatography or gas-liquid chromatography. Sterols are identified with reference to standards, or with the aid of mass spectrometry. The second approach involves using a pulse of a radioactive precursor to indicate which sterols are being synthesised at a given time after exposure to sub-lethal levels of fungicide. The site of action of the compound is indicated by the fact that substrates accumulate, whilst the product is depleted or absent altogether. This approach can be used with intact fungal cells or with subcellular preparations.

2. Inhibition of sterol biosynthesis

The first detailed studies reported were those of the effects on sterol synthesis of triforine (Sherald and Sisler, 1975) and triarimol (Ragsdale, 1975). The field has been reviewed by Buchenauer (1977a), Ragsdale (1977) and Siegel (1981).

It is now established that the principal action of the fungicides in parts A–E of Table 6.4 is to prevent removal of the 14α-methyl group*, so that all the intermediates which accumulate in the presence of inhibitor contain this group. The steps which are inhibited are indicated by the heavy arrows in Fig. 6.6 and this leads to consequent rises in the levels of 24-methylene dihydrolanosterol, obtusifoliol and 14α-methylfecosterol (also referred to in the literature by its systematic name, 14α-methyl-$\Delta^{8,24(28)}$-ergostadienol) (Ragsdale, 1977; Siegel, 1981).

As part of their study on the mechanism of action of buthiobate in yeast cell-free extracts, Kato and Kawase (1976) showed that ^{14}C-14-desmethyl lanosterol (i.e. lanosterol lacking the 14α-methyl group) could undergo 4-demethylation in the presence of buthiobate; thus 4-demethylation itself is not affected. This therefore suggests that, in the studies des-

*The α- denotes the stereochemical orientation of the methyl group.

cribed above, the 4-methyl-14-methyl intermediates (i.e. 24-methylene dihydrolanosterol and obtusifoliol) accumulate because the presence of a 14-methyl group makes removal of the 4-methyl group less efficient.

Corroboration of the site of action of buthiobate was obtained by Tanaka *et al.* (1977), who obtained a good correlation between fungicidal activity against *Sphaerotheca fuliginea* and inhibition of ergosterol biosynthesis in *Monolinia fructigena* in a series of 24 analogues. Thus, for instance, the 2- and 4-pyridyl analogues of buthiobate (which carries a 3-pyridyl substituent – see Table 6.4) had no activity in either system. Further support for the mode of action of compounds of this type has come from the demonstration that a mutant of *Ustilago maydis* deficient in C-14 demethylation was nearly identical to a fenarimol-treated wild-type strain in respect of a number of growth parameters; in addition the sterols of the mutant were the same as those accumulating in wild-type fungal cells treated with fungicides such as fenarimol and etaconazole (Walsh and Sisler, 1982).

Since much of the work described above has involved analysing cells grown in sub-lethal concentrations of chemical, we can take it that the sensitivity of sterol biosynthesis reflects the susceptibility of the fungus. The sensitivity of the sterol demethylation system of, for example, *Aspergillus* to imazalil is illustrated by the fact that Siegel and Ragsdale (1978) found it to be temporarily inhibited when the fungus was treated with concentrations of imazalil in the approximate range 0.01–0.03 µM, levels which are 50–100 times lower than the minimum growth-inhibitory concentration.

Thus we may conclude that the compounds in parts *A-E* of Table 6.4 are fungicidal because they prevent 14-demethylation of sterols. The fungi will be deprived of ergosterol, but it is likely that their growth is impaired because the accumulating methylated sterols find their way into the cell membranes and cause disruption (see Gadher *et al.*, 1983). The compounds do not interfere with a range of other basic cellular functions (Buchenauer, 1977b).

It has been suggested that triadimefon is not active *per se*, but is converted *in vivo* to triadimenol before exerting its effect (Gasztonyi and Josepovits, 1979). The reduction introduces a second asymmetric centre into the molecule and, since more than one enzyme is involved in the reduction, the ratio of the two isomers produced depends on the fungal species. Further, there seems to be a correlation between the susceptibility of a fungus and the extent to which it forms the more fungitoxic isomers of triadimenol (Gasztonyi, 1981). Interestingly, it has recently been shown that triadimefon itself is a good inhibitor of sterol demethylation in a yeast cell-free system (Gadher *et al.*, 1983). However, we cannot be sure that the yeast enzyme is the same as that in other fungi and, in addition, metabolism could have taken place in the *in vitro* system.

3. *Possible additional effects of triarimol and related compounds*

Although the pyrimidine derivatives in part D of Table 6.4 inhibit 14α-demethylation as their major effect, there seems to be an additional effect on the reactions between 14α-methylfecosterol and ergosterol (Ragsdale, 1975, 1977; Buchenauer, 1977c) as indicated below:

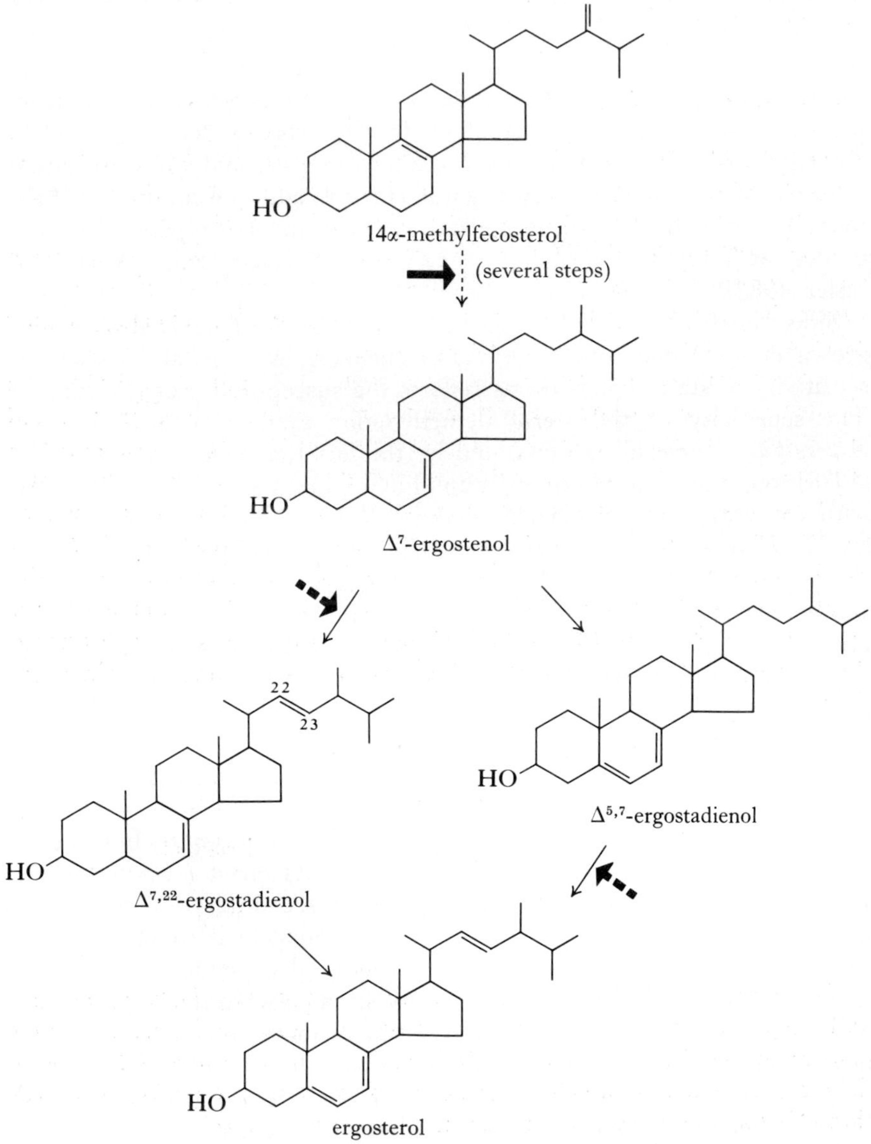

14α-methylfecosterol

(several steps)

Δ⁷-ergostenol

Δ⁷,²²-ergostadienol

Δ⁵,⁷-ergostadienol

ergosterol

The heavy arrow indicates the inhibition of 14α-demethylation. However, the pattern of depletion of subsequent intermediates from treated *Ustilago* cultures suggested that fenarimol, nuarimol and triarimol also prevented the introduction of the C-22 double bond as indicated by the broken arrows above (Ragsdale, 1975, 1977; Buchenauer, 1977c). Nevertheless Ragsdale (1977) allows the possibility that the effects on the side chain are secondary consequences of the inhibition of demethylation of the sterol nucleus.

4. *Morpholines* (Table 6.4)

Available evidence suggests that tridemorph does act by inhibition of sterol biosynthesis, but that it inhibits a different step. By structural analogy, this may also apply to dodemorph and fenpropimorph.

Kerkenaar *et al.* (1979) reviewed earlier work and conducted further investigations leading to the conclusion that inhibition of ergosterol biosynthesis was probably the primary effect since other effects generally took longer and required higher doses of fungicide.

A more detailed investigation was conducted by Kato *et al.* (1980) who compared the effects of buthiobate and tridemorph on the growth and metabolism of *Botrytis cinerea*. Sterol synthesis was the most sensitive process, and buthiobate inhibited 14α-demethylation as indicated above. The pattern of build-up of precursors suggested that tridemorph, on the other hand, prevented the movement of the double bond from the 8- to the 7-position, a reaction which occurs following the removal of the methyl groups from the C-4 and C-14 positions. This conclusion was based on the fact that the main sterol accumulating in tridemorph-treated cultures was considered to be fecosterol, which retains the Δ^8 double bond.

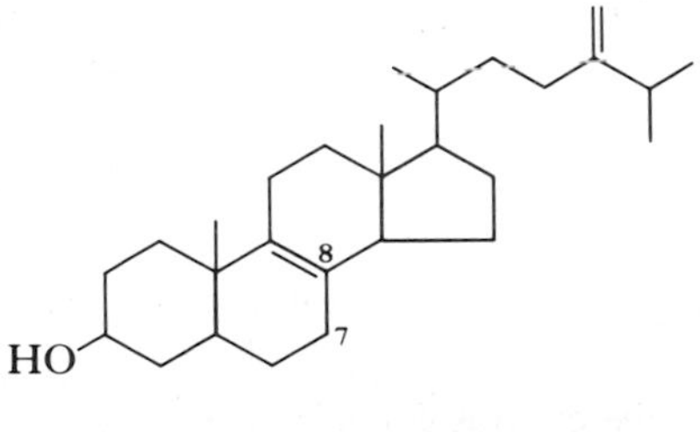

fecosterol

Kerkenaar *et al.* (1981) conducted a similar study using *Ustilago maydis*, and they identified the major accumulating sterol as ignosterol. The conversion of 24-methylenedihydrolanosterol to 4,4-dimethylfecosterol (Fig. 6.6) is in fact a two-stage process and Kerkenaar *et al.* (1981) proposed that it is the second step which is inhibited by tridemorph.

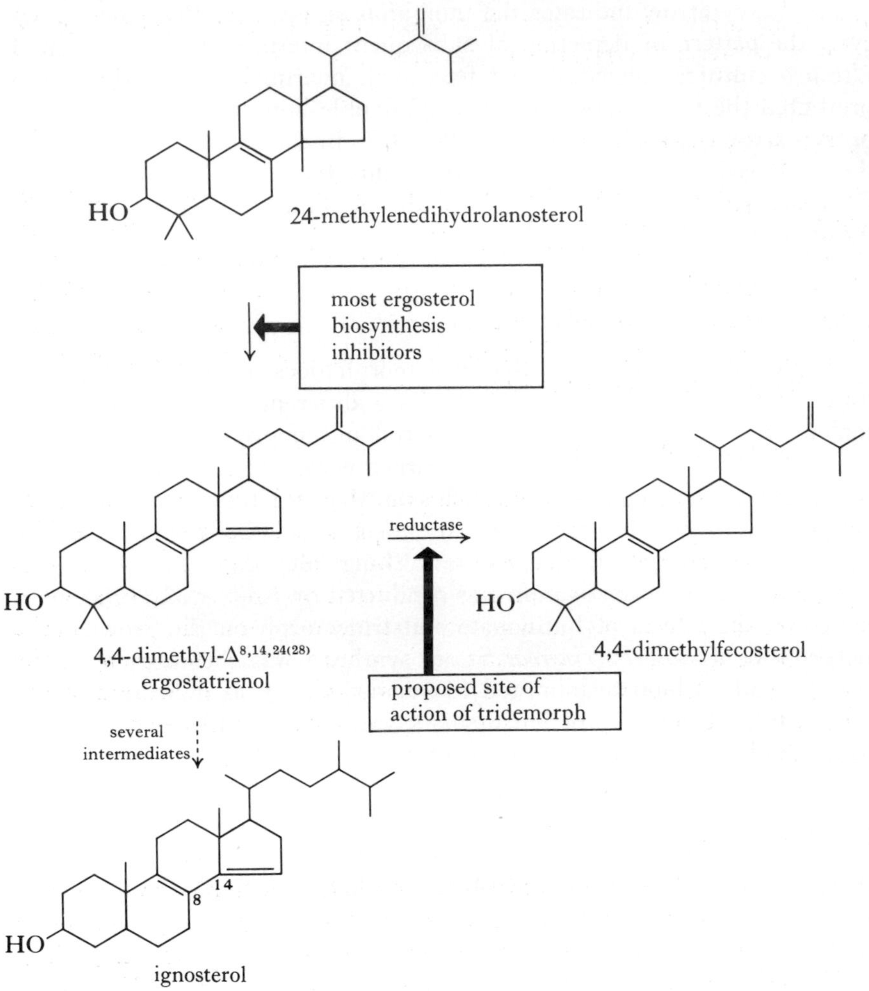

HO 24-methylenedihydrolanosterol

most ergosterol
biosynthesis
inhibitors

reductase

HO

4,4-dimethyl-$\Delta^{8,14,24(28)}$
ergostatrienol

HO

4,4-dimethylfecosterol

proposed site of
action of tridemorph

several
intermediates

HO

8   14

ignosterol

The different conclusions reached by the two groups could be due to technical difficulties in identifying accumulating sterols (see Kerkenaar *et al.*, 1981) or they may simply reflect differences in the responses of the fungi.

In summary, although we cannot specify exactly where tridemorph acts, we can conclude that it inhibits sterol biosynthesis at a different site in the pathway from the one inhibited by the majority of ergosterol biosynthesis inhibitors.

It has been shown that 100 μM dodemorph suppressed the growth of *Verticillium albo-atrum* by 78% while stimulating electrolyte leakage from the fungal cells (Buchenauer, 1975), but whether this is related to inhibition of sterol biosynthesis is not known.

5. *Cross-resistance*

Ragsdale (1977) points out that mutants of *Cladosporium cucumerinum* selected for resistance to triarimol turned out to be cross-resistant to triforine (Sherald *et al.*, 1973) and buthiobate (Sherald and Sisler, 1975). Further examples of cross-resistance are given by Siegel (1981). Whilst cross-resistance could indicate a common site of action, the mechanism of resistance was not established, so that changes in permeability or metabolism could also explain the results.

6. *Molecular mode of action*

A molecular mechanism by which compounds in this section may interfere with sterol biosynthesis has recently been proposed (Gadher *et al.*, 1983). The C-14 demethylation is a cytochrome P-450 dependent reaction (p. 331) and these workers suggest that the compounds occupy the sterol-binding site of the enzyme in such a way that a nitrogen atom of the heterocyclic ring complexes with the iron atom at the centre of the protohaem of the cytochrome. This iron atom normally has a role in the formation of the activated oxygen species involved in the initial step of the 14-methyl group removal.

7. *Other effects*

A common observation in studies on the mode of action of the group of compounds under consideration has been that they also induce rises in free fatty acids and sometimes triglycerides in treated organisms (e.g. Buchenauer 1977a,b; Ragsdale, 1975; Sherald and Sisler, 1975). This is generally accepted to be a secondary effect, perhaps attributable to the lack of a sink for fatty acids once sterol depletion causes a cessation of membrane synthesis. An additional secondary effect is the inhibition of 3-hydroxy-3-methyl-glutaryl-CoA reductase, the enzyme which catalyses the formation of mevalonic acid (see Fig. 5.1, p. 197). This apparently occurs due to feedback inhibition of the enzyme by intermediates accumulating following the blocking of 14α-demethylation (Berg *et al.*, 1981). A further effect at higher concentrations may involve direct attack on the cell membrane and it is possible that this could be the most important action of certain pharmaceutical fungicides which are structurally related to those above (see Walsh and Sisler, 1982, for original references).

8. Inter-relationship with plant growth regulators

Several triazole and pyrimidine ergosterol biosynthesis inhibitors have been shown to affect not only sterol synthesis but also gibberellin formation by higher plants (Buchenauer, 1977a; Buchenauer and Röhner, 1981; Ragsdale, 1977). In a reciprocal way the gibberellin biosynthesis inhibitor ancymidol is weakly fungicidal (Sherald *et al.*, 1973). As we have explained in Chapter 5, ancymidol blocks conversion of *ent*-kaurene to *ent*-kaurenoic acid in plants. Ragsdale (1977) rationalises these observations by pointing out that this is a closely similar operation to the first stages of C-14 demethylation and suggesting that the two compounds are affecting a similar type of mixed-function oxidase in sterol and gibberellin synthesis. It is interesting in this connection that the candidate plant growth retardant paclobutrazol (proposed name), which bears a very close structural resemblance to the fungicide diclobutrazol (Table 6.4), is reported to act by inhibition of gibberellin synthesis (Lever *et al.*, 1982).

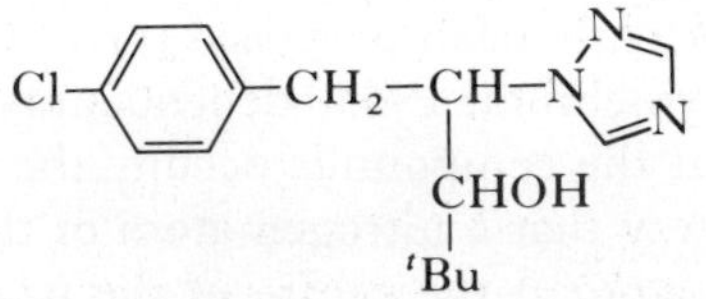

paclobutrazol

Effects of chlormequat chloride and chlorphonium chloride on sterol biosynthesis were briefly dealt with in Chapter 5 (p. 199).

D. Phosphatidyl choline synthesis
1. Introduction

Phosphatidyl choline is a particular member of the class of phospho-glycerides whose general structure is:

where R^1 and R^2 originate from saturated and unsaturated C-16 or C-18 fatty acids, and X is derived from an alcohol, choline ($Me_3\overset{+}{N}CH_2CH_2OH$) in the case of phosphatidyl choline. These phospholipids have a polar 'head group' and a hydrophobic 'tail' and this so-called amphipathic nature is ideally suited to their role in membrane structure. Phosphatidyl choline is one of the most abundant phospholipids found in cellular membranes.

2. Action of inhibitors

S-Benzyl O,O-di-isopropyl phosphorothioate (known in Japan as IBP) and edifenphos are used for the control of Pyricularia oryzae in rice.

$$\text{C}_6\text{H}_5\text{—CH}_2\text{—S—P(OPr}^i)_2 \ (=\text{O}) \qquad (\text{C}_6\text{H}_5\text{—S—})_2 \text{P(=O)—OEt}$$

IBP   edifenphos

Early indications were that IBP might inhibit fungal chitin synthesis *in vivo* since it caused an accumulation of the chitin precursor UDP-N-acetylglucosamine (see p. 272) whereas related, non-fungitoxic, compounds did not (Maeda *et al.*, 1970). However, since chitin synthesis takes place on the outside of the cell membrane the possibility that interference with the membrane could be the primary effect was considered by Maeda *et al.* (1970) and later supported by De Waard (1972). The more specific proposal has now been made that IBP and edifenphos inhibit a particular step in the biosynthesis of phosphatidyl choline (Akatsuka *et al.*, 1977; Kodama *et al.*, 1979, 1980).

Although phosphatidyl choline can be formed from diacylglycerol and choline activated in the form of cytidine-diphosphocholine, the compounds do not affect this pathway. In fact they probably interfere with an alternative route (Fig. 6.7) via the methylation of phosphatidyl ethanolamine; it is this latter pathway which predominates in Pyricularia oryzae (Kodama *et al.*, 1979).

These workers incubated *P. oryzae* mycelia *in vitro* with various radioactive substrates and found that incorporation of label from [S-methyl-^{14}C]-methionine, [S-methyl-^{14}C]-S-adenosylmethionine and [1-^{14}C]-glycerol into phosphatidyl choline was markedly inhibited by 174 μM IBP. This suggested an action on a step after S-adenosyl methionine. Subsequent experiments *in vitro* confirmed that none of the enzymes between glycerol and phosphatidyl ethanolamine were affected by 174 μM IBP. Kodama *et al.* (1979) then prepared a microsomal fraction from *P. oryzae* which was capable of transferring methyl groups from [S-methyl-^{14}C]-S-adenosyl methionine to an endogenous substrate, presumably phosphatidyl ethanolamine, to give phosphatidyl choline. Under their conditions, this activity was half-inhibited by approx. 140 μM IBP.

A similar study by Kodama *et al.* (1980) indicated that edifenphos interfered with phosphatidyl choline synthesis at the same step. In this case the phospholipid N-methyl transferase activity was half inhibited at about 50 μM, the concentration which led to 50% inhibition of fungal growth *in vitro*.

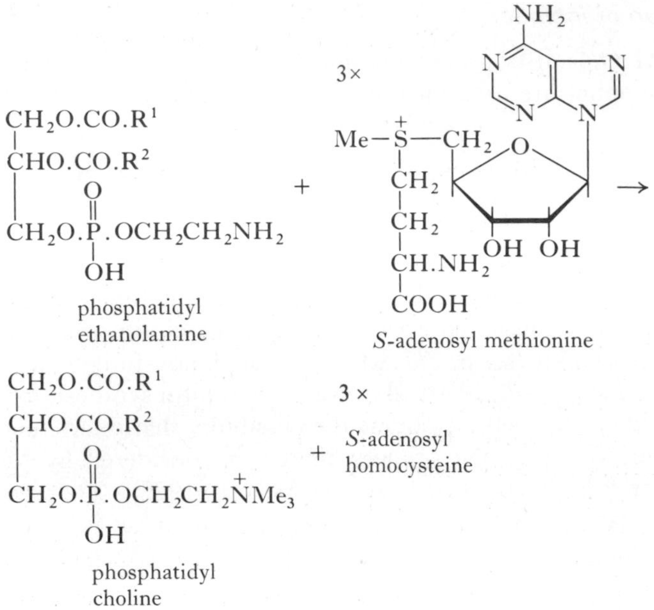

CH₂O.CO.R¹
|
CHO.CO.R²
| O
| ‖
CH₂O.P.OCH₂CH₂NH₂
|
OH

phosphatidyl
ethanolamine

+

3×

Me—S⁺—CH₂
|
CH₂
|
CH₂
|
CH.NH₂
|
COOH

S-adenosyl methionine

→

CH₂O.CO.R¹
|
CHO.CO.R²
| O
| ‖
CH₂O.P.OCH₂CH₂N⁺Me₃
|
OH

phosphatidyl
choline

+   3 ×   *S*-adenosyl
homocysteine

Fig. 6.7 The methylation of phosphatidyl ethanolamine. R¹ and R² are long-chain fatty acids

Thus there is good evidence that the effect of IBP and edifenphos on phospholipid synthesis occurs at the *N*-methyl transferase step although the concentrations used are too high for us to be certain that this is the primary mode of action.

IV. PESTICIDES INTERFERING WITH POLYSACCHARIDE SYNTHESIS
A. Cellulose synthesis
1. *Introduction*

Cellulose is an insoluble (1-4)β-linked glucose polymer which constitutes the essential structural backbone of the plant cell wall (Colvin, 1980). Two herbicides may act through inhibition of cellulose synthesis.

2. *Dichlobenil and chlorthiamid*

Dichlobenil is a pre- and post-emergence herbicide which inhibits meristematic (i.e. actively dividing) tissue (Worthing, 1979), while chlorthiamid is known to be converted to dichlobenil in both plants and soil (Beynon and Wright, 1972) by a process which might involve singlet oxygen (Pillai, 1977).

chlorthiamid   dichlobenil

Hogetsu *et al.* (1974) were the first to suggest that dichlobenil was a cellulose synthesis inhibitor. They fed radiolabelled glucose to sections of adzuki bean epicotyls which had been treated with 100 μM indoleacetic acid to induce elongation. During a 6 h incubation 1 μM dichlobenil reduced incorporation into the cellulose fraction by about 30% compared with controls. In a subsequent experiment the I_{50} for the inhibition was found to be approximately 3 μM.

Subsequently, Umetsu *et al.* (1976) showed that incubation of sus-pension-cultured soybean cells with 0.5–2 μM dichlobenil led to remarkable swelling of cells and modification of cell form, effects also caused by coumarin, a known but considerably less effective cellulose synthesis inhibitor. Meyer and Herth (1978) have extended this work by showing that, after the cell wall was enzymically removed from cultured tobacco cells to give protoplasts, regeneration of the wall was completely inhibited by c. 12 μM dichlobenil.

Montezinos and Delmer (1980) compared a number of compounds as inhibitors of incorporation of radiolabel from glucose into cellulose in cotton fibres growing from *in vitro* cultured ovules (the ovule is the female part of the flower, which will develop into the seed following fertilisation). Dichlobenil was the most effective compound tested, causing specific inhibition of glucose incorporation into cellulosic and to a lesser extent non-cellulosic glucans of the cell wall of the developing cotton fibre. Micromolar concentrations were effective within 10 minutes, and in-hibition could be reversed by washing the herbicide out. The speed of action argues against an indirect effect via inhibition of RNA or protein synthesis (Montezinos and Delmer, 1980). Results obtained in a study of the effect of dichlobenil on cellulose formation by tobacco protoplasts were consistent with an action on the cellulose-synthesising complex at the cell membrane (Galbraith and Shields, 1982).

Montezinos and Delmer (1980) found that chlorthiamid also inhibited glucose incorporation in 2–3 hour incubations suggesting that chlor-thiamid has some activity of its own as well as that due to prior conversion in the soil to dichlobenil.

B. Chitin synthesis

1. *Introduction*

Chitin is a structural component of insect cuticles and many fungal cell walls and is comparable in importance to the cellulose of plants. The nature of fungal cell walls is reviewed by Misato and Kakiki (1977) and that of the insect cuticle by Richards (1978).

Chitin is essentially a polymer of N-acetylglucosamine, and is synthesised by a reaction catalysed by the enzyme chitin-UDP*-N-acetylglucosaminyl-transferase (EC 2.4.1.16, often referred to as chitin synthetase), in which N-acetylglucosamine units are transferred from UDP-N-acetylglucosamine to the growing chitin chain:

$$\text{UDP-}N\text{-acetylglucosamine} \xrightarrow{\substack{\text{chitin-UDP-}N\text{-acetylglucosaminyl-}\\\text{transferase}}} \text{chitin}$$

The pathway from glucose to chitin is given in Fig. 6.8.

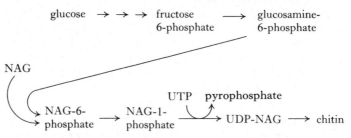

Fig. 6.8 An outline of chitin formation. NAG, N-acetylglucosamine; UDP, uridine 5′-diphosphate; UTP, uridine 5′-triphosphate

2. *Polyoxin fungicides*

The polyoxins are a class of closely inter-related antifungal compounds which have been widely used in Japan. They are all thought to act in the same way, although polyoxin D has been most studied from a biochemical point of view. Work on the biochemical mode of action has been reviewed by Gooday (1979), Misato and Kakiki (1977), and Misato *et al.* (1979). The structures of the polyoxins resemble that of UDP-N-acetylglucosamine:

$$NH_2CO.OCH_2\overset{\underset{\displaystyle OH}{|}}{CH}.\overset{\underset{\displaystyle NH_2}{|}}{CH}.\overset{\underset{\displaystyle OH}{|}}{CH}.CO.NH.CH$$

R = CO$_2$H; polyoxin D
R = CH$_2$OH; polyoxin B

**Abbreviation:* UDP, uridine 5′-diphosphate.

UDP-N-acetylglucosamine

Not surprisingly in view of this similarity, polyoxin D is a competitive inhibitor of chitin synthetase with respect to UDP-N-acetylglucosamine, having a K_i of 1.4 μM, compared with a K_m of 1.4 mM in the case of the *Neurospora crassa* enzyme (Endo *et al.*, 1970). Follow up studies have led to detailed proposals for the interaction of the related polyoxin C and derivatives with the enzyme from *Pyricularia oryzae* (Hori *et al.*, 1974a). This difference between K_m and K_i values seems to be generally found across a range of fungal species although *in vivo* the sensitivity to polyoxins varies. This might in some cases be due to differences in the extent of uptake (Gooday, 1979) though this was not a likely explanation for the lack of sensitivity of *Mucor rouxii* to polyoxin A (Müller *et al.*, 1981). In this context, simple peptides antagonise the fungicidal action of polyoxin *in vivo* (Mitani and Inoué, 1968) but do not protect against chitin synthetase inhibition *in vitro* (Hori *et al.*, 1974b), so the antagonism might occur at the level of the uptake mechanism.

Supporting evidence for the mode of action *in vivo* comes from Ohta *et al.* (1970) who showed that 10 μM polyoxin D inhibited the incorporation of radioactive glucosamine into the cell wall of the fungus *Cochliobolus miyabeanus* by over 85%, while it had no effect on the incorporation of glucose or acetate, though it did inhibit amino acid incorporation to the extent of about 50%. They also showed that UDP-N-acetylglucosamine accumulated in mycelia of fungi treated with polyoxin D. Finally, they found that of seven polyoxin derivatives, the only one that did not inhibit the incorporation of glucosamine into the fungal cell wall was also the only one lacking fungicidal activity.

Thus it may be concluded with some confidence that polyoxin D kills fungi by interfering with chitin synthesis. This mode of action is consistent with the observation that polyoxin D causes swelling of fungal mycelium, as well as that of the germ tubes of treated spores (Dekker, 1971). The recent development of a rapid preparation of chitin synthetase from *Coprinus cinereus* (Adams and Gooday, 1980) may facilitate future work.

3. The insect growth regulator diflubenzuron

Diflubenzuron is an insect growth regulator which is effective against a considerable range of larvae in a variety of situations. The compound disrupts the moulting process and is now clearly established to interfere with chitin synthesis.

The history of the development of the compound and the early mode of action experiments have been reviewed by Post and Mulder (1975). The vagaries of the pesticide discovery process are illustrated by the fact that the compound was synthesised during work designed to exploit the manufacturing process of the herbicide dichlobenil. Some of the reasons for the selection of this particular compound from a range of analogues are given by Verloop and Ferrell (1977).

Early histological work revealed that diflubenzuron interfered with larval cuticle deposition (Mulder and Gijswijt, 1973). Later, Sowa and Marks (1975) studied β-ecdysone-dependent chitin synthesis in regenerating leg tissue of cockroach and found the process to be half inhibited by diflubenzuron at a tissue concentration of 69 μM; similar results were obtained with polyoxin D.

Subsequently Deul et al. (1978) investigated the effect of diflubenzuron on chitin synthesis in fifth-instar larvae of Pieris brassicae. Larvae were fed on treated or control cabbage leaves for 24 hours before being injected with [6-^{14}C]-glucose. After a further 24 hours the incorporation of radioactivity into various cellular fractions was examined, with the result that only the chitin fraction was found to be affected, with incorporation depressed almost to zero. The response was very rapid, so that essentially the same result was obtained when the diflubenzuron was injected (1 μg/larva) at the same time as the ^{14}C-glucose, and the extract prepared 15 minutes later. In this same system, lethal doses of both polyoxin D and diflubenzuron were shown microscopically to cause comparable abnormalities in the cuticle and to inhibit its further growth, as judged by a failure of the cuticle to increase in thickness (Gijswijt et al., 1979).

The work was taken a stage further by Hajjar and Casida (1979) who have developed a system for the study of chitin synthesis in vitro in isolated abdominal walls of the milkweed bug Oncopeltus fasciatus chosen for its size. In this system (cf. the cockroach system mentioned above) the hormone β-ecdysone does not have to be added. Diflubenzuron inhibited incorporation of label from glucose to chitin in this system with an I_{50} of 0.2 μg/g abdominal wall, which seems reasonable considering that the LD_{50}

was 2.4 μg/g fifth-instar nymph. The rapid action was confirmed and significant accumulation of UDP-N-acetylglucosamine was found following additions of [^{14}C]-N-acetylglucosamine to the *in vitro* abdomen system in the presence of 3 μM diflubenzuron. The use of a similar preparation from houseflies gave essentially the same results and conclusions (van Eck, 1979). Further, Hajjar and Casida (1979) demonstrated a correlation between the insecticidal potency of a range of analogues and their ability to inhibit incorporation of label from glucose to chitin in this system.

Taken together, the results described so far indicate that the primary effect of diflubenzuron in insect tissues is to inhibit chitin synthesis. The suggestion that the primary action might be via an inhibition of ecdysone metabolism (Yu and Terriere, 1977) is not consistent with all the evidence (Hajjar and Casida, 1979) and the idea that elevation of chitinase could be the first event (Ishaaya and Casida, 1974) has been ruled out by the timing of the effect (Deul *et al.*, 1978). These and other effects regarded as secondary are tabulated by Verloop and Ferrell (1977).

For some time biochemical limitations precluded an examination of the effects of diflubenzuron on chitin synthetase in cell-free systems but it has now been found that diflubenzuron does not inhibit solubilised preparations from the gut of *Tribolium castaneum* (Cohen and Casida, 1980), pupae of the stable fly *Stomoxys calcitrans* (Mayer *et al.*, 1980) and integument tissue of both the giant silkworm and the cabbage looper (Cohen and Casida, 1982). It has also been shown that fungal chitin synthetase preparations are insensitive to diflubenzuron (Leighton *et al.*, 1981).

The question as to the primary effect of diflubenzuron is therefore still open. One possibility has arisen from the work of DeLoach *et al.* (1981) who showed that diflubenzuron led to a cessation of DNA synthesis only in the cells which give rise to the adult epidermis and cuticle during development of stable fly pupae (Meola and Mayer, 1980). However, the evidence is not yet sufficient to propose that this is the primary effect (DeLoach *et al.*, 1981).

Another suggestion is that diflubenzuron may act by inhibiting a serine protease presumed to be involved in the conversion of an inactive chitinase zymogen to active enzyme (Leighton *et al.*, 1981). However, the evidence for this is very circumstantial, depending on the observations that diflubenzuron could inhibit chymotrypsin at micromolar concentrations, while low levels of known chymotrypsin inhibitors were able to prevent chitin synthesis in the cockroach leg regenerating system described above.

To summarise therefore, chitin synthesis is definitely inhibited following diflubenzuron application, though the effect does not seem to be due to a direct action on the enzyme. The primary target is not yet known.

V. HERBICIDES AND FUNGICIDES INHIBITING MISCELLANEOUS BIOSYNTHETIC REACTIONS

A. Aromatic amino acid biosynthesis

1. *Introduction*

The route of biosynthesis of the aromatic amino acids phenylalanine, tyrosine and tryptophan is given in Fig. 6.9. An account of this pathway in plants is given by Gilchrist and Kosuge (1980). Clearly the amino acids are important as constituents of proteins, but phenylalanine is also the major precursor of the plant lignins which strengthen and protect the cell walls particularly in older (woody) tissue. The phenylpropanoid compounds which derive from phenylalanine and which are intermediates in lignin synthesis also provide part of the structure of the flavonoid materials (such as anthocyanins – see later) many of which are highly coloured. The first step in the formation of both lignin and flavonoids is the cleavage of phenylalanine to cinnamate and ammonia by phenylalanine ammonia lyase (PAL; EC 4.3.1.5):

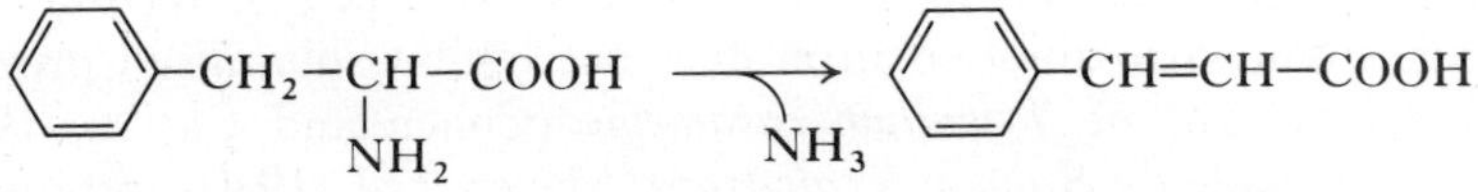

$$
\underset{\text{phenylalanine}}{\bigcirc\!\!-CH_2-\underset{\underset{NH_2}{|}}{CH}-COOH} \quad\xrightarrow{\quad NH_3\quad}\quad \underset{\text{cinnamate}}{\bigcirc\!\!-CH=CH-COOH}
$$

2. *Glyphosate*

This is a slow-acting post-emergence herbicide which is readily translocated within the plant and is therefore particularly useful against perennial weeds which are difficult to control (Franz, 1979; Worthing, 1979).

$$
\underset{\text{glyphosate}}{HOOC-CH_2-NH-CH_2-\overset{\overset{\displaystyle O}{\|}}{P}(OH)_2}
$$

It has now been established that glyphosate has its primary mode of action in the aromatic amino acid biosynthesis pathway. Mode of action work on glyphosate was reviewd by Cole (1982).

(*a*) *Effect on aromatic amino acid formation*

Jaworski (1972) first showed that glyphosate-induced inhibition of the growth of the pond weed *Lemna gibba* and the bacterium *Rhizobium japonicum* could be overcome by various combinations of phenylalanine, tyrosine and tryptophan. The results did not permit a particular enzyme to be identified as the target, but Jaworski favoured an action between

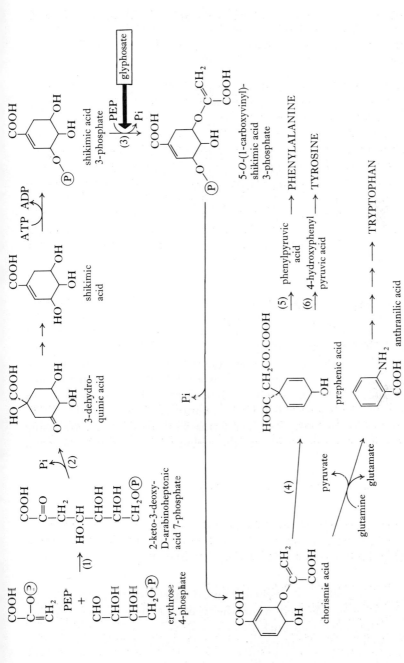

Fig. 6.9 Aromatic amino acid biosynthesis and the site of action of glyphosate. The enzymes catalysing the numbered steps are as follows: (1) phospho-2-keto-3-deoxy-heptonate synthase (EC 4.1.2.15), (2) 3-dehydroquinate synthase (EC 4.6.1.3), (3) 3-phosphoshikimate 1-carboxyvinyltransferase (EC 2.5.1.19), (4) chorismate mutase (EC 5.4.99.5), (5) prephenate dehydratase (EC 4.2.1.51), (6) prephenate dehydrogenase (EC 1.3.1.13). PEP, phosphoenolpyruvate; P$_i$, inorganic phosphate; Ⓟ, phosphate ester

chorismate and the oxo-acid precursors of phenylalanine and tyrosine (i.e. on steps 4, 5 or 6, Fig. 6.9). It has since been shown, in at least one example each of a bacterium, an alga, a higher plant seedling and a higher plant cell culture, that combinations of aromatic amino acids can overcome the effect of glyphosate on growth (Gresshoff, 1979; Haderlie et al., 1977; Roisch and Lingens, 1980). This has not always worked in plants (see Duke and Hoagland, 1978; Gresshoff, 1979) perhaps because exogenous amino acids do not always reach the appropriate cell compartment (Cole et al., 1980).

If glyphosate inhibited synthesis of aromatic amino acids, depletion in the pool sizes could be anticipated and were indeed detected for one or other amino acid in, for instance, bean shoots (Shaner and Lyon, 1980), dark-grown soybean roots (Hoagland et al., 1979) and couch grass rhizomes (Ekanayake et al., 1979), though no changes were measured in cultured carrot cells (Haderlie et al., 1977). Nilsson (1977) found that in glyphosate-treated wheat leaves the amounts of free phenylalanine and tyrosine were not changed, but the total amounts (i.e., including that in protein) were lowered.

Most approaches described above have looked for effects in intact tissues, but the objective of all mode of action studies should be to identify an *in vitro* system which is sensitive to realistically low concentrations of pesticide. With Jaworski's (1972) results in mind Roisch and Lingens (1974) showed that chorismate mutase (step 4, Fig. 6.9) and prephenate dehydratase (step 5, Fig. 6.9) of *E. coli* were not affected by glyphosate, and have since reported inhibition by glyphosate of the first two enzymes (steps 1 and 2, Fig. 6.9) in the aromatic amino acid biosynthesis pathway (Roisch and Lingens, 1980). However, this work was carried out using bacterial enzymes so its relevance to the action of glyphosate in an intact plant cannot be assumed.

Building on the background of results described above, Amrhein and his colleagues have now provided internally consistent and compelling evidence culminating in the proposal that glyphosate inhibits 3-phos-phoshikimate-1-carboxyvinyltransferase (EC 2.5.1.19; PSCV transferase; step 3, Fig. 6.9) formerly known as 5-enolpyruvoyl shikimate phosphate synthase (EPSP synthase). Their evidence for this conclusion is summarised below.

(*i*) Glyphosate inhibited light-enhanced accumulation of phenyl-propanoid substances derived from phenylalanine. For instance, antho-cyanin formation in excised buckwheat hypocotyls was 50% inhibited by 100–200 μM glyphosate. Structurally related compounds such as glyphosine (*N*,*N*-bis(phosphonomethyl)-glycine),   aminomethylphosphonic   acid, methylglycine and iminodiacetic acid gave little or no inhibition (Holländer and Amrhein, 1980), and the effect could be reversed by the provision of phenylalanine.

(*ii*) Phenylalanine levels were decreased in glyphosate-treated buck-wheat hypocotyls, and the rises in phenylalanine brought about by addition of an inhibitor of phenylalanine ammonia lyase (PAL), were effectively reduced by glyphosate. No effect of glyphosate on PAL could be detected (Holländer and Amrhein, 1980).

(*iii*) Glyphosate inhibited incorporation of radioactive shikimic acid into phenylalanine, tyrosine and tryptophan (Holländer and Amrhein, 1980) and caused an accumulation of shikimate; for instance, in buck-wheat hypocotyls treatment with 1 mM glyphosate for 24 hours caused a 20-fold increase in shikimate in the dark, and a 50-fold increase in the light. (Amrhein *et al.*, 1980a). There was a correlation between shikimate accumulation and anthocyanin depletion at various glyphosate concentrations (Amrhein *et al.*, 1980b).

(*iv*) Glyphosate inhibited anthraquinone formation in cultured cells of *Galium mollugo* (Rubiaceae). Since such compounds are derived from chorismate, and since chorismate alleviated the inhibition, a site of action between shikimate and chorismate was suggested (Amrhein *et al.*, 1980a,b).

(*v*) Glyphosate inhibited formation of anthranilate from shikimate in an extract of *Aerobacter aerogenes*, and the dose-response curve of the entire pathway paralleled that of PSCV transferase assayed using radiolabelled substrate (Steinrücken and Amrhein, 1980). The overall pathway was inhibited with an I_{50} of 5–7 µM, though the enzyme was a little less sensitive. When PSCV transferase was assayed in the presence of 1 mM shikimate 3-phosphate and 2 mM phosphoenolpyruvate, the glyphosate I_{50} was 20 µM (Steinrücken and Amrhein, 1980).

(*vi*) The *Aerobacter aerogenes* enzyme was subsequently purified to homogeneity, and it has been shown that glyphosate competes with phosphoenolpyruvate for binding to the enzyme, with a K_i/K_m ratio of about 0.003 (Amrhein *et al.*, 1981). Essentially similar results have since been obtained with a higher plant enzyme. Glyphosate was not a general inhibitor of enzymes catalysing reactions involving phosphoenolpyruvate (Amrhein *et al.*, 1983).

A recent survey of the effects of glyphosate on the enzymes of aromatic amino acid biosynthesis in mung beans has revealed that a particular Co^{2+}-dependent isoenzyme of phospho-2-keto-3-deoxy-heptonate synthase (Fig. 6.9) was also inhibited by glyphosate, to the extent of 86% at 1 mM. Inhibition was competitive with respect to erythrose 4-phosphate. However, a Mn^{2+}-dependent isoenzyme was unaffected by glyphosate and flux through the pathway was not prevented since shikimate accumulated in glyphosate-treated plants (Rubin et al., 1982).

In summary, we may conclude on the basis of current evidence that PSCV transferase is the primary target of glyphosate action.

(b) Other effects

During the search for the primary site of action of glyphosate a number of effects were identified which now seem likely to be secondary or less important compared with the inhibition of PSCV transferase. Most of these are reviewed in detail by Hoagland and Duke (1982) and they include elevation of PAL levels (see Duke et al., 1980; Cole et al., 1980) and inhibitory effects on chloroplast functions such as the formation of the chlorophyll precursor 5-aminolevulinic acid (Kitchen et al., 1981) and photosynthetic electron transport (see Richard et al., 1979). More recently Lee (1982) has suggested that depletion of free IAA levels could be involved in the action of glyphosate.

(c) Why do glyphosate-treated plants die?

Without the benefit of Amrhein's work, Cole et al. (1980) reviewed the available evidence and concluded that glyphosate was herbicidal because it inhibited protein synthesis via depletion of phenylalanine. This fits with the slow onset of phytotoxicity following glyphosate application and the work from Amrhein's group is, of course, fully consistent with this sug-gestion. However, some of the other metabolic perturbations referred to above could also cause or contribute to plant death. It has been shown that various carbon sources can reverse glyphosate inhibition of growth of carrot cell cultures (Killmer et al., 1981). These authors suggest the possibility that the drain on metabolites associated with the rise in shikimate levels (see above) might leave too low a carbon supply to cope with the need to fix ammonia. A consequence of this would be an elevation in the levels of phytotoxic ammonia as well as of glutamate and glutamine, which are closely involved in the incorporation of ammonia into organic mol-ecules; such rises had previously been observed (Ekanayake et al., 1979; Haderlie et al., 1977; Hoagland et al., 1979; Nilsson, 1977).

B. Purine metabolism as the site of action of ethirimol and related compounds

These compounds (Table 6.5) are active against powdery mildew fungi of various crops (Worthing, 1979).

Hollomon has conducted a detailed investigation of the mode of action of ethirimol, and has obtained evidence consistent with the possibility

Table 6.5 Pyrimidine fungicides which may interfere with purine metabolism

Name	Structure
Bupirimate	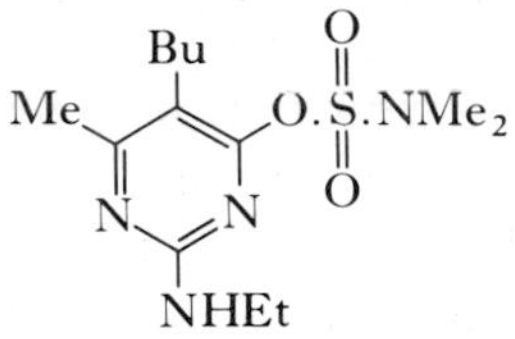 Bu, Me, $O.S.NMe_2$ (with O above and O below the S), pyrimidine ring with N, N, NHEt
Dimethirimol	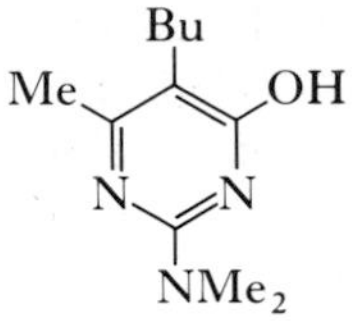 Bu, Me, OH, pyrimidine ring with N, N, NMe_2
Ethirimol	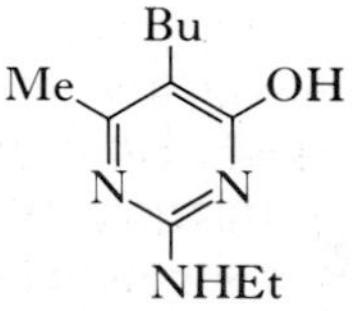 Bu, Me, OH, pyrimidine ring with N, N, NHEt

that the compound acts through inhibition of adenosine deaminase (Hollomon, 1979a,b; Hollomon and Chamberlain, 1981).

At the stage of germination which is inhibited by ethirimol, the fungus does not seem to be making purines *de novo* (Hollomon, 1979a), so these are presumably obtained in other ways. Adenosine deaminase is involved in a salvage pathway which makes use of bases released by nucleic acid breakdown, and, at least in mammals, it is considered that this enables purines synthesised in one tissue to be made available to other tissues (Murray, 1971). An inability to synthesise adenine nucleotides can be confidently predicted to lead to effective growth inhibition, so that for a tissue relying exclusively on purine salvage for its continued growth, adenosine deaminase is likely to be an essential enzyme.

Hollomon (1979b) assayed adenosine deaminase in extracts of conidia and mycelia of various species of mildew and other fungicides, and found that ethirimol sensitivity of the fungi correlated with the susceptibility of their adenosine deaminase to inhibition by the compound. In a series of analogues there was reasonable agreement between the ability of compounds to inhibit adenosine deaminase and their activity as fungicides (Hollomon and Chamberlain, 1981). In the case of the enzyme from conidia of *Erysiphe graminis* (barley powdery mildew), ethirimol was a non-competitive inhibitor, with an apparent K_i of 23 μM (deoxyadenosine as substrate: apparent K_m, approximated from fig. 2 of Hollomon (1979b), 100 μM). Since the inhibition was non-competitive the compound may act at a regulatory site, and this raises the possibility that other enzymes may also be affected (Hollomon, 1979b).

C. Melanin biosynthesis as the possible site of action of tricyclazole

Tricyclazole is a fungicide which controls rice blast disease at concentrations well below those required to inhibit growth of *Pyricularia oryzae in vitro* (Tokousbalides and Sisler, 1978). Although 100 μM tricyclazole is

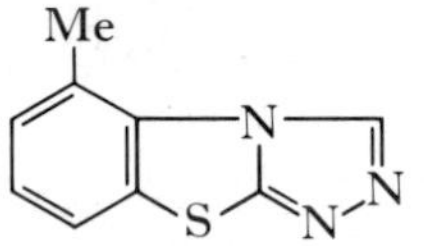

tricyclazole

required to produce measurable effects on fungal growth *in vitro*, as little as 10 nM fungicide leads to detectable inhibition of secondary metabolism involved in melanin formation by this and other fungi (see Woloshuk *et al.*, 1981). The major site of action of tricyclazole in the melanin synthesis pathway of *Verticillium dahliae* has been suggested to be between 1,3,8-trihydroxynaphthalene and vermelone (Tokousbalides and Sisler, 1979).

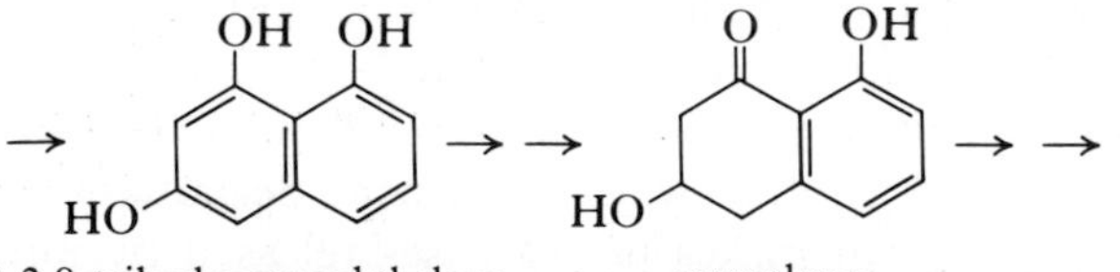

1,3,8-trihydroxynaphthalene   vermelone

Melanin may be important for the pathogenicity of the fungus since buff mutants of *Pyricularia oryzae* which did not contain melanin were not pathogenic (see Woloshuk *et al.*, 1981). The presence of melanin in the pathogenic wild-type strain renders it grey-black, whereas increasing concentrations of tricyclazole cause it to become first tan, then reddish-tan in colour (Woloshuk *et al.*, 1981). These authors also report the selection of mutants which have the same appearance as wild-type strains treated with tricyclazole, and these should help to elucidate the role of melanin in the pathogenic action.

Thus the activity of tricyclazole could be due to its interference with the melanin biosynthesis pathway of the fungus, leading, by some as yet unknown mechanism, to impaired pathogenicity. However, the possibility that the compound influences the metabolism of the host plant still remains a possibility (see Tokousbalides and Sisler, 1979).

D. Glutamine synthetase as the target for bialaphos and glufosinate

Research during the past decade has established that nitrogen in organic molecules of plants is incorporated from the inorganic form by the following reactions (Milflin and Lea, 1980):

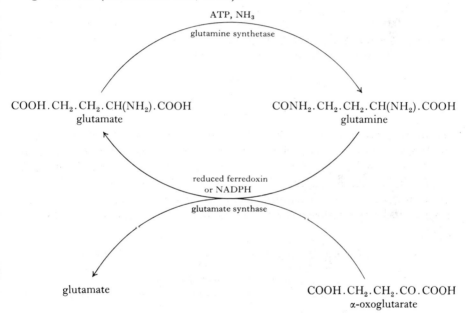

The organic nitrogen is distributed to other amino acids by transamination from glutamate, and some amino acids are used in further biosynthetic reactions.

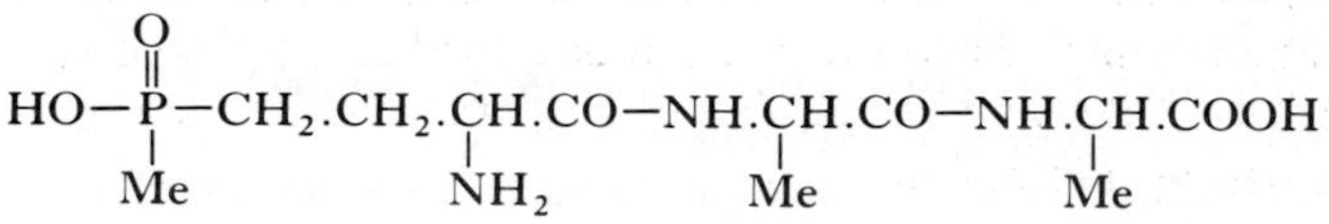

Bialaphos is the name given to L-[2-amino-4-(hydroxymethylphosphinyl)-butyryl]-L-alanyl-L-alanine, a naturally-produced peptide herbicide which is isolated from the broth of *Streptomyces* species (Sekizawa and Takematsu, 1983). It is hydrolysed *in vivo* to yield, as well as the alanine residues, the L-isomer of the compound $HO.PO(Me).CH_2.CH_2.CH(NH_2).COOH$ (2-amino-4-(hydroxymethylphosphinyl)butyric acid) (Tachibana et al., 1982). This material is the active component of glufosinate, also a candidate herbicide (Schwerdtle et al., 1981) and it has been shown to inhibit glutamine synthetase (EC 6.3.1.2) (Bayer et al., 1972; Leason et al., 1982). It is thought that bialaphos-treated plants convert the peptide to glufosinate and are killed by the combination of ammonia accumulation and amino acid depletion which follow inhibition of glutamine synthetase; both these effects can be observed within four hours of treatment (Tachibana et al., 1982).

REFERENCES

Adams, D. J. and Gooday, G. W. (1980). *Biotechnol. Lett.* **2**, 75–78

Akatsuka, T., Kodama, O. and Yamada, H. (1977). *Agric. Biol. Chem.* **41**, 2111–2112

Amrhein, N., Deus, B., Gehrke, P. and Steinrücken, H. C. (1980a). *Plant Physiol.* **66**, 830–834

Amrhein, N., Schab, J. and Steinrücken, H. C. (1980b). *Naturwissenschaften* **67**, 356–357

Amrhein, N., Deus, B., Gehrke, P., Holländer, H., Schab, J., Schulz, A. and Steinrücken, H. C. (1981). *Proc. Plant Growth Regulator Soc. America* **8**, 99–106

Amrhein, N., Holländer-Cyztko, H., Leifeld, J., Schulz, A., Steinrücken, H. C. and Topp, H. (1983). Proc. Journées internationale d'études du Groupe Polyphenols, Toulouse (in press)

Anderson, I. C. and Robertson, D. S. (1960). *Plant Physiol.* **35**, 531–534

Bartels, P. G. and McCullough, C. (1972). *Biochem. Biophys. Res. Commun.* **48**, 16–22

Bartels, P. G. and Watson, C. W. (1978). *Weed Sci.* **26**, 198–203

Bartels, P. G., Matsuda, D., Siegel, A., and Weier, T. E. (1967). *Plant Physiol.* **42**, 736–741

Baxter, R. and McGowan, J. E. (1976). *J. Exp. Bot.* **27**, 525–531

Baxter, R., Knell, V. C., Somerville, H. J., Swain, H. M. and Weeks, D. P. (1973). *Nature (New Biol.)* **243**, 139–142

Bayer, E., Gugel, K. H., Hägele, K., Hagenmaier, H., Jessipow, S., König, W. A. and Zähner, H. (1972). *Helv. Chim. Acta* **55**, 224–239

Ben-Aziz, A. and Koren, E. (1974). *Plant Physiol.* **54**, 916–920

Bent, K. J. and Skidmore, A. M. (1979). *Proc. British Insectic. Fungic. Conf.* Vol. 2, pp. 477–484

Berg, D., Draber, W., von Hugo, H., Hummel, W. and Mayer, D. (1981). *Z. Naturforsch.* **36C**, 798–803

Beyer, P., Kreuz, K. and Kleinig, H. (1980). *Planta* **150**, 435–438

Beynon, K. I. and Wright, A. N. (1972). *Residue Rev.* **43**, 23–53

Bohnen, K., Siegle, H. and Loecher, F. (1979). Proc. British Crop Protection Conf. "Weeds", Vol. 2, pp. 541–548

Bolton, P. and Harwood, J. L. (1976). *Phytochem.* **15**, 1507–1509

Britton, G. (1979). *Z. Naturforsch.* **34C**, 979–985

Buchenauer, H. (1975). Abstract in *Chemical Abstracts* (1978) **88**, 184437n

Buchenauer, H. (1977a). Proc. British Insectic. Fungic. Conf., Vol. 3, pp. 699–711
Buchenauer, H. (1977b). Pestic. Biochem. Physiol. 7, 309–320
Buchenauer, H. (1977c). Z. Pflanzenschutz 84, 286–299
Buchenauer, H. (1977d). Z. Pflanzenkr. Pflanzenschutz 84, 440–450
Buchenauer, H. (1978a). Pestic. Biochem. Physiol 8, 15–25
Buchenauer, H. (1978b). Pest. Sci. 9, 507–512
Buchenauer, H. and Kemper, K. (1981). Meded. Fac. Landbouw. Rijksuniv. Gent 46, 909–921
Buchenauer, H. and Röhner, E. (1981). Pestic. Biochem. Physiol. 15, 58–70
Burns, E. R., Buchanan, G. A. and Carter, M. C. (1971). Plant Physiol. 47, 144–148
Carter, M. C. (1975). In "Herbicides. Chemistry, Degradation and Mode of Action" (P. C. Kearney and D. D. Kaufman, eds) pp. 377–398. Dekker, New York
Casida, J. E., Gray, R. A. and Tilles, H. (1974). Science 184, 573–574
Clarke, I. E., Sandmann, G., Bramley, P. M. and Böger, P. (1982). FEBS Lett. 140, 203–206
Cohen, E. and Casida, J. E. (1980). Pestic. Biochem. Physiol. 13, 129–136
Cohen, E. and Casida, J. E. (1982). Pestic. Biochem. Physiol. 17, 301–306
Cole, D. J. (1982). Proc. British Crop Protection Conf. "Weeds", Vol. 1, pp. 309–315
Cole, D. J., Dodge, A. D. and Caseley, J. C. (1980). J. Exp. Bot. 31, 1665–1674
Colvin, J. R. (1980). In "The Biochemistry of Plants" (J. Preiss, ed.) Vol. 3, pp. 543–570. Academic Press, London and New York
Dekker, J. (1971). World Rev. Pest Control 10, 9–23
DeLoach, J. R., Meola, S. M., Mayer, R. T. and Thompson, J. M. (1981). Pestic. Biochem. Physiol. 15, 172–180
Deul, D. H., De Jong, B. J. and Kortenbach, J. A. M. (1978). Pestic. Biochem. Physiol. 8, 98–105
Devlin, R. M., Kisiel, M. J. and Karczmarczyk, S. J. (1976). Weeds Res. 16, 125–129
Devlin, R. M., Saras, C. N., Kisiel, M. J. and Kostusiak, A. S. (1978). Weed Sci. 26, 432–433
De Waard, M. A. (1972). Neth. J. Plant Pathol. 78, 186–188
Donald, W. W. (1981). Weed Sci. 29, 490–499
Donald, W. W., Fawcett, R. S. and Harvey, R. G. (1979). Weed Sci. 27, 122–127
Duke, S. O. and Hoagland, R. E. (1978). Plant Sci. Lett. 11, 185–190
Duke, S. O., Hoagland, R. E. and Elmore, C. D. (1980). Plant Physiol. 65, 17–21
Eder, F. A. (1979). Z. Naturforsch. 34 C, 1052–1054
Ekanayake, A., Wickremasinghe, R. L. and Liyanage, H. D. S. (1979). Weed Res. 19, 39–43
Ellis, R. J. (1975). Phytochem. 14, 89–93
Endo, A., Kakiki, K. and Misato, T. (1970). J. Bacteriol. 104, 189–196
Ennis, H. L. and Lubin, M. (1964). Science 146, 1474–1476
Eronen, L. and Karunen, P. (1977). Proc. 9th Scand. Symp. Lipids (R. Maccuse, ed.) pp. 206–213. Lipidforum, Sweden
Fang, S. C. (1975). In "Herbicides. Chemistry, Degradation and Mode of Action" (P. C. Kearney and D. D. Kaufman, eds) Vol. 1, pp. 323–348. Dekker, New York
Feierabend, J. and Schubert, B. (1978). Plant Physiol. 61, 1017–1022
Feierabend, J. and Winkelhüsener, T. (1982). Plant Physiol. 70, 1277–1282
Feierabend, J., Schulz, U., Kemmerich, P., and Lowitz, T. (1979). Z. Naturforsch. 34 C, 1036–1039

Fisher, D. J. and Hayes, A. L. (1982). *Pestic. Sci.* **13**, 330–339

Franz, J. E. (1979). In "Advances in Pesticide Science" (H. Geissbühler, ed.) Pt 2, pp. 139–147. Pergamon Press, Oxford

Frasinel, C., Patterson, G. W. and Dutky, S. R. (1978). *Phytochem.* **17**, 1567–1570

Frosch, S., Jabben, M., Bergfeld, R., Kleinig, H. and Mohr, H. (1979). *Planta* **145**, 497–505

Fujii, Y., Kurokawa, T., Inoue, Y., Yamaguchi, I. and Misato, T. (1977). *Nippon Noyaku Gakkaishi* **2**, 431–437

Gadher, P., Mercer, E. I., Baldwin, B. C. and Wiggins, T. E. (1983). *Pestic. Biochem. Physiol.* **19**, 1–10

Galbraith, D. W. and Shields, B. A. (1982). *Physiol. Plant.* **55**, 25–30

Gale, E. F., Cundliffe, E., Reynolds, P. E., Richmond, M. and Waring, M. J. (1981). "The Molecular Basis of Antibiotic Action", 2nd edn. John Wiley and Sons, Chichester

Gasztonyi, M. (1981). *Pestic. Sci.* **12**, 433–438

Gasztonyi, M. and Josepovits, G. (1979). *Pestic. Sci.* **10**, 57–65

Gentner, W. A. (1966). *Weeds* **14**, 27–31

Georgopoulos, S. G. and Vomvoyanni, V. (1972). In "Herbicides, Fungicides, Formulation, Chemistry" (A.S. Tahori, ed.) pp. 337–346. Gordon and Breach, New York

Gilchrist, D. E. and Kosuge, T. (1980). In "The Biochemistry of Plants" (B. J. Milflin, ed.) Vol. 5, pp. 507–531. Academic Press, London and New York

Gijswijt, M. J., Deul, D. H. and De Jong, B. J. (1979). *Pestic. Biochem Physiol.* **12**, 87–94

Gooday, G. W. (1979). In "Systemfungizide" (H. Lyr and C. Polter, eds) pp. 159–166. Akademie-Verlag, Berlin

Goodwin, T. W. (1971). In "Aspects of Terpenoid Chemistry and Biochemistry" (T. W. Goodwin, ed.) pp. 315–356. Academic Press, London and New York

Goodwin, T. W. (1981). "The Biochemistry of the Carotenoids", Vol. 1. Chapman and Hall, London

Gresshoff, P. M. (1979). *Aust. J. Plant Physiol.* **6**, 177–185

Grumbach, K. H. and Bach, T. J. (1979). *Z. Naturforsch.* **34C**, 941–943

Grumbach, K. H. and Drollinger, M. (1980). *Z. Naturforsch.* **35C**, 445–450

Haderlie, L. C., Widholm, J. M. and Slife, F. W. (1977). *Plant Physiol.* **60**, 40–43

Hajjar, N. P. and Casida, J. E. (1979). *Pestic. Biochem. Physiol.* **11**, 33–45

Harvey, B. M. R., Chang, F. Y. and Fletcher, R. A. (1975). *Canad. J. Bot.* **53**, 225–230

Harwood, J. L. (1975). In "Recent Advances in the Chemistry and Biochemistry of Plant Lipids" (T. Galliard and E. I. Mercer, eds) pp. 43–93. Academic Press, London and New York

Harwood, J. L. and Stumpf, P. K. (1971). *Arch. Biochem. Biophys.* **142**, 281–291

Henry, M. J. and Sisler, H. D. (1981). *Pestic. Sci.* **12**, 98–102

Hilton, J. L., (1969). *J. Agric. Food Chem.* **17**, 182–198

Hilton, J. L., Scharen, A. L., St. John, J. B., Moreland, D. E. and Norris, K. H. (1969). *Weed Sci.* **17**, 541–547

Hoagland, R. E. and Duke, S. O. (1982). In "Biochemical Responses Induced by Herbicides" (D. E. Moreland, J. B. St. John and F. D. Hess, eds) pp. 175–205. American Chemical Society, Washington, D.C.

Hoagland, R. E., Duke, S. O. and Elmore, C. D. (1979). *Physiol. Plant.* **46**, 357–366

Hogetsu, T., Shibaoka, H. and Shimokoriyama, M. (1974). *Plant Cell Physiol.* **15**, 389–393

Holländer, H. and Amrhein, N. (1980). *Plant Physiol.* **66**, 823–829
Hollomon, D. W. (1979a). *Pestic. Biochem. Physiol.* **10**, 181–189
Hollomon, D. W. (1979b). Proc. British Crop Protection Conf. "Pests and Diseases", Vol. 1, pp. 251–256
Hollomon, D. W. and Chamberlain, K. (1981). *Pestic. Biochem. Physiol.* **16**, 158–169
Hori, M., Kakiki, K. and Misato, T. (1974a). *Agric. Biol. Chem.* **38**, 699–705
Hori, M., Kakiki, K. and Misato, T. (1974b). *Agric. Biol. Chem.* **38**, 691–698
Horváth, L. and Pulay, Á. (1980). *Pestic. Biochem. Physiol.* **14**, 265–270
Ishaaya, I. and Casida, J. E. (1974). *Pestic. Biochem. Physiol.* **4**, 484–490
Jaworski, E. G. (1972). *J. Agric. Food Chem.* **20**, 1195–1198
Junge, W. (1977). In "Photosynthesis I" (A. Trebst and M. Avron, eds) Encycl. Plant Physiol., New series, Vol. 5, pp. 59–93. Springer-Verlag, Berlin
Kamimura, S., Nishikawa, M. and Takahi, Y. (1976). *Ann. Phytopathol. Soc. Japan* **42**, 242–252
Kato, T. and Kawase, Y. (1976). *Agric. Biol. Chem.* **40**, 2379–2388
Kato, T., Tanaka, S., Ueda, M. and Kawase, Y. (1974). *Agric. Biol. Chem.* **38**, 2377–2384
Kato, T., Shoami, M. and Kawase, Y. (1980). *Nippon Noyaku Gakkaishi* **5**, 69–79
Kerkenaar, A. (1981). *Pestic. Biochem. Physiol.* **16**, 1–13
Kerkenaar, A., Barug, D. and Kaars Sijpesteijn, A. (1979). *Pestic. Biochem. Physiol.* **12**, 195–204
Kerkenaar, A., Uchiyama, M. and Versluis, G. G. (1981). *Pestic. Biochem. Physiol.* **16**, 97–104
Kerr, M. W. and Avery, R. J. (1972). *Biochem. J.* **128**, 132P–133P
Kerridge, D. (1958). *J. Gen. Microbiol.* **19**, 497–506
Khan, M.-U., Lem, N. W., Chandorkar, K. R. and Williams, J. P. (1979). *Plant. Physiol.* **64**, 300–305
Killmer, J., Wildholm, J. and Slife, F. (1981). *Plant Physiol.* **68**, 1299–1302
Kitchen, L. M., Witt, W. W. and Rieck, C. E. (1981). *Weed Sci.* **29**, 571–577
Kleudgen, H. K. (1979). *Pestic. Biochem. Physiol.* **12**, 231–238
Kodama, O., Yamada, H. and Akatsuka, T. (1979). *Agric. Biol. Chem.* **43**, 1719–1725
Kodama, O., Yamashita, K. and Akatsuka, T. (1980). *Agric. Biol. Chem.* **44**, 1015–1021
Kolattukudy, P. E. (1970). *Lipids* **5**, 398–402
Kolattukudy, P. E. (1980). In "The Biochemistry of Plants" (P. K. Stumpf, ed.) Vol. 4, pp. 571–645. Academic Press, London and New York
Kolattukudy, P. E. and Brown, L. (1974). *Plant Physiol.* **53**, 903–906
Kraus, P. (1979). *Pflanzenschutz. Nachr. Bayer* **32**, 17–30
Krinsky, N. I. (1966). In "Biochemistry of Chloroplasts" (T. W. Goodwin, ed.) Vol. I, pp. 423–430. Academic Press, London and New York
Kümmel, H. W. and Grimme, L. H. (1975). *Z. Naturforsch* **30 C**, 333–336
Lay, M.-M. and Casida, J. E. (1976). *Pestic. Biochem. Physiol.* **6**, 442–456
Leason, M., Cunliffe, D., Parkin, D., Lea, P. J. and Miflin, B. J. (1982). *Phytochemistry* **21**, 855–857
Leavitt, J. R. C., Duncan, D. N., Penner, D. and Meggitt, W. F. (1978). *Plant. Physiol.* **61**, 1034–1036
Lee, T. T. (1982). *Physiol. Plant.* **54**, 289–294
Leighton, T., Marks, E. and Leighton, F. (1981). *Science* **213**, 905–907
Leroux, P. and Gredt, M. (1978). *Ann. Phytopathol.* **10**, 45–60

Lever, B. G., Shearing. S. J. and Batch, J. J. (1982). Proc. British. Crop Protection Conf. "Weeds", Vol. 1, pp. 3–10

Lichtenthaler, H. K. and Kleudgen, H. K. (1977) Z. Naturforsch. 32C, 236–240

Lütke-Brinkhaus, F., Liedvogel, B., Kreuz, K. and Kleinig, H. (1982). Planta 156, 176–180.

Maeda, T., Abe, H., Kakiki, K. and Misato, T. (1970). Agric. Biol. Chem. 34, 700–709

Mann, J. D. and Pu, M. (1968). Weed Sci. 16, 197–198

Mann, J. D., Jordan, L. S. and Day, B. E. (1965). Plant Physiol. 40, 840–843

Marriott, M. (1980). J. Gen. Microbiol. 117, 253–255

Martin, J. T. and Juniper, B. E., (1970). "The Cuticles of Plants". Edward Arnold, London

Martin, R. A. and Edgington, L. V. (1982). Pestic. Biochem. Physiol. 17, 1–9

Mayer, R. T., Chen, A. C. and DeLoach, J. R. (1980). Insect Biochem. 10, 549–556

Meola, S. M. and Mayer, R. T. (1980). Science 207, 985–987

Meyer, Y. and Herth, W. (1978). Planta 142, 253–262

Miflin, B. J. and Lea, P. J. (1980). In "The Biochemistry of Plants", Vol. 5 (B. J. Miflin, ed.) pp. 169–202. Academic Press, London and New York

Misato, T. (1969). Residue Rev. 25, 93–106

Misato, T. and Kakiki, K. (1977). In "Antifungal Compounds" (M. R. Siegel and H. D. Sisler, eds) Vol. 2, pp. 277–300. Dekker, New York

Misato, T., Kakiki, K. and Hori, M. (1979). In "Advances in Pesticide Science" (H. Geissbühler, ed.) Pt 3, pp. 458–464. Pergamon Press, Oxford

Mitani, M., and Inoué, Y. (1968). J. Antibiotics 21, 492–496

Mitropoulos, K. A., Gibbons, G. F., Connell, C. M. and Woods, R. A. (1976). Biochem. Biophys. Res. Commun. 71, 892–900

Montezinos, D. and Delmer, D. P. (1980). Planta 148, 305–311

Mulder, R. and Gijswijt, M. J. (1973). Pestic. Sci. 4, 737–745

Müller, H., Furter, R., Zähner, H. and Rast, D. M. (1981). Arch. Microbiol. 130, 195–197

Murray, A. W. (1971). Ann. Rev. Biochem. 40, 811–826

Nilsson, G. (1977). Swed. J. Agric. Res. 7, 153–157

Ohta, N., Kakiki, K. and Misato, T. (1970). Agric. Biol. Chem. 34, 1224–1234

Pappas, A. C. and Fisher, D. J. (1979) Pestic. Sci. 10, 239–246

Pardo, A. D. and Schiff, J. A. (1980). Can. J. Bot. 58, 25–35

Pillai, V. N. R. (1977). Chemosphere 6, 777–782

Poldermans, B., Goosen, N. and Van Knippenberg, P. H. (1979). J. Biol. Chem. 254, 9085–9089

Post, L. C. and Mulder, R. (1975). In "Mechanism of Pesticide Action" (G. K. Kohn, ed.) pp. 136–143. American Chemical Society, Washington, D.C.

Rafii, Z. E., Ashton, F. M. and Glenn, R. K. (1979). Weed Sci. 27, 422–426

Ragsdale, N. N. (1975). Biochim. Biophys. Acta 380, 81–96

Ragsdale, N. N. (1977). In "Antifungal Compounds" (M. R. Siegel and H. D. Sisler, eds) Vol. 2, pp. 333–363. Dekker, New York

Ragsdale, N. N. and Sisler, H. D. (1972). Biochem. Biophys. Res. Commun. 46, 2048–2053.

Richard, E. P., Goss, J. R. and Arntzen, C. J. (1979). Weed Sci. 27, 684–688

Richards, A. G. (1978). In "Biochemistry of Insects" (M. Rockstein, ed.) pp. 205–232, Academic Press, London and New York

Ridley, S. M. and Ridley, J. (1979). Plant Physiol. 63, 392–398

Roisch, U. and Lingens, F. (1974). Ang. Chem. Int. Ed. Engl. 13, 400

Roisch, U. and Lingens, F. (1980). *Hoppe-Seyler's Z. Physiol. Chem.* **361**, 1049–1058
Rubin, J. L., Gaines, C. G. and Jensen, R. A. (1982). *Plant Physiol.* **70**, 833–839
Rüdiger, W. and Benz, J. (1979). *Z. Naturforsch.* **34C**, 1055–1057
Sandmann, G. and Böger, P. (1983). In "Pesticide Chemistry: Human Welfare and the Environment" (J. Miyamoto and P. C. Kearney, eds) Vol. 2, pp. 261–268. Pergamon Press, Oxford
Sandmann, G., Bramley, P. M. and Böger, P. (1980). *Pestic. Biochem. Physiol.* **14**, 185–191
Sandmann, G., Kunert, K.-J. and Böger, P. (1981). *Pestic. Biochem. Physiol.* **15**, 288–293
Schuphan, I. and Casida, J. E. (1979). *J. Agric. Food Chem.* **27**, 1060–1066
Schwerdtle, F., Bieringer, H. and Finke, M. (1981). *Z. PflKrankh. PflSchutz,* Sonderh. IX, 431–440
Sekizawa, Y. and Takematsu, T. (1983). In "Pesticide Chemistry: Human Welfare and the Environment" (J. Miyamoto and P. C. Kearney, eds) Vol. 1, pp. 321–326. Pergamon Press, Oxford
Shaner, D. L. and Lyon, J. L. (1980). *Weed Sci.* **28**, 31–35
Sherald, J. L. and Sisler, H. D. (1975). *Pestic. Biochem. Physiol.* **5**, 477–488
Sherald, J. L., Ragsdale, N. N. and Sisler, H. D. (1973). *Pestic. Sci.* **4**, 719–727
Siegel, M. R. (1977). In "Antifungal Compounds" (M. R. Siegel and H. D. Sisler, eds) Vol. 2, pp. 399–438. Dekker, New York
Siegel, M. R. (1981). *Plant Disease* **65**, 986–989
Siegel, M. R. and Ragsdale, N. N., (1978). *Pestic. Biochem. Physiol.* **9**, 48–56
Siegel, M. R. and Sisler, H. D. (1963). *Nature (Lond.)* **200**, 675–676
Sowa, B. A. and Marks, E. P. (1975). *Insect. Biochem.* **5**, 855–859
Steinrücken, H. C. and Amrhein, N. (1980). *Biochem. Biophys. Res. Commun.* **94**, 1207–1212
Still, G. G., Davis, D. G. and Zander, G. L., (1970). *Plant Physiol.* **46**, 307–314
St. John, J. B. (1982). In "Biochemical Responses Induced by Herbicides" (D. E. Moreland, J. B. St. John and F. D. Hess, eds) pp. 97–109. American Chemical Society, Washington, D.C.
St. John, J. B. and Hilton, J. L. (1976). *Weed Sci.* **24**, 579–582
St. John, J. B., Rittig, F. R., Ashworth, E. N. and Christiansen, M. N. (1979). In "Advances in Pesticide Science" (H. Geissbühler, ed.) Pt 2, pp. 271–273. Pergamon Press, Oxford
Stumpf, P. K. (1980). In "The Biochemistry of Plants" (P. K. Stumpf, ed.) Vol. 4, pp. 177–204. Academic Press, London and New York
Tachibana, K., Watanabe, T., Sekizawa, Y., Konnai, M. and Takematsu, T. (1982). Abstracts 5th Int. Congr. Pestic. Chem., Kyoto, No. IVa-19. International Union of Pure and Applied Chemistry
Tanaka, S., Kato, T., Yamamoto, S. and Yoshioka, H. (1977). *Agric. Biol. Chem.* **41**, 1953 1959
Tokousbalides, M. C. and Sisler, H. D. (1978). *Pestic. Biochem. Physiol.* **8**, 26–32
Tokousbalides, M. C. and Sisler, H. D. (1979). *Pestic. Biochem. Physiol.* **11**, 64–73
Umetsu, N., Satoh, S. and Matsuda, K. (1976). *Plant Cell Physiol.* **17**, 1071–1075
Vaisberg, A. J. and Schiff, J. A. (1976). *Plant Physiol.* **57**, 260–269
van den Bossche, H., Willemsens, G., Cools W., Lauwers, W. F. J. and Le Jeune, L. (1978). *Chem.-Biol. Interact.* **21**, 59–78
van Eck, W. H. (1979). *Insect Biochem.* **9**, 295–300
Vázquez, D. (1979). "Inhibitors of Protein Biosynthesis". Springer-Verlag, Berlin

Verloop, A. and Ferrell, C. D. (1977). In "Pesticide Chemistry in the 20th Century" (J. R. Plimmer, ed.) pp. 237–270. American Chemical Society, Washington, D.C.

Vivekanadan, M. and Gnanam, A. (1975). *Plant Physiol.* **55**, 526–531

Walsh, R. C. and Sisler, H. D. (1982). *Pestic. Biochem. Physiol.* **18**, 122–131

Weeks, D. P. and Baxter, R. (1972). *Biochemistry* **11**, 3060–3064

Weete, J. D. (1973). *Phytochem.* **12**, 1843–1864

Weete, J. D. (1974). "Fungal Lipid Biochemistry". Plenum Press, New York

Weinberg, M. B. and Castelfranco, P. A. (1975). *Weed Sci.*, **23**, 185–187

Wilkinson, R. E. (1978a). In "Chemistry and Action of Herbicide Antidotes" (F. M. Pallos and J. E. Casida, eds) pp. 85–108. Academic Press, London and New York

Wilkinson, R. E. (1978b). *Pestic. Biochem. Physiol.* **8**, 208–214

Wilkinson, R. E. (1983). In "Pesticide Chemistry: Human Welfare and the Environment" (J. Miyamoto and P. C. Kearney, eds) Vol. 3, pp. 233–236. Pergamon Press, Oxford

Wilkinson, R. E. and Ashley, D. (1979). *Weed Sci.* **27**, 270–273

Wilkinson, R. E. and Hardcastle, W. S. (1970). *Weed Sci.* **18**, 125–128

Wilkinson, R. E. and Smith, A. E. (1975). *Weed Sci.* **23**, 100–104

Wilkinson, R. E. and Smith, A. E. (1976). *Phytochem.* **15**, 841–842

Wilkinson, R. E., Smith, A. E. and Michel, B. (1977). *Plant Physiol.* **60**, 86–88

Woloshuk, C. P., Wolkow, P. M. and Sisler, H. D. (1981). *Pestic. Sci.* **12**, 86–90

Worthing, C. R. (ed.) (1979). "The Pesticide Manual", 6th edn. British Crop Protection Council, London

Yu, S. J. and Terriere, L. C. (1977). *Pestic. Biochem. Physiol.* **7**, 48–55

7 | Pesticides with a Non-specific Mode of Action

The preceding chapters have shown that, where the biochemical mode of action of a pesticide has been well defined, there seems to be little doubt that the biological effects of the compound can be attributed to a primary action at one particular site. Other effects seem to be secondary to this, or else to occur only at higher concentrations of the chemical. This chapter, on the other hand, is concerned with a number of pesticides, many of them relatively old, that are thought to have non-specific modes of action.

I. PESTICIDES THOUGHT TO ACT BY INTERFERING WITH MEMBRANES

Most pesticides need to cross one or more cellular membranes in order to reach the target with which they interact, and it is not, therefore, surprising that many pesticides have some effect on membranes even though their primary target is elsewhere. Thus, for instance, comprehensive accounts of the effects that herbicides may have on membrane systems have been given by Morrod (1976) and Rivera and Penner (1979). The compounds which we shall deal with in this section (Table 7.1) are those which may have their primary effect through interfering with membrane structure and function.

291

Table 7.1   Compounds thought to act by interfering with membranes

Name	Structure
Petroleum oils	
Tar oils	
Dodine	$\left[C_{12}H_{25}NH.\underset{\underset{NH_2}{\|}}{C}.NH_2 \right]^{+}$ $[CH_3COO]^{-}$
Guazatine	$H_2N.\underset{\underset{NH}{\|}}{C}.NH(CH_2)_8.NH.(CH_2)_8.NH.\underset{\underset{NH}{\|}}{C}.NH_2$
Pimaricin	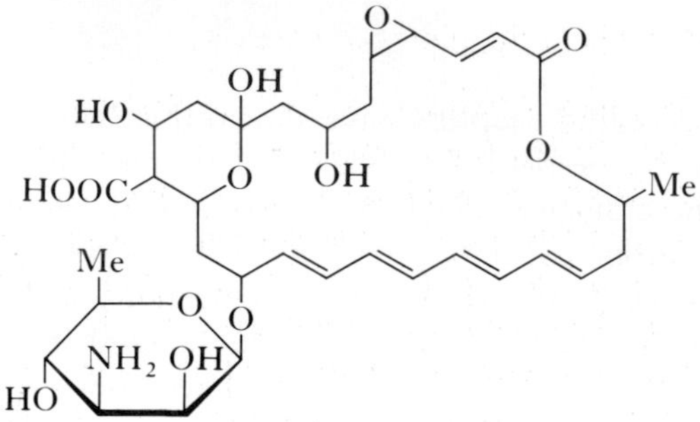

A. Petroleum and tar oils

Petroleum oils, produced by the refinement of crude mineral oils, consist largely of saturated and unsaturated aliphatic hydrocarbons and are used principally to control scale insects and spider mites (especially under glasshouse conditions) and also as insect ovicides on dormant trees (Worthing, 1979). This latter is also the major application of tar oils, which are similar to petroleum oils but, because of their origin, also contain phenols or 'tar acids' which are soluble in aqueous alkali. A general account of the use of oils as pesticides is given by Green *et al.* (1977).

It seems reasonable to attribute the action of oils on plants to disruption of cellular membranes since a plant damaged by oil undergoes darkening and loss of turgor, presumably due to the leakage of the cell contents into the intercellular spaces. Maize coleoptile segments dipped for 2 minutes in hydrocarbon oils, and then rinsed, are unable to elongate normally, presumably because they cannot take up the water required for elongation since their membranes are disrupted (van Overbeek and Blondeau, 1954).

The molecular mechanism by which oils disrupt membranes is not known, though the suggestion of van Overbeek and Blondeau (1954) that they act by solubilising membrane lipids seems likely.

The effect on membranes is probably involved in all the biocidal applications of the petroleum oils but additionally, in insects, the spiracles may become flooded by the oil leading to asphyxiation. Control of mosquito larvae by spreading oil on the surface of the water in which they are developing also relies on asphyxiation, though the oil also prevents the larvae from anchoring themselves at the surface (Green et al., 1977). In support of this Berlin and Micks (1973) found that the pathological changes observed in treated larvae were similar to those caused by anoxia.

B. Long-chain guanidino fungicides

Dodine and guazatine (Table 7.1) are fungicides containing a long alkyl chain attached to a guanidino group. It seems to be generally agreed that dodine, which is used to control apple scab, acts by interfering with membrane structure. Its influence on membranes is like that of a detergent and presumably depends on the ability of the lipophilic chain to dissolve in the lipid portion of the membrane, while the polar guanidino residue will tend to remain in the adjacent aqueous phase where it may be able to interact with the phosphate groups of phospholipids.

Fungal spores take up large amounts of dodine, probably by an indiscriminate non-metabolic process, and levels may reach 1 to 2.4% of the cell weight which is some 35 to 84 millimoles of dodine per kilogramme of fungus. At least some of this becomes immobilised in, for example, the cell wall and may not contribute to the fungitoxic action (Somers and Pring, 1966). Dodine will lyse the external membrane of spores of many fungal species, although, with spores of others, scarcely any lysis occurs at toxic levels of the fungicide (Brown and Sisler, 1960). In a recent study on spores of a *Fusarium* species Miller and Barran (1977) found that dodine caused enhanced water uptake at concentrations below 40 μM, while at concentrations of 50 μM and above, the membranes of the cells lost both their ability to actively transport phenylalanine and their impermeability to divalent cations.

There is no reason to expect that the effects of a detergent such as dodine would be specific to one type of cellular membrane, and mitochondrial membrane functions have also been shown to be susceptible. Pressman (1963) showed that dodine, which was the most active substituted guanidine that he tested, caused 50% inhibition of electron transport in isolated rat liver mitochondria at a concentration of approximately 80 μM. Since the inhibition, or at least that produced by octyl guanidine, was relieved by the uncoupler, 2,4-dinitrophenol, he concluded that alkylguanidines were acting as inhibitors of oxidative phosphorylation. This could have been because the dodine prevented the correct interaction between the membrane phospholipids and the ATP-synthesising complex, leading to inhibition. It could not have been due to general membrane disruption since this would have destroyed the mitochondrial coupling even in the absence of uncoupler.

C. Pimaricin

This fungicidal antibiotic (Table 7.1) is isolated from cultures of *Streptomyces natalensis* and used to control diseases of bulbs, especially basal rot of daffodils (Worthing, 1979). It is a member of the class of polyene antibiotics, many of which have clinical applications (Hamilton-Miller, 1973; Hammond, 1977).

The recent literature relating to the mode of action of the polyenes is succinctly presented by Kotler-Brajtburg *et al.* (1979) who studied the effects of a range of polyene antibiotics on potassium ion leakage and cell mortality in yeast, or haemolysis of mouse red blood cells. They found that, although some polyenes caused potassium leakage or growth inhibition at sub-lethal concentrations, pimaricin was included in a second group which appeared to have an all-or-nothing effect, giving rise to membrane destruction and cell death (to the extent of 50% at about 45 μM in this study) with no noticeable damage at lower levels.

Damage caused by polyene antibiotics results from a molecular association of the antibiotic with the sterols of fungal membranes. The evidence for this, and the detailed models proposed for some other polyenes, are described by Norman *et al.* (1976) and Hammond (1977).

II. AN INDISCRIMINATE ENZYME INACTIVATOR–SODIUM FLUORIDE

Sodium fluoride (NaF) is an old-fashioned cockroach poison, and cryolite (Na_3AlF_6) is an insecticide which was first used in 1929 (Worthing, 1979). Cryolite acts by the release of fluoride ion (O'Brien, 1967).

O'Brien (1967) points out that sodium fluoride is not particularly toxic to insects, the oral LD_{50} to three species lying between 130 and 350 mg kg^{-1}, with the consequence that the molar concentration of fluoride necessary to cause death in the insect will be quite high – very approximately 3 to 8 mM in the examples mentioned. Since fluoride will form complexes with a variety of metal-containing enzymes, including those dependent on iron, calcium and magnesium (Hewitt and Nicholas, 1963), it seems likely that it kills by the simultaneous inhibition of several enzymatic processes in the insect, and not by acting at one specific locus.

III. PESTICIDES WHICH MAY ACT THROUGH INDISCRIMINATE REACTIONS WITH CELLULAR COMPONENTS

A. Herbicidal chloroacetanilides

These herbicides (Table 7.2) contain an α-chloromethyl amide group which will be open to attack by suitably placed nucleophiles present on any macromolecular cell component to which they bind. Such nucleophiles include sulphydryl and amino groups which are, of course, present in most proteins.

Table 7.2   Herbicidal chloroacetanilides

Name	Structure
Alachlor	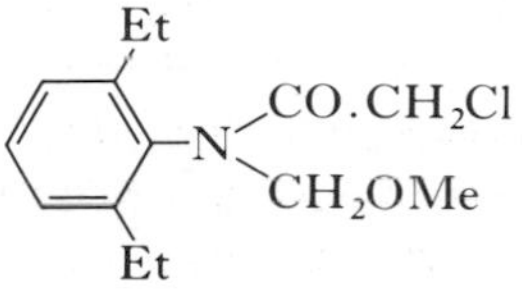 2,6-diethylphenyl group, N bonded to $CO.CH_2Cl$ and CH_2OMe
Butachlor	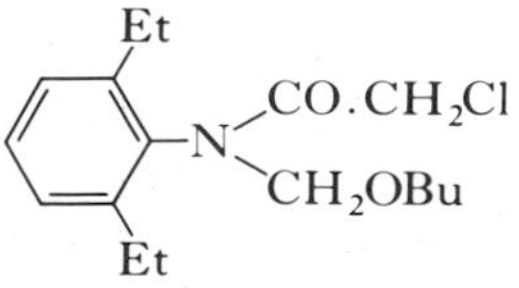 2,6-diethylphenyl group, N bonded to $CO.CH_2Cl$ and CH_2OBu
Diethatyl-ethyl	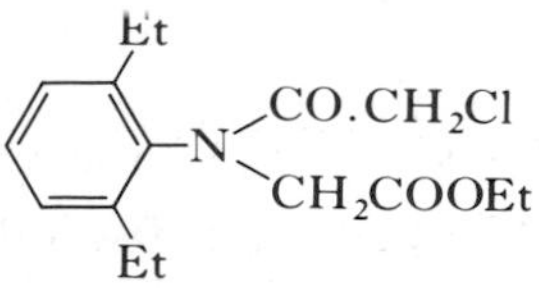 2,6-diethylphenyl group, N bonded to $CO.CH_2Cl$ and CH_2COOEt

Name	Structure

Dimethachlor

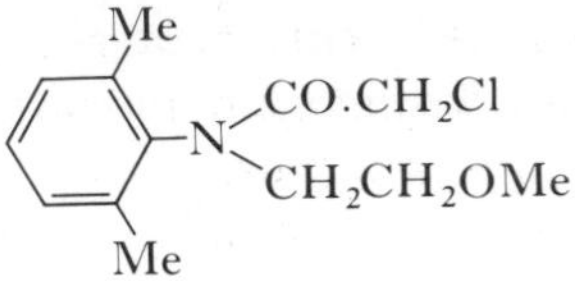

Me

$CO.CH_2Cl$
N
CH_2CH_2OMe

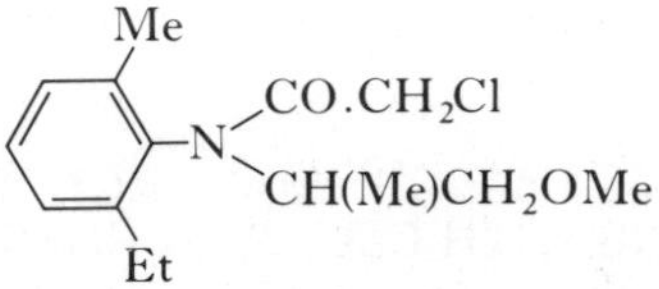

Me

Metolachlor

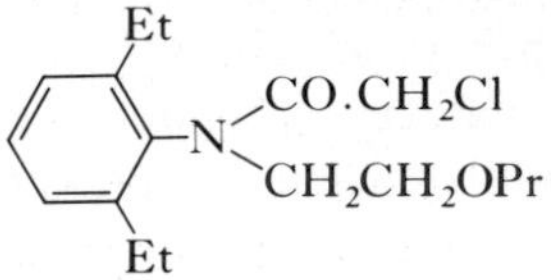

Me

$CO.CH_2Cl$
N
$CH(Me)CH_2OMe$

Et

Pretilachlor

Et

$CO.CH_2Cl$
N
CH_2CH_2OPr

Et

Propachlor

$CO.CH_2Cl$
N
Pr^i

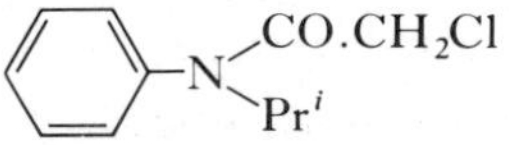

As a result of work conducted with whole plants or excised plant tissue, various groups have suggested that these compounds inhibit protein synthesis (Mann *et al.*, 1965; Jaworski, 1969; Duke *et al.*, 1975). This is unlikely to be due to a primary effect at the level of the ribosome (the cell organelle which catalyses the assembly of amino acids into proteins) since Deal *et al.* (1980) failed to find any effect of 100 μM alachlor, metolachlor, propachlor or the superseded allidochlor (*N,N*-diallylchloroacetamide) in an *in vitro* protein synthesising system prepared from oat roots. Further, protein synthesised *in vitro* by polysomes (p. 236) isolated from herbicide-treated tissue was not distinguishable from that made by control polysome preparations.

Interactions with plant hormones have also been explored. Metolachlor treatment of germinating sorghum seeds causes growth retardation and, amongst other things, ethylene production. However, since the morphological changes accompanying germination inhibition occurred even in the presence of an inhibitor of ethylene biosynthesis, Paradies *et al.* (1981) concluded that ethylene formation was an effect, rather than the cause of growth disruption. Also using germinating sorghum seeds, Wilkinson (1981a) found evidence for an inhibitory effect of metolachlor on terpene synthesis and showed that the terpene-derived gibberellic acid could overcome the inhibitory effect of metolachlor, but only within very specific dose ranges of each compound. More recent studies have indicated an effect of micromolar concentrations of alachlor and metolachlor on formation of gibberellin precursors in an *in vitro* system (Wilkinson, 1981b, 1982). This was proposed to be at least a partial explanation for the herbicidal action of these compounds and to account for effects on cell enlargement seen by Deal and Hess (1980). These latter workers carefully quantified effects of alachlor and metolachlor on cell growth and division in pea and oats, as a result of which they recommended that future work to elucidate the primary mode of action should be done at times and concentrations at which growth of the test species is first inhibited.

In an electron-microscopic study, Ebert (1980) documented the morphological effects occurring in sorghum cells following metolachlor treatment, but could not elucidate the primary effect. Kleudgen (1980) found, using barley leaves, that propachlor-induced depletions in chloroplast components such as chlorophyll and carotenoids were similar to those occurring during senescence. Suggestions have also been made that the compounds may act through a general effect on membranes (Pillai *et al.*, 1976) or membrane transport systems (Marsh *et al.*, 1976).

From the above work it is impossible to reach a conclusion on the primary site of action of the chloroacetanilides. The various effects observed could all be consequences of some as yet unidentified primary action, but in view of the reactivity of the compounds it seems best to conclude at the moment that non-specific interactions at a number of sites may be responsible for the herbicidal effect.

B. Chlorinated aliphatic herbicides

A wide range of biochemical changes can be detected following treatment of plants with the chlorinated aliphatic herbicides shown in Table 7.3 (Ashton and Crafts, 1981; Foy, 1975). However, none has been identified as the primary target, and it is thought instead that they probably act by non-specific combination with cell proteins.

Redemann and Hamaker (1954) showed that a concentration of dalapon (as the acid) of about 1.5 mM precipitated egg albumin from solution in distilled water, and a high concentration of TCA is commonly used as a protein precipitant. Such high concentrations of these herbicides may perhaps be approached in plants since application rates can be up to ten times higher than those used for most herbicides (Worthing, 1979) and,

Table 7.3   Chlorinated aliphatic herbicides

Name	Structure
Chloroacetic acid	$ClCH_2COOH$
Dalapon	$Me.CCl_2.COOH$
TCA	$Cl_3C.COO^-Na^+$

in addition, neither TCA nor dalapon is readily metabolised in higher plants (Foy, 1975). At these high concentrations the herbicides could presumably interfere with the function of both enzymic and non-enzymic (e.g. membrane) proteins.

In a recent survey of the effects of dalapon on a variety of cellular functions of isolated bean leaf cells, Ashton et al. (1977) found that RNA synthesis was the most sensitive of those measured, being 44% inhibited at 1 mM. Other processes were affected to a slightly lesser extent. The results are consistent with the possibility of a non-specific interaction with proteins if one supposes that a protein involved in RNA synthesis was most severely affected under the conditions of the experiment.

If chlorinated aliphatic herbicides do become present in the cytoplasm at high concentrations, it is likely that they will also interfere with the normal function of those cell constituents to which they have some structural resemblance. Examples include the interference with pantoate binding to pantothenate synthetase (Hilton et al., 1963) and the competitive inhibition of pyruvate uptake in mitochondria (Paradies and Papa, 1977).

Both TCA and dalapon applied via the roots will reduce the wax bloom on leaves, as can be seen by electron micrographs of the leaf surfaces of treated peas (Martin and Juniper, 1970). Kolattukudy (1965) showed that 0.05 to 0.1 μmoles of TCA applied per 5 g of broccoli leaf caused approximately 50% inhibition of the synthesis of the surface layer of wax. He also used TCA, at a rather higher concentration, to inhibit alkane synthesis by chopped leaves (Kolattukudy, 1968). It is not known whether inhibition of lipid synthesis contributes to the lethal effect of the chlorinated aliphatic acid herbicides.

C. Pesticides containing mercury

The mercury-containing pesticides listed in Table 7.4 are mostly old compounds that are used mainly as fungicides, particularly for seed dressing. The toxic nature of mercury itself, and the discovery that methyl mercury ($CH_3.Hg^+$), which is more toxic to living organisms than aryl mercurials or inorganic salts of mercury, is synthesised in the biosphere (for review see Vallee and Ulmer, 1972), led to considerable concern over the widespread use of mercury pesticides.

Table 7.4   Mercury-containing pesticides

Name	Structure
Mercuric chloride	$HgCl_2$
Mercuric oxide	HgO
Mercurous chloride	Hg_2Cl_2
2-Methoxyethylmercury chloride	$MeOCH_2CH_2.HgCl$
2-Methoxyethylmercury silicate	$MeOCH_2CH_2.Hg.$silicate
Phenylmercury acetate	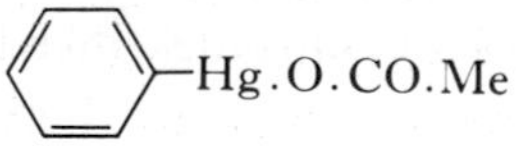 ⟨⟩—Hg.O.CO.Me
Phenylmercury nitrate	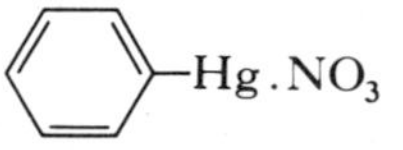 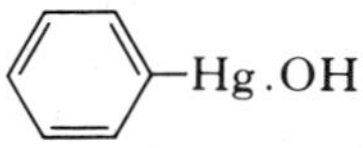 ⟨⟩—Hg.NO₃ ⟨⟩—Hg.OH

In Chapter 1 we classified copper compounds (and copper carriers) as inhibitors of respiration, because of the ability of copper ions to bind to the lipoamide cofactor of pyruvate dehydrogenase. Nevertheless thiol groups of other enzymes could react also, so that the inhibition is not specific. Mercury pesticides will also inhibit glycolysis as a result of binding to essential thiol groups (Lyr, 1977). However, mercury also shows a strong affinity for phosphates, histidine side-chains of proteins, and for purines, pteridines and porphyrins (Vallee and Ulmer, 1972), and is therefore likely to interfere with a large number of cell processes.

The biochemical basis for mercury toxicity is generally thought to be its interaction with the thiol groups of proteins, where Hg^{2+} forms compounds of the following type:

$$R-S-Hg^+, \quad R-S-Hg-S-R, \quad R \overset{S}{\underset{S}{<}} \hspace{-0.3em} \diagdown Hg \quad \text{etc.}$$

Webb (1966) lists over 40 enzymes that are at least 50% inhibited by a concentration of Hg^{2+} of 10 μM or less. This would seem to be more than adequate to explain the fungitoxic action of mercuric chloride since approximately 5 mg of mercury per gramme of spores (or approximately 25 mM, assuming equal distribution throughout the spore) is required to kill half the spores (McCallan and Miller, 1958). In fact, with this degree of enzyme sensitivity it is perhaps surprising that so much compound is required *in vivo*.

It seems likely that mercurous chloride could act after oxidation to the mercuric compound.

The organic compounds in Table 7.4 have the general formula R–Hg–X, where R is an organic moiety and X is a group bound to mercury by a bond that is essentially salt-like in character (Ulfvarson, 1969). Since there is evidence (reviewed by Ulfvarson, 1969) that the toxicity of compounds of the type R–Hg–X is essentially independent of X, it seems that the active species we must consider is R–Hg$^+$.

The differences in chemical reactivity between the organic mercurials and mercuric chloride have been described by Webb (1966). Although R–Hg$^+$ is monofunctional, it is more soluble in lipid than the bifunctional Hg^{2+}, so that it will presumably penetrate more readily into cells and tissues. Webb (1966) concludes that in general Hg^{2+} is a more effective enzyme inhibitor than organic mercurials, though the latter are still powerful toxicants. Thus he lists several examples of phenylmercuric compounds causing over 50% inhibition of enzymes when present at concentrations of 10 μM or less.

D. Alkylene bisdithiocarbamate fungicides

These compounds (Table 7.5) are similar to the dialkyldithiocarbamates which, through their action as copper carriers, affect principally the lipoamide-containing dehydrogenases (p. 14). The biochemical mode of action is, however, different in that no enzyme has been identified as being especially sensitive to the alkylene bisdithiocarbamates, which are thought to work by the reaction of their metabolites with cellular thiol groups generally. A reaction with thiol groups is supported by the fact that the inhibitory action of nabam on fungal spore germination is strongly antagonised by the thiol compounds thioglycollic acid and cysteine, which were presumed to react directly with the fungicide or its metabolites (Kaars Sijpesteijn and van der Kerk, 1954).

Alkylene bisdithiocarbamates are fairly unstable both in non-biological (Hylin, 1973) and biological systems (Kaars Sijpesteijn *et al.*, 1977) and the range of degradation products formed has made identification of the actual toxic species rather uncertain. Ethylene thiuram disulphide was preferred as a candidate by reviewers (Owens, 1969; Lukens, 1971).

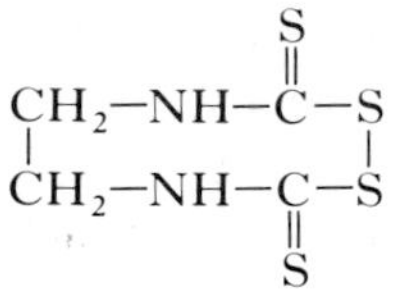

ethylene thiuram disulphide 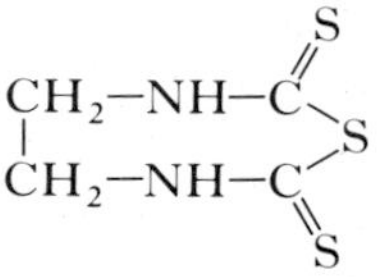   'ethylene thiuram monosulphide'

However, it has more recently been recognised that the structure of the compound referred to as 'ethylene thiuram monosulphide', is not that shown above but is in fact that of etem (Table 7.5) (Pluijgers *et al.*, 1971) a compound which has fungicidal activity itself (Kaars Sijpesteijn and van der Kerk, 1954). Since it seems unlikely that this compound could be converted to the disulphide, the disulphide is probably not the only fungitoxic compound formed from members of this class of fungicide.

Table 7.5   Alkylene bisdithiocarbamate fungicides and a related compound

Name	Structure
Etem	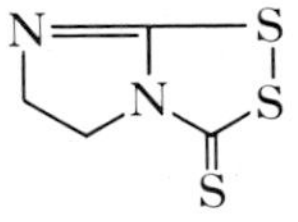
Mancozeb	$\begin{bmatrix} CH_2NH.CS.S.Mn- \\ | \\ CH_2NH.CS.S- \end{bmatrix}_x Zn_y$
Maneb	$\begin{bmatrix} CH_2NH.CS.S.Mn- \\ | \\ CH_2NH.CS.S- \end{bmatrix}_x$

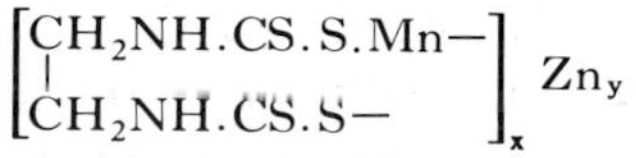

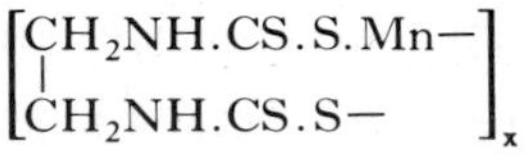

Name	Structure
Nabam	$\begin{aligned}&CH_2NH.CS.S^- \ Na^+\\&\quad\vert\\&CH_2NH.CS.S^- \ Na^+\end{aligned}$
Propineb	$\left[\begin{aligned}&Me{-}CHNH.CS.S.Zn{-}\\&\qquad\vert\\&\quad CH_2NH.CS.S{-}\end{aligned}\right]_x$
Zineb	$\left[\begin{aligned}&CH_2NH.CS.S.Zn{-}\\&\quad\vert\\&CH_2NH.CS.S{-}\end{aligned}\right]_x$
Zineb-ethylenebis-(thiuram disulphide) mixed precipitation	$\left[\begin{aligned}&\left[\left(\begin{aligned}&CH_2NH.CS.S{-}\\&\quad\vert\\&CH_2NH.CS.S{-}\end{aligned}\right)^{2-}\ Zn(NH_3)_2^{2+}\right]_3\\[4pt]&\left(\begin{aligned}&CH_2NH.CS.S{-}\\&\quad\vert\\&CH_2NH.CS.S{-}\end{aligned}\right)\end{aligned}\right]_x \quad x>1$

Kaars Sijpesteijn *et al.* (1977) reviewed the transformations of alkylene bisdithiocarbamates in plants and microorganisms and indicated that the fungitoxic species might be etem or the further degradation product ethylene diisothiocyanate. This compound is also fungicidal in its own right and was first suggested as the reactive metabolite some time ago (see Kaars Sijpesteijn and van der Kerk, 1954).

$$\begin{aligned}&CH_2{-}NCS \qquad \text{ethylene}\\&\quad\vert\\&CH_2{-}NCS \qquad \text{diisothiocyanate}\end{aligned}$$

To summarise, it cannot be said with certainty whether there is a single important fungitoxic principle *in vivo*, though a number of compounds with suitable properties have been identified. It may be that more than one of these compounds reacts with vital thiol groups to kill the fungus; the combination need not necessarily be the same with different alkylene bisdithiocarbamates or different fungi.

E. *N*-Trichloromethylthio fungicides and related compounds

Compounds of this class (Table 7.6) are particularly susceptible to attack by electron-rich groups, including the hydroxide ion (Kohn, 1977) and, in fact, this author emphasises that the compounds are only protected from hydrolysis by their low water-solubility.

Table 7.6 *N*-Trichloromethylthio fungicides and related compounds

Name	Structure
Captafol	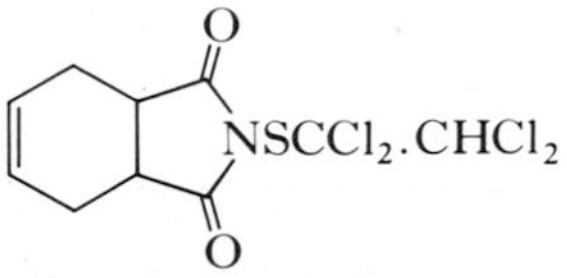 ![structure] NSCCl$_2$.CHCl$_2$
Captan	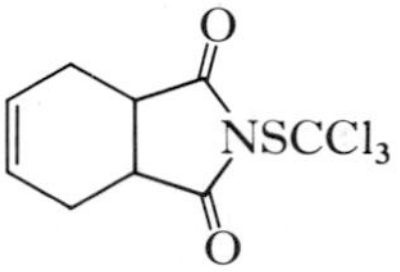 ![structure] NSCCl$_3$
Folpet	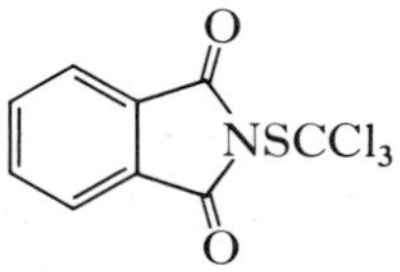 ![structure] NSCCl$_3$
Dichlofluanid	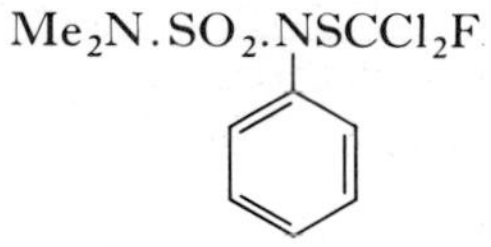 Me$_2$N.SO$_2$.NSCCl$_2$F
Tolylfluanid	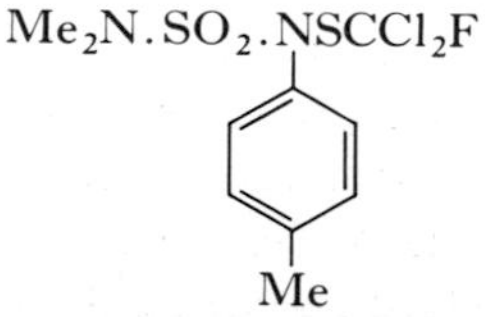 Me$_2$N.SO$_2$.NSCCl$_2$F Me

The *N*-trichloromethylthio fungicides probably work through their non-specific reactions with cell components, particularly thiols, though the exact mechanism of the reactions has been uncertain for some time (Lukens, 1971; Kohn, 1977). Reaction of the fungicide with thiols produces thiophosgene (Cl.CS.Cl) as a product, and this highly reactive chemical is also able to combine with –SH groups (Lukens, 1971). As an example of the susceptibility of thiols, 20 μM folpet caused 80% inhibition of glyceraldehyde-phosphate dehydrogenase in 60 minutes, and the effect could be prevented if dithiothreitol was added before the fungicide (Siegel, 1971).

The non-specific nature of the action of compounds of this group is illustrated by the breadth of biochemical effects reported. These embrace the inhibition of thiol enzymes, interactions with membranes, and disruption of mitochondrial reactions including oxidative phosphorylation and NADH oxidation (see Kumar *et al.*, 1975 and Kohn, 1977, for original references).

F. Miscellaneous pesticides

Table 7.7 gives the structures of miscellaneous compounds which may react non-specifically with cellular components.

Table 7.7   Miscellaneous pesticides

Name and use	Structure
Acrolein (aquatic herbicide and algicide)	$CH_2{=}CH.CHO$
Allyl alcohol (herbicide)	$CH_2{=}CH.CH_2OH$
Chlorothalonil (fungicide, algicide)	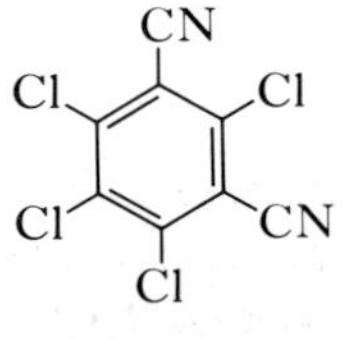
2,3-Dichloro-*N*-(4-fluorophenyl)maleimide (fungicide)	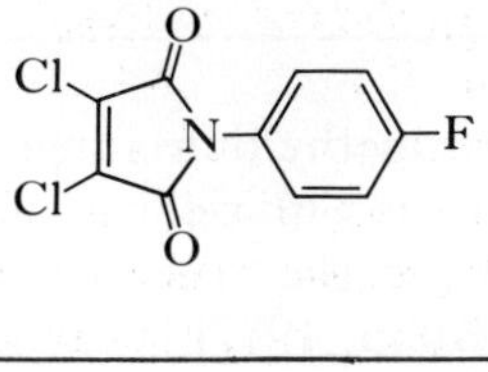

Name	Structure
Dithianon (fungicide)	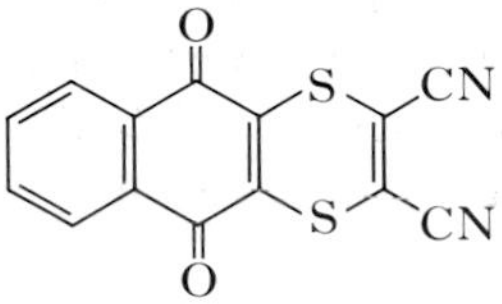 (naphthoquinone-dithiin structure with two CN groups)
Quinomethionate	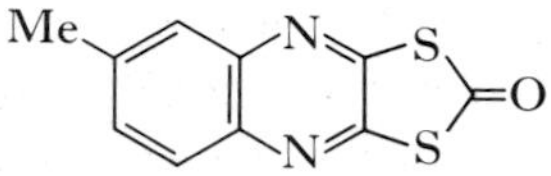 (Me-substituted quinoxaline dithiole structure with =O)
Sodium chlorate	$NaClO_3$

1. Acrolein and allyl alcohol

The reasons for the herbicidal activity of acrolein and allyl alcohol do not seem to have been sought specifically, but acrolein is a generally reactive chemical, capable of undergoing addition reactions with nucleophilic groups within the cell such as –SH and –NH_2.

Allyl alcohol is not reactive *per se*, but it can be converted to acrolein, at least by alcohol dehydrogenase (Rando, 1974). In the case of the yeast enzyme, acrolein formation occurs much more quickly than enzyme inhibition, which suggests that the inhibition does not occur immediately the acrolein is formed at the active site. Instead Rando (1974) suggests that the acrolein is first released from the active site and then returns from solution to bind these irreversibly.

2. Chlorothalonil

The fungicide and algicide chlorothalonil probably works by reaction with cellular thiol groups since thiol compounds reverse the fungistatic action. In addition, levels of the protective thiol compound glutathione fell rapidly in treated cells, and cell mortality did not begin until the protective system had been exhausted so that inhibition of thiol-dependent enzymes became significant (Vincent and Sisler, 1968; Tillman *et al.*, 1973).

Effects on individual enzymes *in vitro* fit with this mode of action (Long and Siegel, 1975). Inhibition of the thiol-dependent glyceraldehyde-phosphate dehydrogenase from yeast could be prevented by dithiothreitol, and treatment with the thiol reagent DTNB (5,5'-dithiobis(2-nitrobenzoic acid)) greatly diminished binding of ^{14}C-labelled chlorothalonil to the enzyme. Chlorothalonil did not bind to non-thiol groups of either glyceraldehyde-phosphate dehydrogenase or α-chymotrypsin (Long and Siegel, 1975).

3. 2,3-Dichloro-N-(4-fluorophenyl)maleimide

This fungicide was the subject of an investigation by Tsuda et al. (1976). Spore germination of *Pyricularia oryzae* was half-inhibited by 11 μM fungicide and several biochemical processes were inhibited to the extents of 30–55% by 40 μM compound. All the effects were greatly alleviated by sulphydryl compounds such as glutathione and dithiothreitol and the authors therefore concluded that the compound acts by combination with cellular thiol groups.

4. Dithianon

Dithianon probably acts by general reaction with thiol groups, a conclusion based on studies with model thiols, thiol enzymes, and biochemical systems dependent on reduced thiols (see Sturdik and Drobnica, 1980). These authors found, for instance, that the growth of yeast cells was half-inhibited by about 1.4 μM fungicide, that radioactive leucine incorporation was similarly sensitive, and that treatment of yeast protoplasts and yeast cell homogenates with 300 μM dithianon led to strong inhibition of a selection of glycolytic enzymes.

5. Quinomethionate

Aziz and Knowles (1973) examined the reasons for the efficacy of this acaricide against *Tetranychus urticae*. They found that radioactive compound became irreversibly bound to proteins in a range of tissue homogenates, including one from spider mites, and that sulphydryl reagents blocked this binding. The dithiol formed by loss of the carbonyl group (i.e. 6-methyl-2,3-quinoxalinedithiol) was a significant metabolite of quinomethionate in mites, and since it was shown to be toxic in its own right this could perhaps have been binding to proteins as well. Rojakovick and March (1976) found that (for instance) cockroach brain phosphodiesterase was very sensitive to quinomethionate (I_{50} c. 0.9 μM), but they regarded this as an example of the susceptibility of a thiol enzyme rather than indicative of a primary site of action.

6. Sodium chlorate

Sodium chlorate probably owes its phytotoxicity to conversion to the oxidising chlorite ion (ClO_2^-), formed from chlorate by the plant enzyme nitrate reductase which normally reduces nitrate (NO_3^-) to nitrite (NO_2^-) (see Hofstra, 1977).

REFERENCES

Ashton, F. M. and Crafts, A. S. (1981). "Mode of Action of Herbicides", 2nd edn. Wiley-Interscience, New York

Ashton, F. M., De Villiers, O. T., Glenn, R. K. and Duke, W. B. (1977). *Pestic. Biochem. Physiol.* **7**, 122–141

Aziz, S. A. and Knowles, C. O. (1973). *J. Econ. Ent.* **66**, 1041–1045

Berlin, J. A. and Micks, D. W. (1973). *Ann. Entomol. Soc. Amer.* **66**, 775–780

Brown, I. F. and Sisler, H. D. (1960). *Phytopathol.* **50**, 830–839

Deal, L. M. and Hess, F. D. (1980). *Weed Sci.* **28**, 168–175

Deal, L. M., Reeves, J. T., Larkins, B. A. and Hess, F. D. (1980). *Weed Sci.* **28**, 334–340

Duke, W. B., Slife, F. W., Hanson, J. B. and Butler, H. S. (1975). *Weed Sci.* **23**, 142–147

Ebert, E. (1980). *Pestic. Biochem. Physiol.* **13**, 227–236

Foy, C. L. (1975). In "Herbicides. Chemistry, Degradation, and Mode of Action" (P. C. Kearney and D. D. Kaufman, eds) Vol. 1, pp. 399–452. Dekker, New York

Green, M. B., Hartley, G. S. and West, T. F. (1977). "Chemicals for Crop Protection and Pest Control". Pergamon Press, Oxford

Hamilton-Miller, J. M. T. (1973). *Bacteriol. Rev.* **37**, 166–196

Hammond, S. M. (1977). *Prog. Med. Chem.* **14**, 105–179

Hewitt, E. J. and Nicholas, D. J. D. (1963). In "Metabolic Inhibitors" (R. M. Hochster and J. H. Quastel, eds) Vol. II, pp. 311–436. Academic Press, London and New York

Hilton, J. L., Jansen, L. L. and Hull, H. M. (1963). *Ann. Rev. Plant Physiol.* **14**, 353–384

Hofstra, J. J. (1977). *Physiol. Plant.* **41**, 65–69

Hylin, J. W. (1973). *Bull. Environ. Contam. Toxicol.* **10**, 227–233

Jaworski, E. G. (1969). *J. Agric. Food Chem.* **17**, 165–170

Kaars Sijpesteijn, A. and van der Kerk, G. J. M. (1954). *Biochim. Biophys. Acta* **13**, 545–552

Kaars Sijpesteijn, A., Dekhuijzen, H. M. and Vonk, J. W. (1977). In "Antifungal Compounds" (M. R. Siegel and H. D. Sisler, eds) Vol. 2, pp. 91–147. Dekker, New York

Kleudgen, H. K. (1980). *Weed Res.* **20**, 41–46

Kohn, G. K. (1977). In "Pesticide Chemistry in the 20th Century" (J. R. Plimmer, ed.) pp. 153–169. American Chemical Society, Washington, D.C.

Kolattukudy, P. E. (1965). *Biochemistry* **4**, 1844–1855

Kolattukudy, P. E. (1968). *Plant Physiol.* **43**, 375–383

Kotler-Brajtburg, J., Medoff, G., Kobayashi, G. S., Boggs, S., Schlessinger, D., Pandey, C. and Rinehart, K. L. (1979). *Antimicrob. Agents Chemother.* **15**, 716–722

Kumar, S. S., Sikka, H. C., Saxena, J. and Zweig, G. (1975). *Pestic. Biochem. Physiol.* **5**, 338–347

Long, J. W. and Siegel, M. R. (1975). *Chem.-Biol. Interact.* **10**, 383–394

Lukens, R. J. (1971). "Chemistry of Fungicidal Action". Chapman and Hall, London

Lyr, H. (1977). In "Antifungal Compounds" (M. R. Siegel and H. D. Sisler, eds) Vol. 2, pp. 301–332. Dekker, New York

Mann, J. D., Jordan, L. S. and Day, B. E. (1965). *Plant Physiol.* **40**, 840–843

Marsh, H. V., Bates, J. and Downs, S. (1976). *Plant Physiol.* **57**, 61

Martin, J. T. and Juniper, B. E. (1970). "The Cuticles of Plants". Edward Arnold, London

McCallan, S. E. A. and Miller, L. P. (1958). *Adv. Pest. Control Res.* **2**, 107–134.

Miller, R. W. and Barran, L. R. (1977). *Can. J. Microbiol.* **23**, 1373–1383

Morrod, R. S. (1976). In "Herbicides. Physiology, Biochemistry, Ecology" (L. J. Audus, ed.) Vol. 1, pp. 281–304. Academic Press, London and New York

Norman, W. A., Spielvogel, A. M. and Wong, R. G. (1976). *Adv. Lipid Res.* **14**, 127–170

O'Brien, R. D. (1967). "Insecticides: Action and Metabolism". Academic Press, London and New York

Owens, R. G. (1969). In "Fungicides: An Advanced Treatise" (D. C. Torgeson, ed.) Vol. 2, pp. 147–301. Academic Press, London and New York

Paradies, G. and Papa, S. (1977). *Biochim. Biophys. Acta* **462**, 333–346

Paradies, I., Ebert, E. and Elstner, E. F. (1981). *Pestic. Biochem. Physiol.* **15**, 209–212

Pillai, C. G. P., Davis, D. E. and Truelove, B. (1976). *Proc. South. Weed Sci. Soc.* **29**, 403

Pluijgers, C. W., Vonk, J. W. and Thorn, G. D. (1971). *Tetrahedron Lett.* 1317–1318.

Pressman, B. C. (1963). *J. Biol. Chem.* **238**, 401–409

Rando, R. R. (1974). *Biochem. Pharmacol.* **23**, 2328–2331

Redemann, C. T. and Hamaker, J. (1954). *Weeds* **3**, 387–388

Rivera, C. M. and Penner, D. (1979). *Residue Rev.* **70**, 45–76

Rojakovick, A. S. and March, R. B. (1976). *Pestic. Biochem. Physiol.* **6**, 10–19

Siegel, M. R. (1971). *Pestic. Biochem. Physiol.* **1**, 225–233

Somers, E. and Pring, R. J. (1966). *Ann. Appl. Biol.* **58**, 457–466

Sturdik, E. and Drobnica, L. (1980). *Chem.-Biol. Interact.* **30**, 105–114

Tillman, R. W., Siegel, M. R. and Long, J. W. (1973). *Pestic. Biochem. Physiol.* **3**, 160–167

Tsuda, M., Nakajima, T., Kasugai, H., Kawada, S., Yamaguchi, I. and Misato, T. (1976). *Nippon Noyaku Gakkaishi* **1**, 101–106

Ulfvarson, U. (1969). In "Fungicides: An Advanced Treatise" (D. C. Torgeson, ed.) Vol. 2, pp. 303–329. Academic Press, London and New York

Vallee, B. L. and Ulmer, D. D. (1972). In "Fungicides: An Advanced Treatise" (D. C. Torgeson, ed.) Vol. 2, pp. 303–329. Academic Press, London and New York

van Overbeek, J. and Blondeau, R. (1954). *Weeds* **3**, 55–65

Vincent, P. G. and Sisler, H. D. (1968). *Physiol. Plant.* **21**, 1249–1264

Webb, J. L. (1966). "Enzyme and Metabolic Inhibitors", Vol. II. Academic Press, London and New York

Wilkinson, R. E. (1981a). *Pestic. Biochem. Physiol.* **16**, 63–71

Wilkinson, R. E. (1981b). *Pestic. Biochem. Physiol.* **16**, 199–205
Wilkinson, R. E. (1982). *Pestic. Biochem. Physiol.* **17**, 177–184
Worthing, C. R. (ed.) (1979). "The Pesticide Manual", 6th edn. British Crop
Protection Council, London

8 | Pesticides whose Mode of Action is Unknown

I. INTRODUCTION

Apart from the exceptions stated in the Preface, the aim of this book has been to cover all pesticides considered to be in current use by their inclusion in the *Pesticide Manual* (Worthing, 1979). In this chapter we have listed those pesticides whose mode of action appears to be unknown.

It must be noted that many compounds whose mode of action is not clear have been included in previous chapters, if the indications that are presently available suggest a particular mode of action. In our view this cannot be said of the compounds which follow, though it should be appreciated that a dividing line between these two categories is necessarily difficult to draw.

The inclusion of a compound in this chapter does not necessarily mean that no work has been carried out on the biochemical effects of the compound, but it does suggest that the results have so far been inconclusive.

II. ACARICIDES

There appears to be no published information on the modes of action of the unrelated acaricides benzoximate, 3,6-bis(2-chlorophenyl)-1,2,4,5-tetrazine (proposed name clofentezine) (Bryan *et al.*, 1981) and propargite (Table 8.1).

Table 8.1 Miscellaneous acaricides

Name	Structure
Benzoximate	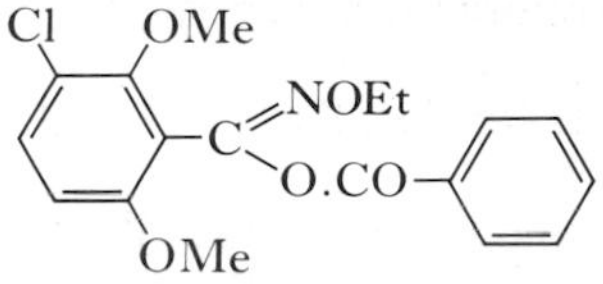 Cl OMe
3,6-Bis(2-chlorophenyl)-1,2,4,5-tetrazine (clofentezine proposed)	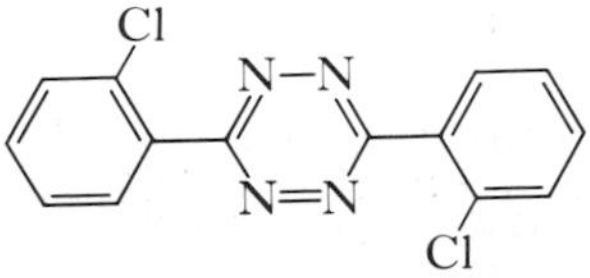
Propargite	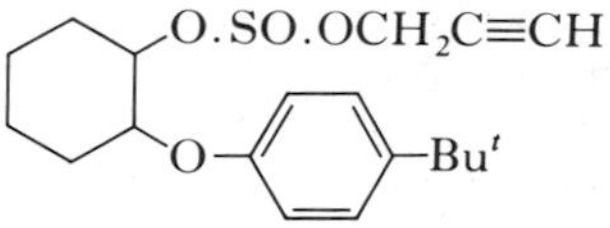

Table 8.2 Organophosphorus fungicides

Name	Structure
Ditalimfos	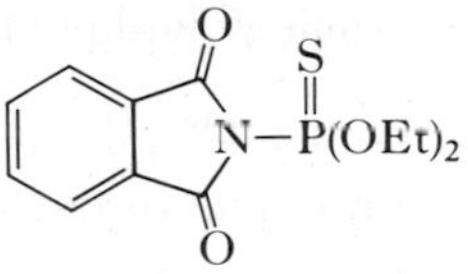
Fosetyl-aluminium	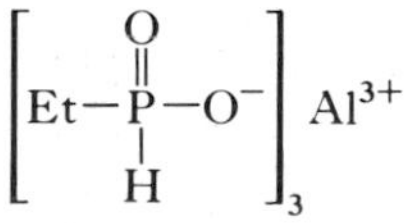
Pyrazophos	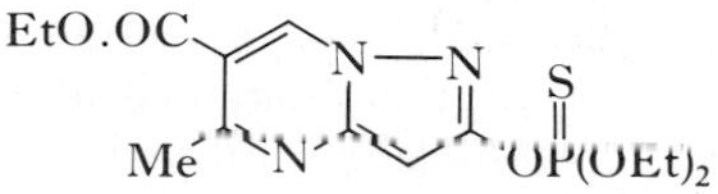
Triamiphos	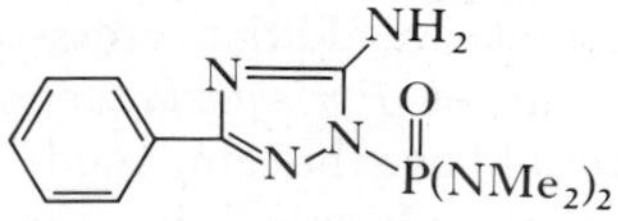

III. FUNGICIDES

A. Organophosphorus fungicides

The modes of action of the organophosphorus fungicides shown in Table 8.2 are not yet clear.

Investigating pyrazophos, De Waard (1974) found that cultures of *Pyricularia oryzae* converted pyrazophos to 2-hydroxy-5-methyl-6-ethoxycarbonylpyrazolo-(1,5-a)-pyrimidine. Since this compound was found to be more fungitoxic than pyrazophos itself the conversion was interpreted as lethal synthesis, a conclusion supported by the demonstration that a resistant fungal strain did not form this metabolite (De Waard and van Nistelrooy, 1980).

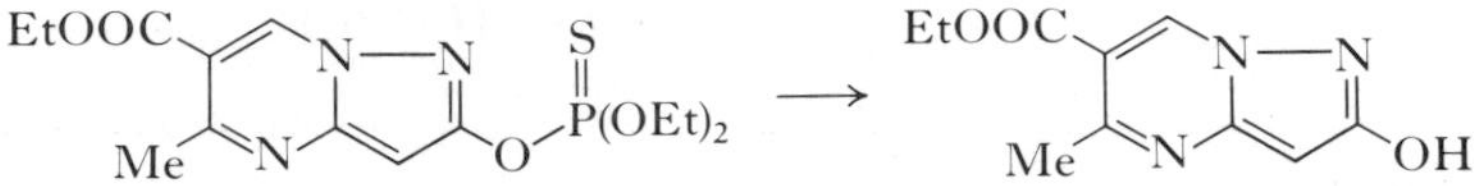

pyrazophos   →   2-hydroxy-5-methyl-6-ethoxycarbonyl-pyrazolo(1,5-a)-pyrimidine

De Waard (1974) found that growth inhibitory concentrations of the active metabolite (of the order of 100 μM) caused inhibition of oxygen uptake as well as of DNA, RNA and protein synthesis in cultures of *Pyricularia oryzae*. This suggested a non-specific mode of action.

The action of fosetyl-aluminium may depend on its ability to cause the production of fungitoxic polyphenols in plants (Bompeix *et al.*, 1981).

B. Miscellaneous fungicides

The modes of action of the fungicides in Table 8.3 have not been established, but relevant observations have been made in some cases.

It has been shown (Despreaux *et al.*, 1981) that RNA formation from [^{14}C]-uridine was more sensitive to cymoxanil than DNA synthesis, protein synthesis or respiration in *Botrytis cinerea*.

A 17 μM concentration of the systemic rice blast (*Pyricularia oryzae*) fungicide isoprothiolane strongly inhibited mycelial growth *in vitro* and after 3 days about 50% of the mycelial tips were swollen. Growth was not inhibited during the first 24 h, however, allowing the possibility that a metabolite could be the active compound (Kakiki and Misato, 1979a). Further results using this fungus (Kakiki and Misato, 1979a,b) suggested that 170 μM isoprothiolane inhibited sugar uptake (and thereby cell wall formation) and acetate incorporation into fatty acids (and thereby triglyceride and phospholipid formation). Extracted fatty acid synthetase was not inhibited. In addition cross-resistance studies using laboratory-induced mutants of *Pyricularia oryzae* raised the possibility that isoprothiolane could have the same mode of action as IBP, which may inhibit biosynthesis of the phospholipid, phosphatidyl choline (p. 269) (Katagiri and Uesugi, 1977).

Table 8.3   Miscellaneous fungicides

Name	Structure
Cymoxanil	EtNH.CO.NH.CO.C=N.OMe with CN below
Dehydroacetic acid	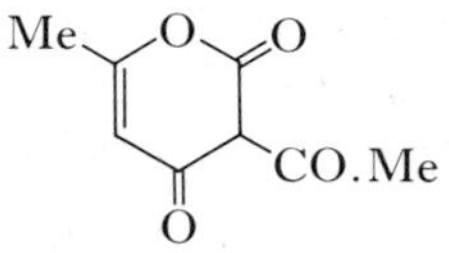 (ring structure) Me, O, O, CO.Me, O
Isoprothiolane	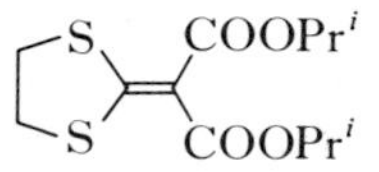 (ring structure) S, S, COOPr^i, COOPr^i
Lime sulphur	CaS_x
Nitrothal-isopropyl	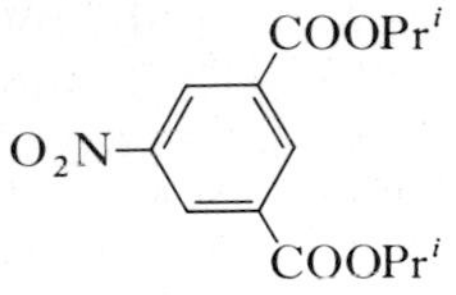 (ring structure) O_2N—, COOPr^i, COOPr^i
Propamocarb (as the hydrochloride)	$Me_2N(CH_2)_3NH.COOPr.HCl$
Prothiocarb (as the hydrochloride)	$Me_2N(CH_2)_3NH.COSEt.HCl$
Sulphur	S_x

Name	Structure
4,5,6,7-Tetrachlorophthalide	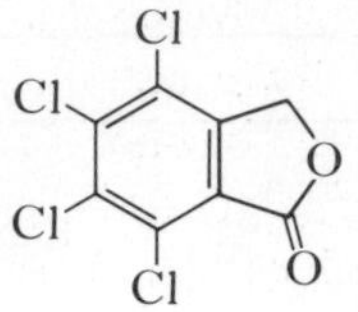

| Validamycin A | 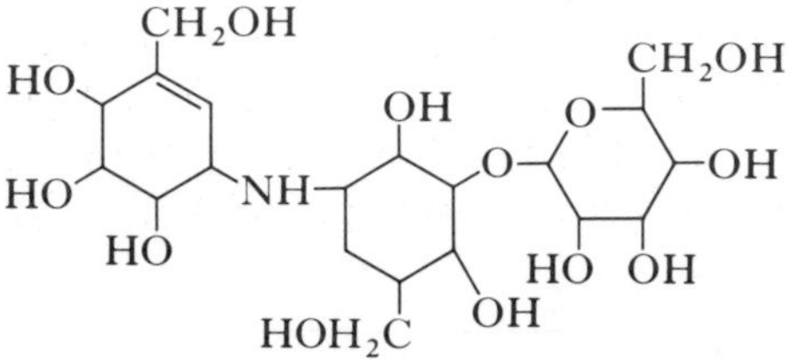 |

Propamocarb and prothiocarb are closely related structurally and pre-sumably have the same mode of action. Kaars Sijpesteijn *et al.* (1974) noted that only fungi possessing cellulose walls were sensitive to prothio-carb, but more recent work suggests that the compound may interfere indirectly with membrane structure or function. Papavizas *et al.* (1978) found that cell constituents such as phosphate, carbohydrate and protein leaked into the medium of *Pythium ultimum* grown in the presence of propamocarb (in the concentration range 1–10 μM). Leakage did not occur if the propamocarb was added after the mycelium had developed, so the effect could not have been on the completed membrane. In addition, the leakage was largely prevented by supplying sterols such as cholesterol to the fungus (*Pythium* species do not have sterols in their membrane). The significance of this is not entirely clear, but it is possible that sterol-fungicide complex formation rendered the pesticide unavailable (Papavizas *et al.*, 1978).

Kerkenaar and Kaars Sijpesteijn (1977) found that in the case of a few fungi (e.g. *Achyla radiosa*) prothiocarb was active because it released ethyl mercaptan, assumed to be toxic after conversion to ethionine, an analogue and antimetabolite of the amino acid methionine. However, this was not the case with other fungi, including *Pythium* species against which prothiocarb is used in the field.

The compound 4,5,6,7-tetrachlorophthalide, used against *Pyricularia oryzae*, does not prevent growth of the fungus, but it interferes with the invasion of the plant by the organism (Aoki and Yamada, 1979). Recent work suggests that the compound may act, like tricyclazole, through inhibition of melanin biosynthesis (p. 282) (Woloshuk and Sisler, 1982).

The mode of action of the fungistatic agent validamycin A is not established but Akechi and Matsuura (1982) report that it inhibited inositol formation in *Rhizoctonia solani* and that pathogenicity of *R. solani* was restored by supplying inositol to validamycin A-treated fungus. As the authors point out, this could be connected with a structural resemblance of part of the validamycin A molecule to inositol (hexahydroxycyclohexane). The structure of validamycin A has recently been revised (Suami *et al.*, 1980) and it is this revised structure which is included in Table 8.3.

The mode of action of the other fungicides in Table 8.3 does not seem to be known. Information on the general biochemistry of dehydroacetic acid is reviewed in detail by Webb (1966), and a concise summary of the properties and uses of sulphur and lime sulphur is given by Green *et al.* (1977).

IV. HERBICIDES

A. Benzoylprop-ethyl and related compounds

It has been clearly established that these wild oat herbicides (Table 8.4) owe their activity to conversion to the parent acids, and their selectivity to the fact that this conversion takes place sufficiently slowly in wheat and barley for the acid to be removed by further transformation to inactive polar conjugates (Jeffcoat and Harries, 1973; Venis, 1982). There is,

Table 8.4 Benzoylprop-ethyl and related compounds

Name	Structure
Benzoylprop-ethyl	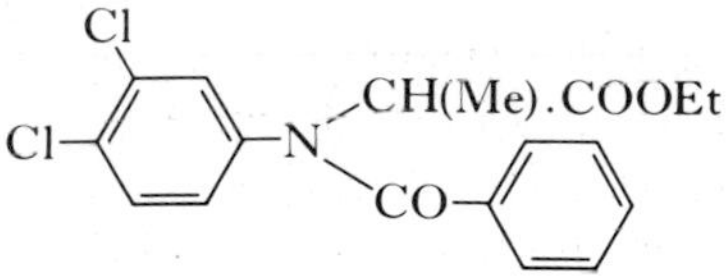
Flamprop-isopropyl (racemate and *R*-enantiomer)	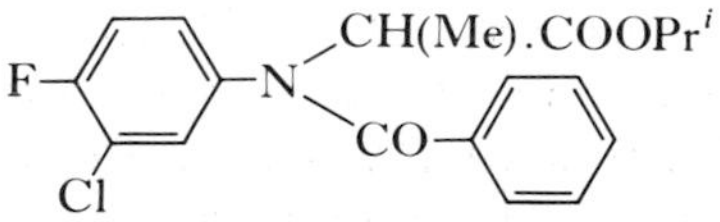
Flamprop-methyl	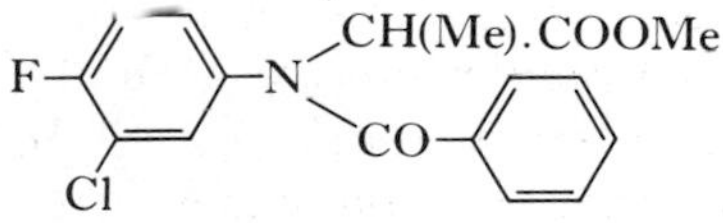

however, much less work bearing on the primary mode of action of the compounds. Jeffcoat and Harries (1975) found that the expansion of oat coleoptile sections could be reduced to about half that of control sections by 10 μM flamprop and 33 μM benzoylprop, and also observed effects on treated plants which suggested that cell expansion was the major process affected. In this context Morrison *et al.* (1979) found that 2,4-D was antagonistic to effects of benzoylprop-ethyl and flamprop-methyl, so some interference with auxin action might be inferred. Since they found that wild oat internode length was more depressed (as a percentage of control values) than the lengths of the individual cells, Morrison *et al.* (1979) concluded that there must also be an effect on cell division.

B. Dichlofop-methyl and related compounds

Compounds such as diclofop-methyl and its relatives (Table 8.5) are used for post-emergence grass control in a variety of crops. Both the ester and the parent acid may be toxic to plant cells (Shimabukuro *et al.*, 1978) and the insensitivity of some plant species is due to their ability to metabolise the herbicide by ring hydroxylation and conjugation (Shimabukuro *et al.*, 1979; Gorecka *et al.*, 1981). Although we cannot yet be certain about the primary site of action, two suggestions have been made.

Firstly, an examination of the effects of diclofop-methyl and clofop-isobutyl on auxin(IAA)-stimulated extension of coleoptiles of wheat, oat and maize has led to the suggestion that the compounds may be antiauxins

Table 8.5   Diclofop-methyl and related compounds

Name	Structure
Diclofop-methyl	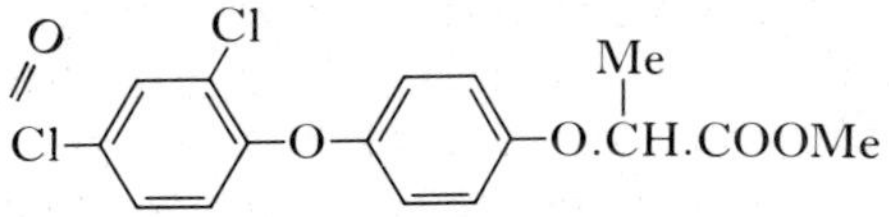
Clofop-isobutyl	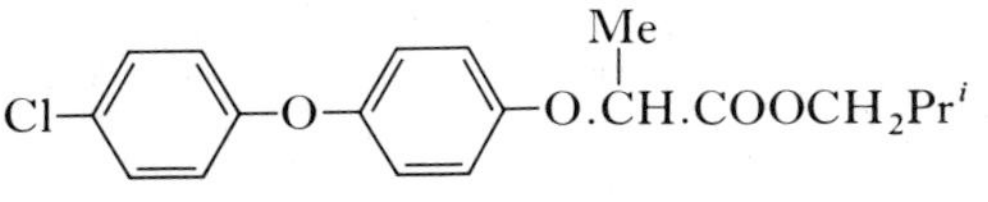
Fluazifop-butyl	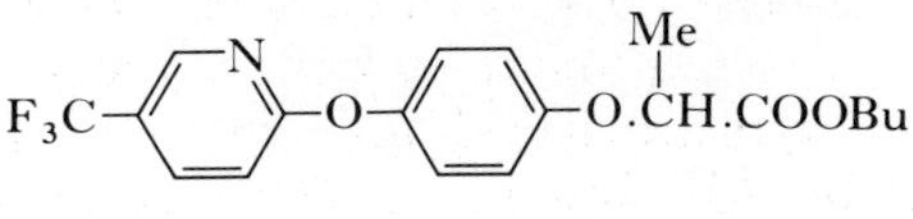

(Shimabukuro *et al.*, 1978; Gorecka *et al.*, 1981). Diclofop-methyl at 10 μM caused 51% inhibition of growth stimulated in oat coleoptiles by 10 μM IAA whereas wheat coleoptiles were less sensitive. This fits with its use as a post-emergence wild oat herbicide in wheat. The growth inhibition was partly overcome by the inclusion of more IAA (Shimabukuro *et al.*, 1978) but it was not demonstrated that IAA could completely overcome the effects, as would be expected if the diclofop-methyl was competing for an IAA binding site.

Wheat and oat are tolerant to clofop-isobutyl, and the effects of the compound on IAA-induced coleoptile extension *in vitro* reflected this. Maize, on the other hand, is a sensitive species and extension of maize coleoptiles *in vitro* was reduced to about half by clofop-isobutyl, though the effect was independent of concentration, occurring at 0.1, 1 and 10 μM (Gorecka *et al.*, 1981). Dose-dependence would be expected from a straight-forward auxin antagonist.

The second suggestion is that diclofop-methyl inhibits fatty acid bio-synthesis in a manner which might be direct or indirect (Hoppe, 1981; Hoppe and Zacher, 1982). This suggestion followed a survey of the effects of the herbicide on various biosynthetic processes of maize roots. Lipid synthesis was the most strongly affected; the rate of incorporation of radiolabelled acetate into a lipid fraction was reduced by half following treatment of the roots with 0.5–5 μM diclofop-methyl. Uptake of the acetate into the tissue was affected, but to a lesser extent. The chemical is in fact used post-emergence and only a very small proportion of diclofop-methyl applied to leaves is translocated to the roots (Brezeanu *et al.*, 1976). However, run-off from the leaves would allow the chemical to reach the roots through the soil, and roots have in fact been shown to be very sensitive to diclofop-methyl (Morrison *et al.*, 1981). Recent preliminary results using the related candidate herbicide ethyl 2-[4-(6-chloro-2-benzox-azolyloxy)phenoxy]propanoate (proposed name, fenoxaprop-ethyl) suggest that lipid synthesis in leaf tissue is also affected (Köcher *et al.*, 1982).

fenoxaprop-ethyl

Structural investigations of the effect of diclofop-methyl on plant tissue have been carried out (Brezeanu *et al.*, 1976; Morrison *et al.*, 1981) and the effects of relatively high concentrations on mitochondrial functions have also been examined (Cohen and Morrison, 1981) but these studies have not brought forward alternative candidates for the primary site of action of these chemicals.

C. Chlorfenprop-methyl

There is some evidence that the wild oat herbicide chlorfenprop-methyl may exert its action through interference with membrane function. Both chlorfenprop-methyl and its parent acid chlorfenprop are herbicidal, and it is thought that the ester is active after hydrolysis to the parent acid, a reaction which occurs very quickly in plant tissue; further, only the (–) enantiomer of chlorfenprop-methyl is effective (Fedtke and Schmidt, 1977), which supports an interaction with a specific biological receptor.

$$Cl-\langle\bigcirc\rangle-CH_2-\underset{\underset{Cl}{|}}{CH}-COOMe \qquad \text{chlorfenprop-methyl}$$

Fedtke (1972) conducted a broad survey of biochemical processes affected by chlorfenprop-methyl, concluding that it acted by inducing autolysis of cells through the release of degrading enzymes such as ribonucleases and proteases from membrane-bound cell compartments in which they are normally isolated. This could have come about via an effect on the membranes of these compartments.

Andreev and Amrhein (1976) found that micromolar concentrations of chlorfenprop-methyl inhibited a number of auxin-mediated processes such as *Avena* coleoptile elongation and proton secretion, as well as inhibiting auxin uptake, transport and metabolism in excised coleoptile tissue. They also used radiolabelled compounds to show that chlorfenprop (and its methyl ester) and the synthetic auxin naphthyl-1-acetic acid (NAA) could interfere with each other's binding to cellular membrane fractions, though not in a competitive way. Andreev and Amrhein's experiments did not support the interpretation that chlorfenprop (or its ester) was acting through competition for an auxin-binding site so as to mimic or antagonise the effect of the natural hormone. However, the possibility remains that chlorfenprop could act at a specific site on the cell membrane so as to impair the membrane's function(s), including the ability to recognise when auxin is occupying its receptor.

In summary the effects described seem to indicate an effect on membranes, presumably as a result of a specific interaction at a so far unidentified site which accepts only one enantiomer of the compound.

Chlorfenprop-methyl has a structural resemblance to diclofop-methyl and benzoylprop-ethyl (and their related compounds) and in each case there has been a suggestion that they antagonise auxin effects. On this basis Fedtke (1982) classifies the compounds together as auxin-inhibitor herbicides.

D. Inorganic herbicides

The inorganic herbicides shown in Table 8.6 have largely been superseded by organic herbicides introduced since 1945.

Brian (1976) considers inorganic compounds in his review of herbicide classification, concluding that little is known of their biochemical action. Reference may also be made to the article by Hewitt and Nicholas (1963) which covers the metabolic effects of cations and anions in general.

Table 8.6   Inorganic herbicides

Name	Structure
Ammonium sulphamate	$H_2N.SO_2.ONH_4$
Borax	$Na_2B_4O_7.10\,H_2O$
Disodium octaborate	approximate composition $Na_2B_8O_{13}.4H_2O$
Sodium metaborate	$(NaBO_2)_x$
Sulphuric acid	H_2SO_4

E. Miscellaneous organic herbicides

The compounds to be considered in this section are given in Table 8.7. Iwataki and Hirono (1979) have conducted a structure-activity study and described the symptoms which develop in alloxydim-sodium treated plants, prominent among which are growth inhibition and chlorosis. Active meristematic tissues seem especially sensitive, so the compound may affect some aspect of cell division.

The biochemical site of action of the post-emergence wild oat herbicide difenzoquat has not been established, though some progress has been made (Pallett and Caseley, 1980). These workers studied the difference in herbicide sensitivity of two wheat cultivars, and found that it could not be attributed to differences in retention, penetration and translocation of chemical, the development of chlorosis, or to effects on cell elongation. However, 100 μM herbicide depressed incorporation of [^{14}C]-thymidine into DNA by meristematic tissue of the sensitive and tolerant cultivars

Table 8.7 Miscellaneous organic herbicides

Name	Structure
Alloxydim-sodium	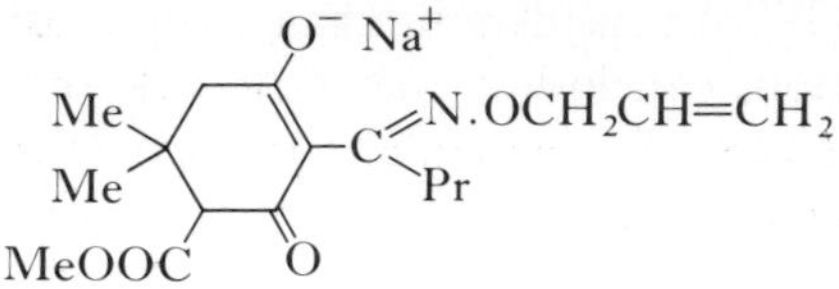
Bensulide	
Butam	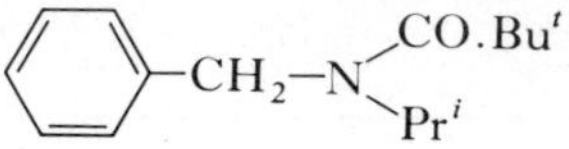
Difenzoquat	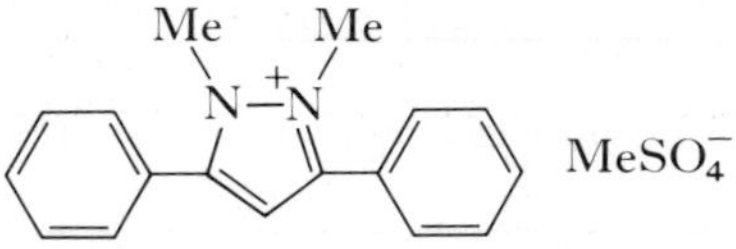
Diphenamid	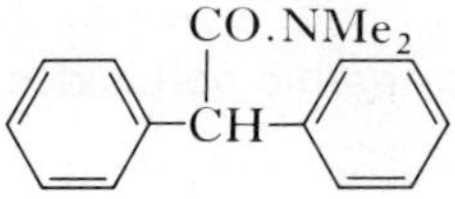
2-[1-(Ethoxyimino) butyl]-5-[2-(ethylthio) propyl]-3-hydroxy-2-cyclohexen-1-one (sethoxydim)	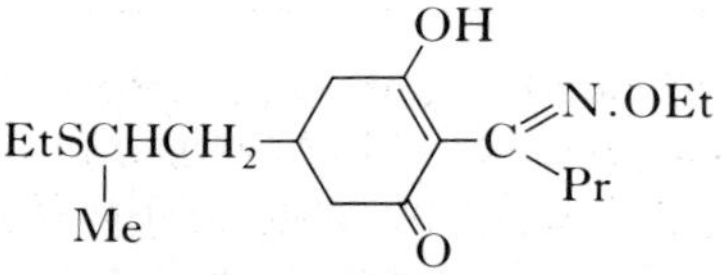
Fosamine-ammonium	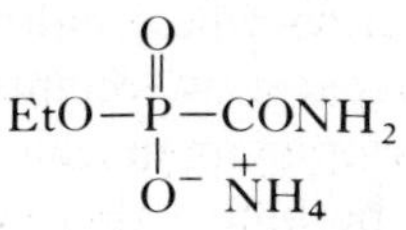

Name	Structure

Mefluidide

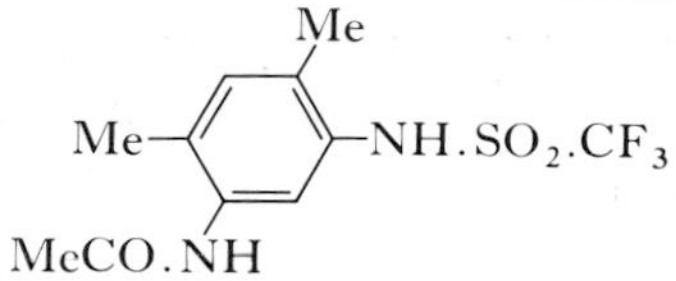

$$Me-\text{(ring)}-NH.SO_2.CF_3$$

with Me at top and McCO.NH at bottom of ring

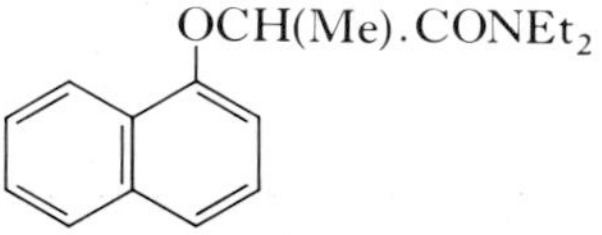

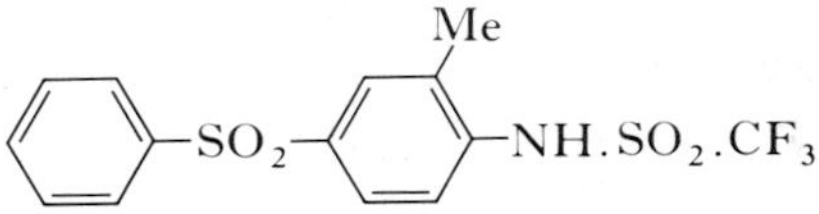

Napropamide

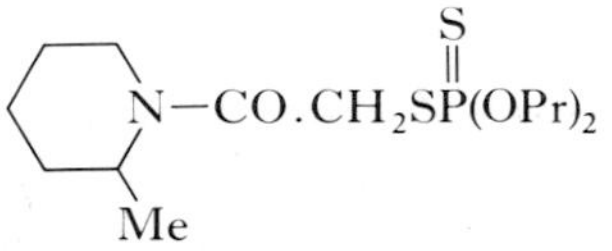

$$OCH(Me).CONEt_2$$

(naphthalene ring system)

Perfluidone

$$\text{(phenyl)}-SO_2-\text{(ring)}-NH.SO_2.CF_3$$

with Me on second ring

Piperophos

$$N-CO.CH_2SP(OPr)_2$$

with S double bonded to P, Me on piperidine ring

Pyridate

$$O.CO.SC_8H_{17}$$

(phenyl-pyridazine ring with Cl, N–N)

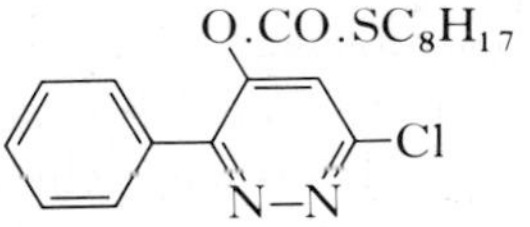

3-o-Tolyloxypyridazine

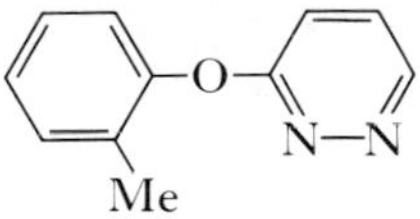

(phenyl)–O–(pyridazine ring, N–N), with Me on phenyl

to the extent of 50% and 9% respectively. This could not be explained solely on the basis of inhibition of uptake of the radiolabelled precursor, though this did occur to some extent. These data do not establish DNA synthesis as the primary site of action of difenzoquat, particularly since the concentrations needed were fairly high. Nevertheless they provide a basis for further study.

Fosamine-ammonium is an unusual compound in that the effect of summer or autumn application to susceptible woody species is most noticeable during the following growing season, when retardation of shoot growth (or development of abnormal shoots) is seen. The mode of action is not known at a biochemical level though it was found in a cytological examination of the effects of the compound on tissues of mesquite (*Prosopis juliflora* var. *glandulosa*), that fosamine strongly inhibited the number of mitoses seen in dividing regions of the plant (Morey and Dahl, 1980). The authors suggest, on the basis of a comparison of the observed effects with those caused by the cell division inhibitor trifluralin (p. 214), that fosamine may not inhibit the process of cell division itself, but rather some process essential for it.

Mefluidide (a herbicide and plant growth regulator) and perfluidone (a herbicide) are both trifluoromethanesulphonanilides, whose history and structure-activity relationships have been reviewed by Fridinger (1979). Although it is not possible to draw conclusions about their primary site of action (if indeed they act at the same site), they have nevertheless been the subject of several investigations.

Probably the most detailed examination of the effects of perfluidone on biochemical systems has been carried out by Moreland and his colleagues (Moreland, 1981). At concentrations in the range of about 40–100 μM perfluidone uncoupled oxidative phosphorylation. Photophosphorylation by chloroplast preparations was also affected, being half-uncoupled by about 60 μM perfluidone with noticeable effects being seen down to 10 μM. These results indicate that perfluidone is able to interfere with ATP synthesis driven by vectorial flow of protons across a membrane. In addition, inhibitory effects on mitochondrial electron transport were seen at higher concentrations, and a further effect at relatively high levels was the induction of mitochondrial swelling in the presence of isotonic potassium chloride, suggesting a direct effect on the membrane leading to a greater permeability to potassium ions (Moreland, 1981).

A second reported effect is the inhibition *in vitro* of a microsomal membrane fraction from spinach, which was capable of incorporating radiolabel from [^{14}C]-L-glycerol-3-phosphate into glycerides; 50% inhibition was caused by 50 μM perfluidone (St. John and Hilton, 1973). This is consistent with the further observation by these workers that 100 μM perfluidone in the water supplied to growing wheat seedlings led to a decrease in the shoot levels of neutral and polar lipids, but an increase in the amount of the fatty acid precursors. Hence, there is the possibility of a longer-term effect on the synthesis of membrane lipids as well as the direct effect on membrane-bound organelles mentioned above.

A further report of a biochemical effect of perfluidone is that of Tafuri *et al.* (1977) who report that an IAA oxidase preparation from *Lens culinaris* roots was stimulated *in vitro* by, e.g., about 40% at 1 μM perfluidone. It is not known, however, whether this contributes to the herbicidal action of the compound.

Davis and Dusbabek (1975) compared the effects of perfluidone on cotton (tolerant) and nutsedge (sensitive). The study was a physiological rather than a biochemical one, but they noted that almost no dividing cells were seen in nutsedge roots, whereas cell divisions did take place in roots of cotton. Finally, Bendixen (1975) noted that, as well as its phytotoxic effects, perfluidone mimicked cytokinins in its effect on nutsedge, leading to breaking of bud dormancy and apical dominance, induction of basal bulb formation and dwarfing of shoot growth. Some of these responses would involve increased cell division, and Bendixen suggests that such morphological effects might pre-dispose the plants to the toxic properties of the compound.

Turning now to mefluidide, Truelove et al. (1977) noted, on the basis of effects on incorporation of radiolabelled leucine into protein by cucumber cotyledon discs, that mefluidide stimulated protein synthesis at 29 μM, but inhibited it at 290 μM. However, since these effects depended on an 8 h pre-incubation with herbicide, the authors regarded the effect as a secondary one. They found no effect on mitochondrial functions or photosynthesis. Glenn (1979) also found that mefluidide stimulated protein as well as RNA and DNA synthesis of corn coleoptiles, effects which he took to be a secondary consequence of the auxin activity possessed by mefluidide; 0.01, 0.1 and 1 μM mefluidide increased corn coleoptile elongation to the same extent as did 1 μM IAA (though it seems curious that concentrations two orders of magnitude apart should give rise to the same response), and also increased IAA transport through the coleoptile sections (Glenn, 1979).

Considering perfluidone and mefluidide together, the results described above seem to have little in common. In view of their structural similarity, it might be rewarding to compare the effects of mefluidide and perfluidone in various biochemical systems.

There does not seem to have been much work on the biochemical mode of action of 3-o-tolyloxypyridazine though it is said to inhibit both auxin-induced growth and cell division (Takeuchi et al., 1973).

V. PLANT GROWTH REGULATORS
A. Dikegulac

Zilkah and Gressel (1979, 1980) have looked critically for the primary site of action of dikegulac (Table 8.8) using both cell suspension cultures and protoplasts, and studying the effect of the compound on amino acid uptake and incorporation into proteins, leakage of dye from pre-loaded cells, and the morphological appearance of the protoplasts.

Table 8.8   Miscellaneous plant growth regulators

Name	Structure
Dikegulac-sodium	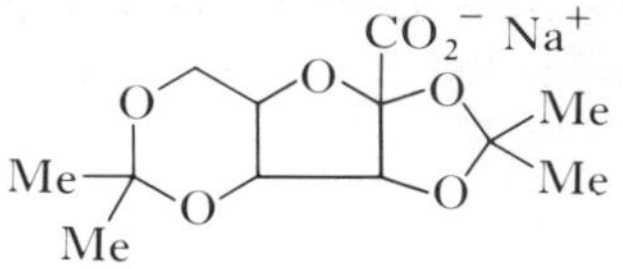
Glyphosine	
Hymexazol	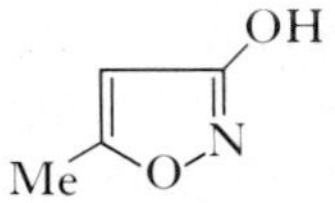
Thidiazuron	
Tributyl phosphorotrithioite	$(BuS)_3P$
$S,S,S,$-Tributyl phosphorotrithioate	$(BuS)_3PO$

Cell division of protoplasts was the most sensitive process, being reduced to half by 10–30 μM dikegulac; this fits with the use of the compound at low rates to stop hedge growth (De Silva *et al.*, 1976). At higher (millimolar) concentrations membrane damage of dividing cells was found, but cells which had stopped dividing were considerably less sensitive; this is consistent with the fact that the compound can be translocated from the leaves of plants (where cells remain unharmed) to the growing points where the dividing cells are killed. Thus the compound can be used as a chemical 'pinching' agent (Arzee *et al.*, 1977). The effects of dikegulac on amino acid uptake and incorporation into proteins were considered to be consequent upon membrane damage.

Zilkah and Gressel (1980) supposed that cell division is stopped by interaction of the compound at an unidentified intracellular site which has a relatively high affinity for dikegulac. Cell damage, however, is probably caused by a direct effect of higher concentrations of compound on the cell membrane.

B. Glyphosine

Glyphosine (Table 8.8) is used to increase the sugar content and hasten the ripening of sugar cane (Worthing, 1979). It is structurally related to glyphosate (p. 276) and causes some of the same effects, though it is much less active. The effects include stimulation of phenylalanine ammonia lyase, and decrease of formation of phenylalanine-derived secondary products (Hoagland, 1980).

As described on p. 279, glyphosate inhibited the conversion of shikimate to anthranilate in cell-free preparations from *Aerobacter aerogenes* with an I_{50} of 5–7 μM; glyphosine caused only about 35% inhibition at 1 mM (Amrhein *et al.*, 1980). Glyphosine was also much less effective than glyphosate at inhibiting anthocyanin formation in excised buckwheat hypocotyls (Holländer and Amrhein, 1980). It is possible that glyphosine could act through a weak effect at the glyphosate site (PSCV transferase – p. 278) or, perhaps, after cleavage to yield glyphosate (which does not seem to have been reported), but it is not obvious how this might lead to sucrose increase.

It has been reported that glyphosine is an inhibitor of chloroplast ribosome formation (Croft *et al.*, 1974), though, in contrast, Slovin and Tobin (1981) found that ribosomes were both present and functional in plastids of treated duckweed (*Lemna gibba*, L). The latter authors suggest, on the basis of a study of the proteins being synthesised by treated and untreated duckweed, that glyphosine differentially affects the synthesis and/or processing of soluble proteins and some chloroplast membrane proteins. This does not, as the authors point out, establish the primary mode of action.

C. Others

The structures of the compounds mentioned below are given in Table 8.8.

Hymexazol is used both as a growth regulator and as a fungicide; the latter activity may be due to inhibition of DNA synthesis (p. 237) but the reason for the plant growth regulating activity does not seem to be known.

Although certain derivatives of thidiazuron carrying chloro substituents in the phenyl ring were found to interfere with various aspects of oxidative and photosynthetic phosphorylation, thidiazuron itself was only effective at concentrations in excess of 100 μM (Hauska *et al.*, 1975). Also, Mok *et al.* (1980) reported that thidiazuron could substitute for compounds with cytokinin activity in promoting the growth of bean tissue cultures. However, it is not clear whether there is a connection between either of these effects and the use of the compound as a cotton defoliant (Worthing, 1979).

It seems reasonable to suppose that tributyl phosphorotrithioite is oxidised to tributyl phosphorotrithioate *in vivo* but, although biochemical changes caused by this latter compound have been studied (see, e.g., Cathey *et al.*, 1981) the primary mode of action of these defoliants does not seem to have been established.

VI. MISCELLANEOUS COMPOUNDS

The mode of action of the molluscicide metaldehyde has not been established.

metaldehyde

REFERENCES

Akechi, K. and Matsuura, K. (1982). Abstracts 5th Int. Congr. Pestic Chem., Kyoto, No. IVc-11, International Union of Pure and Applied Chemistry

Amrhein, N., Deus, B., Gehrke, P. and Steinrücken, H. C. (1980). *Plant Physiol.* **66**, 830–834

Andreev, G. K. and Amrhein, N. (1976). *Physiol. Plant.* **37**, 175–182

Aoki, K. and Yamada, M. (1979). *Japan Pestic. Information* No. 36, 32–35

Arzee, T., Langenauer, H. and Gressel, J. (1977). *Bot. Gaz.* **138**, 18–28

Bendixen, L. E. (1975). *Weed Sci.* **23**, 445–447

Bompeix, G., Fettouche, F. and Saindrenan, P. (1981). *Phytiatr. Phytopharm.* **30**, 257–272

Brezeanu, A. G., Davis, D. G. and Shimabukuro, R. M. (1976). *Can. J. Bot.* **54**, 2038–2048

Brian, R. C. (1976). In "Herbicides. Physiology, Biochemistry, Ecology" (L. J. Audus, ed.) pp. 1–54. Academic Press, London and New York

Bryan, K. M. G., Geering, Q. A. and Reid, J. (1981). Proc. British Crop Protection Conf. "Pests and Diseases", Vol. 1, pp. 67–74

Cathey, G. W., Elmore, C. D. and McMichael, B. L. (1981). *Physiol. Plant.* **51**, 140–144

Cohen, A. S. and Morrison, I. N. (1981). *Pestic. Biochem. Physiol.* **16**, 110–119

Croft, S. M., Arntzen, C. J., Vanderhoef, L. N. and Zettinger, C. S. (1974). *Biochim. Biophys. Acta* **335**, 211–217

Davis, D. G. and Dusbabek, K. E. (1975). *Weed Sci.* **23**, 81–86

De Silva, W. H., Graf, H. R. and Walter, H. R. (1976). Proc. British Crop Protection Conf. "Weeds", Vol. 1, pp. 349–356

Despreaux, D., Fritz, R. and Leroux, P. (1981). *Phytiatr. Phytopharm.* **30**, 245–255

De Waard, M. A. (1974). *Meded. LandbHogesch. Wageningen* **74**, (14), 1–98

De Waard, M. A. and van Nistelrooy, J. G. M. (1980). *Neth. J. Plant Pathol.* **86**, 251–258

Fedtke, C. (1972). *Weed Res.* **12**, 325–336

Fedtke, C. (1982). "Biochemistry and Physiology of Herbicide Action". Springer-Verlag, Berlin

Fedtke, C. and Schmidt, R. R. (1977). *Weed. Res.* **17**, 233–239

Fridinger, T. L. (1979). In "Advances in Pesticide Science" (H. Geissbühler, ed.) Pt 2., pp. 261–270. Pergamon Press, Oxford

Glenn, D. S. (1979). Ph.D. Thesis, University of Kentucky, Lexington, U.S.A.

Gorecka, K., Shimabukuro, R. H. and Walsh, W. C. (1981). *Physiol. Plant.* **53**, 55–63

Green, M. B., Hartley, G. S. and West, T. F. (1977). "Chemicals for Crop Protection and Pest Control". Pergamon Press, Oxford

Hauska, G., Trebst, A., Kötter, C. and Schulz, H. (1975). *Z. Naturforsch.* **30 C**, 805–810

Hewitt, E. J. and Nicholas, D. J. D. (1963). In "Metabolic Inhibitors" (R. M. Hochster and J. H. Quastel, eds) pp. 311–436. Academic Press, London and New York

Hoagland, R. E. (1980). *Weed Sci.* **28**, 393–400

Holländer, H. and Amrhein, N. (1980). *Plant Physiol.* **66**, 823–829

Hoppe, H. H. (1981). *Z. Pflanzenphysiol.* **102**, 189–197

Hoppe, H. H. and Zacher, H. (1982). *Z. Pflanzenphysiol.* **106**, 287–298

Iwataki, I. and Hirono, Y. (1979). In "Advances in Pesticide Science" (H. Geissbühler, ed.) Pt 2, pp. 235–243. Pergamon Press, Oxford

Jeffcoat, B. and Harries, W. N. (1973). *Pestic. Sci.* **4**, 891–899

Jeffcoat, B. and Harries, W. N. (1975). *Pestic. Sci.* **6**, 283–296

Kaars Sijpesteijn, A., Kerkenaar, A. and Overeem, J. C. (1974). *Meded. Fac. Landbouw.* **39**, 1027–1034

Kakiki, K. and Misato, T. (1979a). *J. Pesticide Sci.* **4**, 129–135

Kakiki, K. and Misato, T. (1979b). *J. Pesticide Sci.* **4**, 305–313

Katagiri, M. and Uesugi, Y. (1977). *Phytopathol.* **67**, 1415–1417

Kerkenaar, A. and Kaars Sijpesteijn, A. (1977). *Neth. J. Plant Pathol.* **83**, (Suppl. 1), 145–152

Köcher, H., Kellner, H. M., Lötzsch, K., Dorn, E. and Wink, O. (1982). Proc. British Crop Protection Conf. "Weeds", Vol. 1, pp. 341–347

Mok, D. W. S., Mok, M. C. and Armstrong, D. J. (1980). *Plant Physiol.* **65**, (Suppl.), 24 (Abstract 121)

Moreland, D. E. (1981). *Pestic. Biochem. Physiol.* **15**, 21–31

Morey, P. R. and Dahl, B. E. (1980). *Weed Sci.* **28**, 251–255

Morrison, I. N., Hill, B. D. and Dushnicky, L. G. (1979). *Weed Res.* **19**, 385–393

Morrison, I. N., Owino, M. G. and Stobbe, E. H. (1981). *Weed Sci.* **29**, 426–432

Pallett, K. E. and Caseley, J. C. (1980). *Pestic. Biochem. Physiol.* **14**, 144–152

Papavizas, G. C., O'Neill, N. R. and Lewis, J. A. (1978). *Phytopathol.* **68**, 1667–1671

Shimabukuro, M. A., Shimabukuro, R. H., Nord, W. S. and Hoerauf, R. A. (1978). *Pestic. Biochem. Physiol.* **8**, 199–207

Shimabukuro, R. H., Walsh, W. C. and Hoerauf, R. A. (1979). *J. Agric. Food Chem.* **27**, 615–623

Slovin, J. P. and Tobin, E. M. (1981). *Biochim. Biophys. Acta* **637**, 177–184

St. John, J. B. and Hilton, J. L. (1973). *Weed Sci.* **21**, 477–480

Suami, T., Ogawa, S. and Chida, N. (1980). *J. Antibiot.* **33**, 98–99

Tafuri, F., Businelli, M., Scarponi, L. and Marucchini, C. (1977). *J. Sci. Food Agric.* **28**, 180–184

Takeuchi, Y., Konnai, M. and Takematsu, T. (1973). *Weed Res.* (*Japan*) **16**, 32–37 [*Weed Abstracts* **23**, No. 1341, p. 147]

Truelove, B., Davis, D. E. and Pillai, C. G. P. (1977). *Weed Sci.* **25**, 360–363

Venis, M. A. (1982). *Pestic Sci.* **13**, 309–317

Webb, J. L. (1966). "Enzyme and Metabolic Inhibitors", Vol. II. Academic Press, London and New York

Woloshuk, C. P. and Sisler, H. D. (1982). *J. Pestic. Sci.* **7**, 161–166

Worthing, C. R. (1979). "The Pesticide Manual", 6th edn. British Crop Protection Council, London

Zilkah, S. and Gressel, J. (1979). *Planta* **145**, 273–278

Zilkah, S. and Gressel, J. (1980). *Planta* **147**, 274–276

9 | Resistance, Synergists and Safeners

I. RESISTANCE

Resistance to a pesticide occurs when some individuals of a sensitive species are no longer killed by the chemical at dose rates that were previously active, and various examples have been given throughout the book. Resistance may ultimately render a pesticide totally ineffective against the species in question. It has most commonly been found to insecticides, with over 432 species showing resistance to a range of compounds, but is now also significant with fungicides, where 67 species are resistant to 52 fungicides and bactericides (Conway, 1982). Herbicide resistance is still uncommon, but is known to the triazines (p. 73) and a few other classes of herbicides (LeBaron and Gressel, 1982). Although these figures appear large, they refer in many cases to resistance detected in the laboratory, and the level of control in the field may still be satisfactory.

The development of resistance is an example of Darwinian selection in favour of individuals that possess by chance some pre-existing mechanism of avoiding or otherwise dealing with the toxicant. Such individuals will possess an overwhelming advantage over other members of the species that encounter the chemical, and they will be able to exploit an ecological niche rendered vacant by the death of the sensitive majority of the population of that species.

Putting aside behavioural resistance, where the pest avoids or minimises contact with the chemical, resistant individuals may differ from those that are susceptible because of differences in the way in which they take up, store (possibly at an inert site), excrete, or metabolise the toxicant. In addition the sensitivity of the target site itself may differ. From the information available on the biochemical mechanisms underlying resistance, it seems that the commonest involve uptake, metabolism, and sensitivity of the target site. Examples of the latter are collated from elsewhere in

329

this book in Table 9.1. From the biochemical point of view these resistant mutant strains are valuable since they provide excellent evidence on the site of action of the compound concerned.

Table 9.1   Instances of resistance to a pesticide at the site of action

Compound (or class)	Site	Page
Oxathiin carboxanilides	succinate dehydrogenase complex	28
Uncouplers	uncoupler binding protein – ATPase?	35
Triazines	herbicide binding protein of photosystem II	73
Organophosphorus compounds and carbamates	acetylcholinesterase	137
Pyrethroids, DDT	Na$^+$ channel (?)	152, 157
Carbendazim	tubulin	201

Resistance is a major problem with chemical control of pests, and it is important to learn more about the biochemistry involved. It is useful, for instance, to know of the occurrence of cross-resistance, in which selection of resistant organisms by one agent is accompanied by resistance to another compound that may or may not be chemically related. Knowledge of resistance mechanisms involved in a particular case should allow a more rational choice of a follow-up compound, by avoiding one likely to be affected by the resistance mechanism operating.

It should be noted that the biochemical mechanisms underlying selectivity of pesticide action between different species are the same as those which cause resistance. It can be conferred chemically, at least to herbicides, by the use of safeners (see p. 333).

II. SYNERGISTS

In toxicology, synergism is defined as the case where the toxicity of two compounds applied together is greater than would be expected from the sum of their individual effects. Insecticide synergists usually work by inhibiting the microsomal breakdown of the compound.

A. Microsomal oxidation

One of the commonest ways in which pesticides can be metabolised involves the action of the mixed function oxidase system. This is located on the cell's endoplasmic reticulum, which consists of a network of intracellular membranes that are thought to channel and compartmentalise materials inside the cell, as well as to provide a structural framework for enzyme reactions. In addition to the insecticide, the substrates are NADPH and molecular oxygen, and the system contains a flavoprotein (NADPH-cytochrome c reductase) and a unique cytochrome, cytochrome P-450. Many of the reactions catalysed are hydroxylations (though the products may be unstable or may participate in further reactions) but other oxidations also occur. The role of microsomal oxidation in insecticide metabolism has been reviewed (Nakatsugawa and Morelli, 1976).

B. Action of insecticide synergists

The use of synergists originally arose from the observation that sesame oil would potentiate the action of natural pyrethrins (reviewed by Hodgson and Tate, 1976; Wilkinson, 1976). Many compounds have been studied

Table 9.2   Insecticide synergists which interfere with cytochrome P-450

Name	Structure
N-(2-Ethylhexyl)-8,9,10-trinorborn-5-ene-2,3-dicarboximide	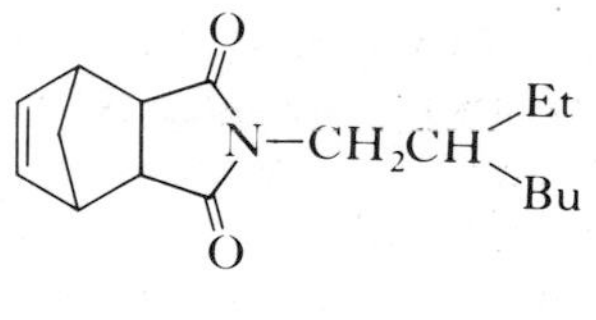
Piperonyl butoxide	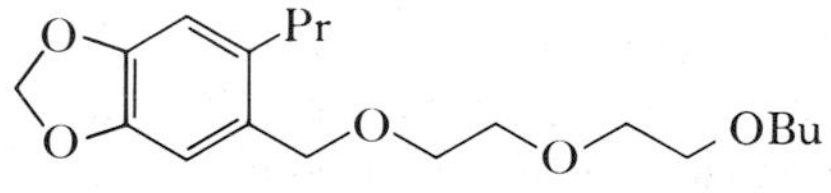
Sesamex	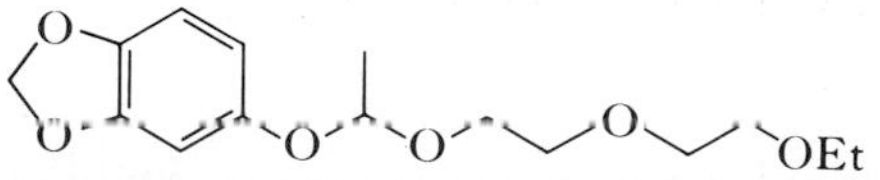
Sulfoxide	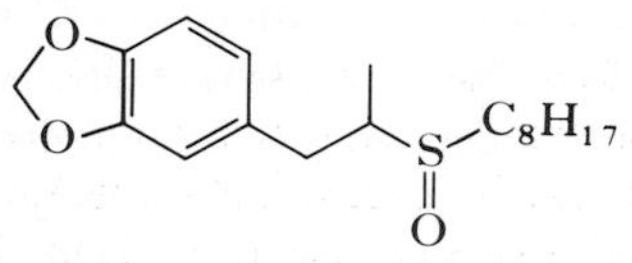

(see Casida, 1970; Wilkinson, 1971) but four which are currently of commercial interest (Worthing, 1979) are given in Table 9.2. They are used mainly in conjunction with relatively expensive and metabolically-sensitive pyrethroids.

It has been shown that compounds of the methylenedioxyphenyl type (e.g. piperonyl butoxide) work by interfering with the oxidative break-down of the insecticide catalysed by cytochrome P-450, with which they form complexes that can be recognised by their spectral characteristics (Kulkarni and Hodgson, 1976). This view is upheld by broad supportive evidence for a correlation between the ability of compounds to form a particular type of spectral complex and their capacity to inhibit micro-somal oxidation; however, the correlation is not absolute and there are some anomalies still to be explained (Hodgson and Tate, 1976; Chang *et al.*, 1981).

It is generally agreed that the methylenedioxyphenyl synergists are effective because of the stability of the inhibitory complex with cytochrome P-450 and this is thought to be formed by interaction of the methylene group at the 2-position with a cytochrome P-450 activated oxygen complex (Hodgson and Tate, 1976). The exact structure of the complex is still not certain, though a cytochrome P-450-iron(II)-carbene structure has been proposed (Mansuy *et al.*, 1979), as indicated in the scheme below:

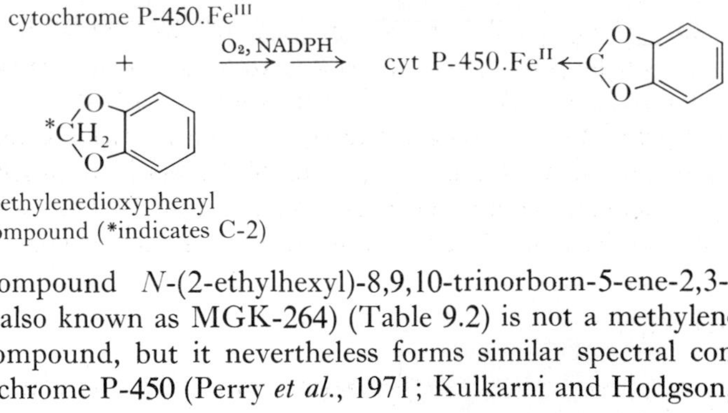

$$\text{cytochrome P-450.Fe}^{III} + \overset{O_2, \text{NADPH}}{\longrightarrow} \text{cyt P-450.Fe}^{II} \leftarrow C$$

methylenedioxyphenyl
compound (*indicates C-2)

The compound *N*-(2-ethylhexyl)-8,9,10-trinorborn-5-ene-2,3-dicarb-oximide (also known as MGK-264) (Table 9.2) is not a methylenedioxy-phenyl compound, but it nevertheless forms similar spectral complexes with cytochrome P-450 (Perry *et al.*, 1971; Kulkarni and Hodgson, 1976).

C. Herbicide synergists

Herbicide synergists are rare, but the addition of ammonium thiocyanate (NH_4SCN) to aminotriazole increases its effectiveness (Ashton and Crafts, 1981). There have been reports using various species that ammonium thiocyanate either does or does not increase absorption of aminotriazole and that it either enhances or diminishes aminotriazole movement about the plant (see Ashton and Crafts (1981) and Cook and Duncan (1979) for original references). It therefore seems likely that the explanation for its action lies elsewhere.

The compound 3-(3-amino-1,2,4-triazol-1-yl)alanine is considered to be a less mobile detoxification product of aminotriazole (Carter, 1975)

$$\begin{array}{c} \text{N—CH}_2.\text{CH.COOH} \\ \text{N}\diagdown \text{N} \qquad \text{NH}_2 \\ \text{NH}_2 \end{array} \qquad \text{3-(3-amino-1,2,4-triazol-1-yl)alanine}$$

which might perhaps be formed by a mechanism involving free radicals (Castelfranco and Brown, 1963). Formation of this derivative in plants is strongly inhibited by ammonium thiocyanate (see Carter, 1975). Cook and Duncan (1979) showed that ammonium thiocyanate largely prevented oxidation of aminotriazole by free radical generating systems and therefore proposed that the synergist acts *in vivo* as a free radical inhibitor to reduce the formation of the less toxic alanine derivative. If this explanation is correct then this herbicide synergist, like the insecticide synergists mentioned above, works by inhibition of a detoxification reaction.

As another example of herbicide synergism, the insecticide carbaryl prevents hydrolysis by the plant of the herbicide propanil. The mixture of propanil and carbaryl can be used commercially for weed control (Matsunaka, 1969).

The compound O,O-diethyl-O-phenyl phosphorothioate (PhO.PS (OEt)$_2$) is able to extend soil persistence and biological activity of thiocarbamates such as EPTC and is therefore described as a herbicide 'extender' (Miaullis et al., 1982). These workers propose that the chemical is effective because it interferes with microbial breakdown of EPTC in the soil; the action can therefore be seen as a particular example of synergism.

No commercially significant example of a fungicide synergist is known.

III. SAFENERS

Safeners provide a chemical method of improving the selectivity of a herbicide. They selectively reduce the activity of the pesticide, whereas synergists increase activity. Three compounds are currently commercially available as herbicide safeners, while two more are close to the market (Table 9.3). Safeners are also called protectants, or antidotes, though the latter term is inappropriate since (by analogy with the medical use of the term) it implies, incorrectly, that the chemical can reverse damage which has already occurred. In fact most safeners are applied to the seed of the crop plant, but N,N-diallyl-2,2-dichloroacetamide (R-25788) is mixed with the herbicide (EPTC) and applied to the soil. The safener reduces

Table 9.3   Herbicide safeners

Name	Structure
Benzyl 2-chloro-4-(trifluoromethyl)-5-thiazolecarboxylate[a]	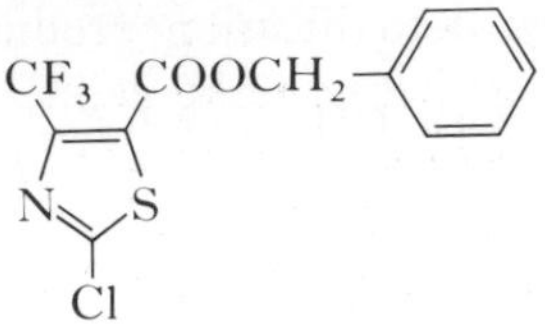
Cyometrinil	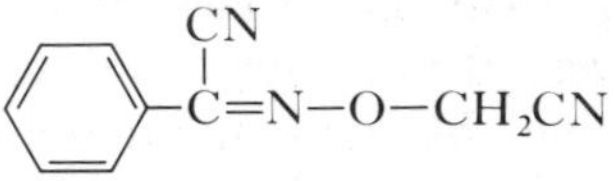
N,N-Diallyl-2,2-dichloroacetamide (R-25788)	
N-(1,3-Dioxolan-2-yl-methoxy)imino-benzeneacetonitrile[b] (oxolabentril proposed)	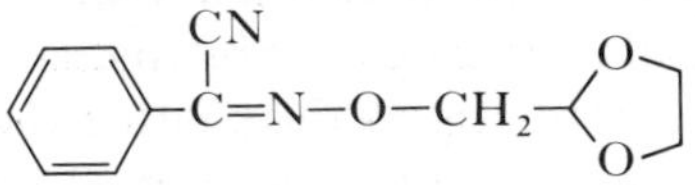
Naphthalic anhydride	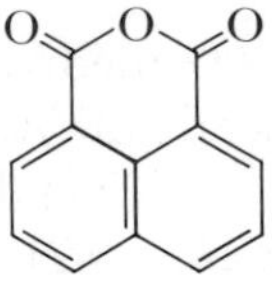

[a]Brinker et al. (1982); [b]Rufener et al. (1982).

the toxicity of the herbicide to the crop plant, while leaving unimpaired its ability to kill weeds.

Many aspects of the chemistry and action of herbicide safeners are collated in the book edited by Pallos and Casida (1978) and the subject has been reviewed by Blair et al. (1976) and by Stephenson and Ezra (1982). Ideas for the biochemical basis of the activity have been put forward only in the case of R-25788.

R-25788 is applied to the soil in mixture with EPTC in order to prevent this herbicide from damaging the maize crop. Lay and Casida (1976, 1978) demonstrated that increased levels of glutathione (GSH) and GSH-S-transferase resulted from treatment of maize seedlings with R-25788. Since these components are involved in EPTC detoxification, this was suggested as the means by which the safener brought about its effect and, indeed, there was a correlation between safener activity and the ability to elevate levels of GSH and its transferase in a series of R-25788 analogues (Lay and Casida, 1978). This view of the biochemical basis of safener action has been accepted or corroborated by a number of subsequent workers (e.g. Rennenberg et al., 1982).

The subject was further examined by the use of maize cell cultures (Ezra and Gressel 1982; Ezra et al., 1982, 1983). R-25788 did cause an elevation of glutathione levels (the transferase was not examined) but not until 12 hours after addition. On the other hand the inhibition by EPTC of acetate incorporation into lipids was evident much earlier and, indeed, cells treated with EPTC began to recover in about 6 hours. Further, R-25788 exhibited competitive inhibition of EPTC uptake, stimulated neutral lipid synthesis itself, and gave partial reversal of the inhibitory effects on acetate incorporation into lipids caused by EPTC. These workers placed considerable emphasis on the fact that they applied EPTC and R-25788 simultaneously (as they are in practice), whereas Lay and Casida pre-incubated their maize seedlings with safener for 24 hours. Nevertheless it ought not to be assumed that chemicals applied together reach the inside of the target cells simultaneously. Ezra and Gressel (1982) do not discount an involvement of glutathione in the action of R-25788 but suggest that additional earlier effects on EPTC uptake and lipid metabolism could contribute to the safening action.

Reviewing the mode of action of safeners, Stephenson and Ezra (1982) detail a number of other relevant if inconclusive studies. They observe that the one effect correlating with safener activity which has been observed by most workers is the elevation of glutathione levels.

REFERENCES

Ashton, F. M. and Crafts, A. S. (1981). "Mode of Action of Herbicides", 2nd edn. Wiley-Interscience, New York

Blair, A. M., Parker, C. and Kasasian, L. (1976). *PANS* **22**, 65–74

Brinker, R., Schafer, D., Radke, R., Boeken, G. and Frazier, H. (1982). Proc. British Crop Protection Conf. "Weeds", Vol. 2, pp. 469–473

Carter, M. C. (1975). In "Herbicides. Chemistry, Degradation, and Mode of Action" (P. C. Kearney and D. D. Kaufman, eds) Vol. 1, pp. 377–398. Dekker, New York

Casida, J. E. (1970). *J. Agric. Food Chem.* **18**, 753–772

Castelfranco, P. and Brown, M. S. (1963). *Weeds* **11**, 116–124

Chang, K.-M., Wilkinson, C. F. and Hetnarski, K. (1981). *Pestic. Biochem. Physiol.* **15**, 32–42

Conway, G. (ed.) (1982). "Pesticide Resistance and World Food Production". Imperial College Centre for Environmental Technology, London

Cook, G. T. and Duncan, H. J. (1979). *Pestic. Sci.* **10**, 281–290

Ezra, G. and Gressel, J. (1982). *Pestic. Biochem. Physiol.* **17**, 48–58

Ezra, G., Krochmal, E. and Gressel, J. (1982). *Pestic. Biochem. Physiol.* **18**, 107–112

Ezra, G., Flowers, H. M. and Gressel, J. (1983). In "Pesticide Chemistry: Human Welfare and the Environment" (J. Miyamoto and P. C. Kearney, eds) Vol. 3, pp. 225–231. Pergamon Press, Oxford

Hodgson, E. and Tate, L. G. (1976). In "Insecticide Biochemistry and Physiology" (C. F. Wilkinson, ed.) pp. 115–148. Heyden, London

Kulkarni, A. P. and Hodgson, E. (1976). *Pestic. Biochem. Physiol.* **6**, 183–191

Lay, M.-M. and Casida, J. E. (1976). *Pestic. Biochem. Physiol.* **6**, 442–456

Lay, M.-M. and Casida, J. E. (1978). In "Chemistry and Action of Herbicide Antidotes" (F. M. Pallos and J. E. Casida, eds) pp. 151–160. Academic Press, London and New York

LeBaron, H. M. and Gressel, J. (eds) (1982). "Herbicide Resistance in Plants". Wiley-Interscience, New York

Mansuy, D., Battioni, J.-P., Chottard, J.-C. and Ullrich, V. (1979). *J. Am. Chem. Soc.* **101**, 3971–3973

Matsunaka, S. (1969). *Residue Reviews* **25**, 45–58

Miaullis, B., Nohynek, G. J. and Pereiro, F. (1982). Proc. British Crop Protection Conf. "Weeds", Vol. 1, pp. 205–210

Nakatsugawa, T. and Morelli, M. A. (1976). In "Insecticide Biochemistry and Physiology" (C. F. Wilkinson ed.) pp. 61–114. Heyden, London

Pallos, F. M. and Casida, J. E. (eds) (1978). "Chemistry and Action of Herbicide Antidotes". Academic Press, London and New York

Perry, A. S., Dale, W. E. and Buckner, A. J. (1971). *Pestic. Biochem. Physiol.* **1**, 131–142

Rennenberg, H., Birk, C. and Schaer, B. (1982). *Phytochemistry* **21**, 5–8

Rufener, J., Nyffeler, A. and Peek, J. W. (1982). Proc. British Crop Protection Conf. "Weeds", Vol. 2, pp. 461–467

Stephenson, G. R. and Ezra, G. (1982). Proc. British Crop Protection Conf. "Weeds", Vol. 2, pp. 451–459

Wilkinson, C. F. (1971). *Bull. Wld. Hlth. Org.* **44**, 171–190

Wilkinson, C. F. (1976). In "The Future for Insecticides" (R. L. Metcalf and J. J. McKelvey, eds) pp. 195–218. Wiley-Interscience, New York

Worthing, C. R. (1979). "The Pesticide Manual", 6th edn. British Crop Protection Council, London

10 | Conclusions

I. SUMMARY OF MODES OF ACTION

A. Diagrammatic summary

The flow diagram (Table 10.1) is intended to act both as a convenient summary of the biochemical mode of action of most pesticides (though those that act non-specifically are excluded) and to demonstrate inter-relationships between pesticide actions which may not be obvious from the preceding text.

The diagram is, of course, oversimplified. For example, it might give the impression that the primary site of action of the pesticides included is absolutely established. Although this is the case for many of them, the text should be consulted before conclusions are drawn about individual compounds. Another oversimplification is that, for the sake of clarity, many possible interconnections between the four 'major functions' that are subject to disruption are not shown.

The biochemical mode of action of many pesticides is well understood. The obvious exceptions are those pesticides which affect processes whose detailed biochemistry is itself complicated and incompletely known. A good example of such a process is cell division; on the basis of symptomology it is not uncommon for a chemical to be described as a cell division inhibitor. It may well be true that the compound has this action, but this

337

Table 10.1 A summary diagram of the mode of action of pesticides (p. 337)

Pesticide	Likely primary action	Major function modified or disrupted
Organophosphorus and carbamate insecticides	Acetylcholinesterase inhibited	NERVOUS → COORDINATION → DEATH
Pyrethroids: DDT and its analogues	Axonal transmission disrupted since Na^+ channel kept open	
Nicotine; cartap; thiocyclam	Combination with acetylcholine receptors	
Cyclodienes; gamma-HCH	Excess acetylcholine released	
Chlordimeform and its analogues	Combination with octopamine receptors	
Trifenmorph; ryanodine; avermectins	Various possible targets in nerve/muscle function	
Dichlobenil (chlorthiamid)	Cellulose synthesis inhibited	STRUCTURAL ORGANISATION → DEATH
Polyoxins; diflubenzuron (indirectly)	Chitin synthesis inhibited	
Thiocarbamates; ethofumesate	Fatty acid synthesis inhibited	
Imidazoles; triazoles; pyrimidines; morpholines; miscellaneous compounds	Ergosterol synthesis inhibited	
IBP; edifenphos	Phosphatidyl choline synthesis inhibited	
Pyridazinones; aminotriazole; fluridone; methoxyphenone	Carotenoid synthesis inhibited	destruction of chlorophyll
Petroleum and tar oils; dodine; guazatine; pimaricin		membranes disrupted
Nitrodiphenylethers	Diversion of electrons (from photosynthetic electron transport?) gives radical species	
Bipyridiniums	Photosynthetic electron transport diverted	toxic oxygen species
Ureas; triazines; acylanilides; uracils; some phenylcarbamates; triazinones; miscellaneous compounds	Photosynthetic electron transport interrupted	

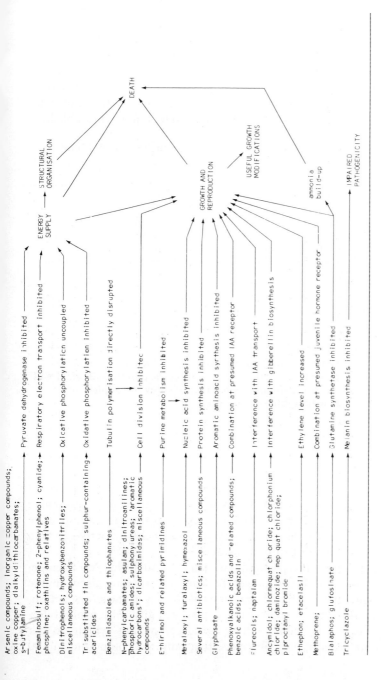

could result from a number of primary effects and more detailed biochemical explanations are needed in such cases.

The types of site with which pesticides interfere are also of interest.
Table 10.2 shows that compounds can inhibit the proper functioning of
enzymes, natural receptors and other less well defined sites of action, and
we have illustrated this using examples where the evidence for the mode
of action is particularly good. For other pesticides the mode of action,
though known in general terms, is not yet understood at the molecular
level, so that the involvement of a site, although likely, is not yet established.

Table 10.2   Examples of the types of primary site to which pesticides bind

Type of site	Example
Enzyme	Ancymidol; arsenicals; copper compounds; cyanide; ethirimol; fungicides inhibiting ergosterol biosynthesis; glufosinate and bialaphos; glyphosate; organophosphorus and carbamate insecticides; oxathiins and related compounds; polyoxin D; precocenes; trisubstituted tins
Natural receptor	Chlordimeform and its analogues; flurecols and naptalam; methoprene; nicotine, cartap and thiocyclam; phenoxyacetic acids and related herbicides
Other sites	Carbendazim and its relatives; herbicidal N-phenyl-carbamates; Hill reaction inhibitors; pyrethroids, and DDT and its analogues; rotenone; possibly uncouplers of oxidative phosphorylation

B. Primary, secondary and minor effects

It will be convenient to define and discuss these terms as part of one general
discussion. We will define the *primary site* of action as 'that site at which
the compound acts sufficiently powerfully to give a useful effect in practice'.
It is implicit in this definition that lack of activity at this site, irrespective
of other effects, would render the compound insufficiently active *in vivo*
for commercial development to be considered.

The odds operating in the pesticide industry are such that, on average,
15–20,000 new compounds are presently made in order to discover one
chemical possessing the combined attributes of efficacy at the site of action,
ability to reach the site *in vivo*, safety, marketability and profitability
required for commercialisation. Setting aside compounds which do not
meet the last four criteria, let us suppose that only one compound in 100

is active enough at a target site to be a development candidate. Assuming that its ability to interfere at another site is independent of its capacity to interfere at the first one, then the chances that the compound has activity at a second site sufficient to justify consideration for further development is also one in 100. On this basis there is a ten thousand to one chance that a single pesticide has two equally significant sites of action. Obvious exceptions are compounds whose site of action is itself non-specific, e.g. membrane-disrupting agents or compounds reacting with cell thiol groups.

The comments above do not mean that compounds with more than one primary site of action do not exist, but it does suggest that they are rare. In particular it should not be assumed that a compound has a dual or multiple mode of action simply because the currently available evidence is insufficient to distinguish between more than one candidate site.

It is, of course, quite likely that a commercial compound acting with great efficacy against its primary target will have *minor effects* on other unrelated processes. These minor effects could contribute to the perturbations which eventually result in the death of the organism without being sufficiently damaging in themselves. It is also conceivable that a compound may have no primary effect but act through a combination of a few minor ones. Nevertheless it should be emphasised that, from the point of view of understanding the action of pesticides and designing improved compounds, it is vital to attempt to identify a primary target.

Secondary effects are those which occur consequent upon the action of a pesticide at its primary target, e.g. the membrane damaging effects which follow the action of Hill reaction inhibitors at their binding site. The flow diagram (Table 10.1) illustrates that essentially the same secondary effect can follow different primary actions. For instance Hill inhibitors, electron divertors (such as the bipyridinium and nitrodiphenylether herbicides) and carotenoid synthesis inhibitors all lead to widespread disruption of chloroplasts.

C. Levels of pesticides required *in vivo* and *in vitro*

There have been few attempts to determine the amount of a pesticide required at the site of action to cause the desired effect. Even when we know the amount of pesticide required to kill a given organism, much of this may not reach the site of action because, for example, it does not penetrate well into the pest, or because it is readily metabolised.

Devonshire (1980) estimated the proportion of the lethal dose of an acetylcholinesterase inhibitor which actually became bound to the enzyme. Starting with the calculated molar specific activity of fly head acetylcholinesterase and the rate of acetylcholine hydrolysis per fly, Devonshire calculated that each fly had about 0.4 pmoles of acetylcholinesterase. Since the LD_{50} values for malaoxon and omethoate were 1.0 and 0.034 nmoles/fly respectively, this meant that only 0.04% (malaoxon) and 1.2% (omethoate) of the LD_{50} became covalently attached to the enzyme to kill the insect. The rest had to be accounted for by losses due to poor uptake, metabolism and excretion.

With older fungicides the action is usually non-specific so that rather high concentrations (in the millimolar range) are present in fungal spores (Chapter 7), though new fungicides with a more specific action are likely to act at lower concentrations *in vivo*.

There is little information about the *in vivo* levels of herbicides required to cause a lethal effect, but Davis *et al.* (1972) established that an approximately 10 μM concentration of picloram, which is not readily metabolised, prevented the normal development of half the leaves in cotton. Chlorsulfuron is a new, unusually active herbicide. Ray (1982) showed that when maize plants were grown hydroponically in solutions containing chlorsulfuron, root growth was half-inhibited at a concentration in the range 2.8–28 nM, while shoot growth was reduced by half at 28–280 nM. However, the level at the site of action would be difficult to quantify and cannot be deduced from this information alone.

The situation *in vitro* is clearly less complicated than the *in vivo* one, but it is still difficult to establish accurately the true concentration of the inhibitor at the site of action except in those cases, such as experiments with soluble enzyme preparations, where a true solution is used. If intact cells, or even organelles are concerned there is, besides the possibility of metabolism, the chance that high local concentrations of the pesticide may occur at the site of action, or, alternatively, that much of the pesticide is bound away from the site.

However, bearing these limitations of the data in mind, it seems to be generally the case that where a specific mode of action is concerned, and in some cases of non-specific action as well, pesticides cause marked inhibition or other interference *in vitro* at concentrations of about 10 μM, and frequently much lower. This can be confirmed from *in vitro* inhibition data given throughout this book. If a pesticide interferes with a reaction *in vitro* at a concentration much in excess of 10 μM it would seem a wise precaution to obtain additional evidence before concluding that this was likely to be the primary target.

As we have seen (Table 10.2) some pesticides act by inhibiting enzymes, many of which are present at 1–10 μmoles per kilogramme of tissue (Srere, 1967, 1970). It therefore seems likely that any pesticide present at a concentration of 10 μM in the tissues of a pest would be present at a concentration similar to, or higher than, that of most enzymes, with the result that severe interference with the functioning of the target enzyme might be expected.

D. A possible lower limit for herbicide application rates

As mentioned above there are now examples of especially low pesticide application rates and it is of interest to try to estimate a lower limit for future chemicals. This is done below for a post-emergence wild oat herbicide.

We will take the following data as the basis for our calculations:

(i) The total number of cells in a square centimetre of the first leaves of *Avena sativa* is 286,000 (Jellings and Leech, 1982). We will assume that this is true of all the leaves of wild oats.

(ii) Three-week-old wild oat plants at the $2\frac{1}{2}$-leaf stage (at which stage they might be sprayed in the field) have, according to our own measurements, a leaf area of about 11 cm².

It is possible to proceed in a number of different ways but we will suppose that the weeds could, in the limit, be killed by one molecule of a herbicide binding to a single site in each cell in the green part of the plant.

Using the data above we can see that the green part of a typical wild oat plant contains 286,000 × 11 cells, say about 3×10^6. Thus, to fulfil the condition above, 3 million molecules need to gain entry to each plant. For a herbicide of molecular mass 250 daltons, 250 g contains 6.023×10^{23} (Avogadro's number) molecules, so that 3×10^6 molecules are contained in

$$\frac{3 \times 10^6 \times 250}{6.023 \times 10^{23}} \text{ or about } 1.25 \times 10^{-15} \text{g,}$$

i.e. 1.25×10^{-15} g herbicide needs to gain entry to each plant. If we further assume that (a) 10% of the chemical falling on a plant actually gains entry and (b) only 10% of the chemical gaining entry survives inactivation by metabolism or storage at an inert site, then 1.25×10^{-13} g herbicide must be applied to each plant.

As stated above, the leaf area of a typical plant is 11 cm². Because of the habit of the plant, let us suppose that only half of this, say 5 cm², is available to intercept chemical from a sprayer positioned above the crop. Hence, 1.25×10^{-13} g must be applied to 5 cm² and this defines the lower limit under the stated conditions as

$$\frac{1.25 \times 10^{-13} \text{g}}{5 \text{ cm}^2} \text{ or } 2.5 \times 10^{-6} \text{g/ha.}$$

The assumptions made in the above calculations are, of course, debatable. Probably the most implausible assumption made is that a herbicidal effect could result from one molecule of herbicide per cell. On the other hand there is the reasonable possibility that plant kill might follow occupation of only a proportion of the cells in a plant. Even if the amount of herbicide required in the field has been under-estimated by 10^3 (e.g. because a herbicidal effect might require 1000 molecules/cell, because less than 10% of applied compound gains entry to the plant, or because less than 10% of the chemical gaining entry survives inactivation) there remains the possibility that herbicides could one day be effective in the field at rates of the order of milligrammes/hectare.

E. Natural ligands

To finish this section on a speculative note, it is worth considering the case where a pesticide binds to a site with no known function (Table 10.2). Although many pesticides are known to compete with natural ligands for binding sites, this is not so for all. It is therefore suggested that a binding site for a toxicant may possibly be a binding site for a natural ligand which has not yet been identified.

II. MODE OF ACTION STUDIES
A. Introduction

Dispersed throughout this volume are examples of various approaches to determining the primary site of action of a pesticide. It seems worthwhile to bring these together in a general way, and to point out the approaches that are available in biochemical mode of action work.

B. Experimental approaches

There are essentially two phases to determining the primary biochemical mode of action of a pesticide. Firstly, it is necessary to establish one or a number of candidate biochemical effects, and secondly, this (or these) must be critically examined to see whether the effect can in fact be regarded as primary.

1. *Candidate effects*

Firstly, these can be identified with respect to the symptoms given by the pesticide on the intact organism, perhaps with reference to effects caused by chemicals of known mode of action. A similar approach can be taken using model systems such as the alga *Chlorella* (Fedtke, 1982). Secondly, some kind of 'metabolic survey' study could be conducted, examining perhaps respiration and photosynthesis as well as nucleic acid, protein and lipid synthesis (Ashton *et al.*, 1977) or perhaps quantifying effects on cell division which have been deduced by observations on intact organisms (Sumida and Yoshida, 1982).

   In addition to these two approaches there will be the opportunity to test some speculative ideas based usually on the chemical structure of the pesticide and a knowledge of the biochemistry of the target organism.

2. *Is it the primary effect?*

Several approaches are available with which to examine whether an effect on a biochemical system *in vitro* is likely to be the basis of the pesticidal effect.

(a) In vitro/in vivo *activity correlations*

When close analogues of the test material are available, a good correlation between the activities of the chemicals against the biochemical system and the activities against the intact organism lends confidence to the view that the former is the cause of the latter. The correlation need not be (and cannot be expected to be) absolute, since differential penetration and metabolism between individual compounds in a series may often lead to anomalous results. However, if a compound which is active *in vivo* does not have the corresponding *in vitro* activity this is a clear warning that the biochemical system being tested may not be the primary target. Conversely, if a close, pesticidally inactive analogue of an active compound nevertheless displays *in vitro* activity, this suggests that the biochemical function under study may not be related to the primary action. Lack of penetration and/or increased metabolism are relatively common explanations for the lack of *in vivo* activity in a compound of high *in vitro* activity. However, it is rare for poor *in vitro* activity to nevertheless be the basis for good *in vivo* performance as a result of conversion of the compound to an active molecule or perhaps its concentration at the site of action. It is to be expected, of course, that protected derivatives (amides, esters, etc.) of active materials will often be inactive *in vitro* but effective *in vivo*, and the possibility of a less predictable activation of a molecule *in vivo* should also be considered.

(b) Consequences of action at the proposed site

A second approach is to consider the consequences of action at the proposed site, and to examine whether any of these consequences occur *in vivo*. Consider the pathway:

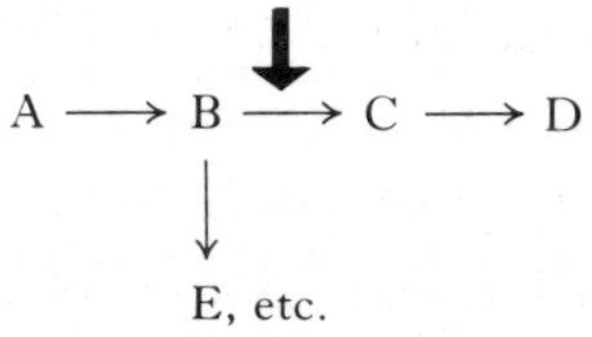

$$A \longrightarrow B \overset{\displaystyle\downarrow}{\longrightarrow} C \longrightarrow D$$
$$\downarrow$$
$$E, \text{ etc.}$$

If a compound were to inhibit the conversion of B to C, falls in the levels of C and D would be expected and rises in the levels of A and B could also occur. In addition the block between B and C could lead to the accumulation of products in a branch pathway, possibly products which are not normally found (E, etc).

A special case occurs when the accumulating substrates themselves (rather than the depletion of product) are the cause of the pesticidal activity. This might be revealed by the timing of the effects, e.g. accumulation of toxic levels of A, B or E is likely to occur more quickly than depletion of D to a critical level. In all cases an attempt should be made to reconcile the kinetics of changes in the levels of metabolites with the timing of the onset of visible symptoms of damage.

(c) Can the action be overcome?

Although it is not always feasible, it is reasonable to ask whether the effect caused by a pesticide on a process can be overcome by adding back to the system a component whose depletion is proposed (e.g. D in the scheme above), or perhaps by adding components capable of neutralising the damaging effects of a proposed accumulating toxicant.

(d) Neurophysiological studies

Aspects of the points outlined in (a)–(c) above are employed in a specific way in electrophysiological studies aimed at elucidating effects on the nervous system which may be suspected from symptoms exhibited by treated insects. In such studies a compound may be identified as an agonist at a particular receptor by comparing the effects caused with those induced by compounds having known stimulatory activity at this receptor.

Additionally, the provision of an antagonist at the receptor under study would be expected to block the effects of the compound.

(e) Mutation at the proposed target site

If strains exist, or can be induced, that are resistant to the pesticide due to differences in a proposed target site, this provides powerful evidence that the primary target has been identified. (A table of examples was given in the previous chapter – Table 9.1.) An active search for such mutants may be carried out under laboratory conditions, particularly in the case of fungi, mites and other readily cultured and fast-breeding organisms.

C. General remarks

It is worth noting that a knowledge of the distribution of applied chemical within the pest could well be of use in interpreting the results of some of the experiments suggested above. The distribution of the chemical within the target organism needs to correspond with the location(s) of the proposed primary site.

The technique of photoaffinity labelling has been mentioned earlier in the book, and it can provide detailed information about pesticide binding sites. Although it has proved of great value in the later stages of an investigation, it is unlikely to be applicable in the early stages when the primary target has yet to be established.

The comments above indicate a number of ways in which information can be sought which will establish the primary biochemical site of action of a pesticide. Obviously such a treatment cannot be complete and it is to be expected that in particular cases further experiments can be devised which will help distinguish primary from secondary and minor effects.

III. DESIGN
A. Approaches to pesticide discovery

As noted above 15–20,000 new compounds are presently made for every pesticide discovered which is capable of development. This means that a chemist working in the industry for forty years and making compounds at the rate of, say, 125 per year, will make a total of 5000 compounds, insufficient for him to have more than a small statistical chance of being associated with a successful invention.

Clearly any technique which holds promise of improving such odds is of great interest. With this in mind we shall examine the three possible ways in which pesticides can be sought. It should be stressed that, in all cases, it is vital that the biological screen accurately reflects market needs and opportunities and is also capable of revealing unexpected activities.

1. Random synthesis and screening

The aim of this approach is to devise and make large numbers of novel compounds. The compounds to be synthesised are chosen on a chemical basis alone, for example because the chemist has an interest in a particular type of compound or a specific kind of reaction. In general the synthetic routes employed should be short and of wide application to allow examination of a wide range of new structures.

2. Analogue synthesis

Information on the structure and biological activity of new compounds is available from two main sources. The first of these is the patent literature which discloses discoveries from industry. The second is the academic literature which provides access to the structure and biological activity of novel natural products and other active materials. In principle either source may offer a good starting point for the synthesis of further novel, patentable analogues.

3. Biochemical design

In this approach an attempt is made to identify biochemical processes vital to the survival of the pest and then to design and synthesise compounds which could interfere with them. Despite a number of attempts (Baillie *et al.*, 1972, 1975; Schroeder *et al.*, 1978; Hammock *et al.*, 1982) no commercial compound known to the authors has yet been obtained by this means. Nevertheless we will consider it here for completeness and because the method is beginning to find use in the pharmaceutical industry. Much of this work involves the identification of key enzymes as targets for inhibition, and the design and synthesis of potential inhibitors; however, non-enzymatic sites have also been explored (see, for example, Wright *et al.*, 1980).

An early approach to enzyme inhibition (Baker, 1967) involved the synthesis of substrate analogues containing a reactive group (such as an epoxide or a halomethyl ketone) which could react with a nucleophile at the active site. However, the groups chosen were usually so reactive as to combine indiscriminately with many cell components. In addition it is usually not known in advance if the target enzyme contains a suitably positioned nucleophile.

An improvement on this technique has been to present the enzyme with a stable analogue of the substrate, which is then converted by the enzyme to an active species. This species is usually one which will react readily with nucleophiles, so again a nucleophilic amino acid must be present at the active site for inhibition to occur. This approach, the k_{cat} approach, has been used increasingly in the drug industry in recent years (see, for example, Kalman, 1979).

Finally, we should note that the potential inhibitor may bind non-covalently to the enzyme and still bring about good inhibition. As an example Ondetti *et al*. (1977) have described the design and pharmaceutical properties of compounds having this mode of inhibition. In particular the potency of compounds which resemble the transition state (or a reactive intermediate) of an enzyme reaction has been stressed (Wolfenden, 1976).

B. Origin of existing pesticides

It appears from the literature that all of the compounds currently used as pesticides have arisen as a result of the first two approaches. In many cases there was no reason, *a priori*, for expecting the first compound of a new series to have biological activity of any sort. In other cases, notably pyrethroids, biological activity was found to reside in specific natural products and this acted as a spur for the effort subsequently expended in making analogues of these lead compounds. Analogue making can also be stimulated by fundamental research. An example is provided by the development of insect growth regulators following the discovery and identification of the insect juvenile hormones (p. 224).

C. Structure-activity correlations

The first two approaches outlined above usually involve the synthesis of a large number of derivatives of the lead compound. Clearly one would wish to choose derivatives in such a way that a minimum number of compounds would provide the maximum amount of information on the series. Various approaches have been developed to meet this objective and the subject has been reviewed by Martin (1978).

One of the most common approaches is that of Hansch (see, for example, Hansch, 1969), which attempts to define the relationship between structure and biological activity in a quantitative manner. The biological activity of a compound is considered to be due to a combination of its lipophilicity, electronic properties and size. Hansch's method examines the way in which substituents influence the properties of a parent compound and attempts to relate these changes to alterations in biological activity.

The hydrophobic, electronic and steric effects of many common substituents are known or can be experimentally determined (see Hansch and Leo, 1979) and computing techniques allow the derivation of an equation showing the contribution of each property to biological activity (an example of this type of correlation has already been described; see p. 128). This equation can then be used to predict new active structures. It should be emphasised that although it is often possible to obtain good correlations within a series when the biological effect is assessed on an isolated biochemical system (e.g., acetylcholinesterase, the Hill reaction) it is more difficult to obtain good correlations using *in vivo* data (Draber *et al.*, 1974).

The question of which analogues to make initially when exploring a series has been addressed in a number of ways (see Martin, 1978). Hansch and his colleagues have identified groups of substituents of closely similar properties (see Hansch and Leo, 1979) and suggested the selection of a substituent from each group so that the compounds in the initial batch have a wide variation in their properties. Alternatively Topliss (1972) has described a procedure in which members of a series are synthesised in a stepwise fashion, each new analogue being chosen on the basis of results with its predecessor.

Another approach to structure-activity correlations and one which makes no reference to substituent constants at all is the statistical method of Free and Wilson (1964). It is assumed in this technique that a substituent in a given position always has the same effect on activity, irrespective of other substituents. The aim of the method is to identify those substituents which increase or decrease the biological activity compared with the mean value of the series, so that useful new combinations can be predicted.

An approach to structure-activity correlations which is in its infancy but will certainly be useful in the future is the use of the computer to try to determine the shape of an active molecule. This shape can then be compared with those calculated for closely-related molecules in an attempt to correlate shape with pesticidal activity. A recent example of such a study is provided by Heritage (1982).

In some chemical series attempts to correlate pesticidal activity with *in vitro* data (e.g. on the Hill reaction) are unsuccessful because members of the series vary in their ability to reach the site of action in the pest. Such discrepancies may sometimes be elucidated by considering the distribution of chemical between the target organism and the surrounding environment (Briggs, 1981). This distribution depends on the physical properties of the pesticide, and a range of these (such as the octanol-water partition coefficient and the vapour pressure) can be estimated from a knowledge of the structure of the chemical alone, or a combination of its structure and melting point (Briggs, 1981).

D. Future pesticides

The world is heavily dependent on pesticides, not only to protect its (inadequate) food supply, but also to control a variety of insect-borne diseases. A detailed technical examination of alternative non-chemical methods of pest control suggests that there is no likelihood of pesticides being displaced within the foreseeable future (Corbett, 1978). New compounds are required not only to tackle the relatively few totally unsolved pest problems, but also to improve upon existing materials. It is important to continue the search for improved pesticides since the living world itself will change and man will always require up-to-date weapons to deal with new pest situations. What form will these new pesticides take?

Two trends have become apparent that will doubtless continue to influence the discovery and development of new pesticides. The first of these is the emergence of compounds which have high biological activity per unit weight. This should allow the application of less chemical per unit area, an aspect which should be beneficial from an environmental standpoint. Examples include the herbicide chlorsulfuron and the pyrethroid insecticides. These compounds are applied at a rate of tens of grammes per hectare rather than the more normal kilogramme or so per hectare. As already discussed above, it is theoretically possible that compounds active at rates of milligrammes per hectare could be discovered. Such developments will encourage the exploration of more expensive and more complex structures than might have been considered otherwise; for example, it is possible that natural products will become more important either as active compounds themselves, or as synthetic leads. Existing natural products used as pesticides include several antibiotic fungicides, the new herbicide bialaphos, the pyrethrins and the avermectins.

Improved activity may also arise from the use of chiral pesticides prepared either by stereospecific synthesis or the separation of individual enantiomers. The normally high cost of production of these may in some cases be justified by the increased activity obtained. It is also conceivable that genetic engineering techniques could be used to provide complex intermediates for new pesticides, or even the final product itself.

The second trend, already seen in the pharmaceutical industry, is towards an increased adoption of rational approaches to pesticide discovery. These will include computer-aided techniques for correlating structure with activity as well as analysis of the molecular shape required to fit receptor sites in the pest. The pesticide industry may also make more use of predictive biochemical design in pesticide discovery. Such an approach should ultimately reduce the cost of discovery and allow a more rational appreciation of the toxicological and environmental side effects that the new pesticide may possess. Knowledge of the biochemical mode of action of a new pesticide should also allow the adoption of a more rational strategy of use in order to avoid or minimise the development of resistance. The authors hope that this book will aid those who are interested in applying biochemical techniques to pesticide discovery.

352 THE BIOCHEMICAL MODE OF ACTION OF PESTICIDES

REFERENCES

Ashton, F. M., de Villiers, O. T., Glenn, R. K. and Duke, W. B. (1977). *Pestic. Biochem. Physiol.* **7**, 122–141

Baker, B. R. (1967). "Design of Active-Site-Directed Irreversible Enzyme Inhibitors". Wiley, New York

Baillie, A. C., Corbett, J. R., Dowsett, J. R. and McCloskey, P. (1972). *Pestic. Sci.* **3**, 113–120

Baillie, A. C., Corbett, J. R., Dowsett, J. R., Sattelle, D. B. and Callec, J.-J. (1975). *Pestic. Sci.* **6**, 645–653

Briggs, G. G. (1981). Proc. British Crop Protection Conf. "Pests and Diseases", Vol. 3, pp. 701–710

Corbett, J. R. (1978). In "Applied Biology" (T. H. Coaker, ed.) Vol. 3, pp. 224–330. Academic Press, London and New York

Davis, F. S., Villareal, A., Baur, J. R. and Goldstein, I. S. (1972). *Weed Sci.* **20**, 185–188

Devonshire, A. L. (1980). In "Insect Neurobiology and Pesticide Action", pp. 473–480. Society of Chemical Industry, London

Draber, W., Büchel, K. H., Timmler, H. and Trebst, A. (1974). In "Mechanism of Pesticide Action" (G. K. Kohn, ed.) pp. 110–116. American Chemical Society, Washington, D.C.

Fedtke, C. (1982). In "Biochemical Responses Induced by Herbicides" (D. E. Moreland, J. B. St. John and F. D. Hess, eds) pp. 231–250. American Chemical Society, Washington, D.C.

Free, S. M. and Wilson, J. (1964). *J. Med. Chem.* **7**, 395–399

Hammock, B. D., Wing, K. D., McLaughlin, J., Lovell, V. M. and Sparks, T. C. (1982). *Pestic. Biochem. Physiol.* **17**, 76–88

Hansch, C. (1969). *Acc. Chem. Res.* **2**, 232–239

Hansch, C. and Leo, A. J. (1979). "Substituent Constants for Correlation Analysis in Chemistry and Biology". Wiley, New York

Heritage, K. J. (1982). *Biochem. Soc. Trans.* **10**, 310–312

Jellings, A. J. and Leech, R. M. (1982). *New Phytol.* **92**, 39–48

Kalman, T. I. (1979). (ed.). "Drug Action and Design: Mechanism-Based Enzyme Inhibitors". Elsevier/North-Holland, New York

Martin, Y. (1978). "Quantitative Drug Design". Marcel Dekker, New York

Ondetti, M. A., Rubin, B. and Cushman, D. W. (1977). *Science* **196**, 441–444

Ray, T. B. (1982). *Pestic. Biochem. Physiol.* **17**, 10–17

Schroeder, M. E., Boyer, A. C., Flattum, R. F. and Sundelin, K. G. R. (1978). In "Pesticide and Venom Neurotoxicity" (D. L. Shankland, R. M. Hollingworth and T. Smyth, Jr., eds) pp. 63–82. Plenum Press, New York

Srere, A. (1967). *Science* **158**, 936–937

Srere, A. (1970). *Biochem. Med.* **4**, 43–46

Sumida, S. and Yoshida, R. (1982). In "Biochemical Responses Induced by Herbicides" (D. E. Moreland, J. B. St. John and F. D. Hess, eds) pp. 251–260. American Chemical Society, Washington, D.C.

Topliss, J. G. (1972). *J. Med. Chem.* **15**, 1006–1011

Wolfenden, R. (1976). *Ann. Rev. Biophys. Bioenerg.* **5**, 271–306

Wright, B. J., Baillie, A. C., Wright, K., Dowsett, J. R. and Sharpe, T. M. (1980). *Phytochemistry* **19**, 61–65

Subject Index

3-Hydroxy-3-methylglutaryl CoA reductase, 243, 267
8-Hydroxyquinoline, 11
Hydroxyquinoline fungicides, 9, 11–12
  structures, 26
Hymexazol, 237, 324, 325
Hyperlipemic hormone, 169

I_{50} definition, 20
IAA (see Indol-3-yl-acetic acid)
IBP (see S-Benzyl O,O-di-isopropyl phosphorothioate)
Ignosterol, 265, 266
Imazalil
  mode of action, 263
  structure, 257
Imidazole fungicides
  mode of action, 262
  structures, 257
Iminodiacetic acid, 279
Indiscriminate pesticides (see Nonspecific pesticides)
Indol-3-yl-acetic acid, 186, 189, 193, 271, 280
  antiauxin activity, 316–317
  auxin activity, 185–7, 323
  herbicidal activity 186–187
  mode of action 180–181
  structure, 180
  structure and activity, 190–192
  synthetic mimics, 181–192
4-(Indol-3-yl) butyric acid, 181
  effect on growth, 187
  mode of action, 186
  structure, 183
Indophenyl acetate, 120
Inhibitory uncouplers, 39
Inorganic herbicides, 319
Inositol, 315
Insect hormone action, interference with, 223–227
Insecticidal esters, 143
2-Iodo-4-nitro-6-isobutylphenol, 71
Iodofenphos, 107
Ioxynil, 81
  binding to chloroplast, 71, 76
  effect on Hill reaction, 65, 69, 73
  structure, 31, 65
  uncoupling activity, 37, 39
Ioxynil octanoate, 65
  structure, 32

Iprodione, 222
Iron-sulphur protein, 53, 54
Isocarbamid, 78
  structure, 59
Isofenphos, 107
Isomethiozin, 63
Isopentenyl pyrophosphate, 244
Isoprocarb, 129
  structure, 114
Isopropalin, 210
3'-Isopropoxy-2-methyl-benzanilide, 26
Isoprothiolane
  mode of action, 312
  structure, 313
Isoproturon, 57
Isothioate, 107

Juvenile hormones, 224, 349
  and action of precocenes, 226–227
  synthetic analogues, 224–225

K_i
  definition of, 15
  in kinetics of acetylcholine inhibition, 121–125
  value for
    aldicarb, 131
    sec-butylamine, 15
    carbamates, 121–125, 127–131, 134
    carboxins, 25, 27
    chlorobenzilate, 159
    ethirimol, 282
    glyphosate, 279
    isoprocarb, 129
    nicotine, 160
    organophosphorus compounds, 121–125, 131–134
    phenyl N-methylcarbamates, 128–131
    phosphine, 25
    polyoxin D, 273
    promecarb, 129
    propoxur, 129
    m-tolyl methylcarbamate, 129
    3,4-xylyl methylcarbamate, 129
    3,5-xylyl methylcarbamate, 129
K_m, definition of, 15
Kasugamycin
  mode of action, 239
  structure, 238
ent-Kaurene, 198, 200, 254, 268